Birkhäuser

Trends in Mathematics

Research Perspectives Ghent Analysis and PDE Center

Volume 2

Research Perspectives Ghent Analysis and PDE Center is a book series devoted to the publication of extended abstracts of seminars, conferences, workshops, and other scientific events related to the Ghent Analysis and PDE Center. The extended abstracts are published in the subseries Research Perspectives Ghent Analysis and PDE Center within the book series Trends in Mathematics. All contributions undergo a peer-review process to meet the highest standard of scientific literature.

Volumes in the subseries will include a collection of revised written versions of the communications or short research announcements or summaries, grouped by events or by topics. Contributing authors to the extended abstracts volumes remain free to use their own material as in these publications for other purposes (for example a revised and enlarged paper) without prior consent from the publisher, provided it is not identical in form and content with the original publication and provided the original source is appropriately credited.

Michael Ruzhansky • Karel Van Bockstal
Editors

Extended Abstracts 2021/2022

Ghent Analysis and PDE Seminar

Birkhäuser

Editors
Michael Ruzhansky
Ghent Analysis and PDE Center,
Department of Mathematics: Analysis,
Logic and Discrete Mathematics
Ghent University
Ghent, Belgium

School of Mathematical Sciences
Queen Mary University of London
London, UK

Karel Van Bockstal
Ghent Analysis and PDE Center,
Department of Mathematics: Analysis,
Logic and Discrete Mathematics
Ghent University
Ghent, Belgium

ISSN 2297-0215 ISSN 2297-024X (electronic)
Trends in Mathematics
ISSN 2948-1724 ISSN 2948-1732 (electronic)
Research Perspectives Ghent Analysis and PDE Center
ISBN 978-3-031-42541-7 ISBN 978-3-031-42539-4 (eBook)
https://doi.org/10.1007/978-3-031-42539-4

This book is published under the imprint Birkhäuser, www.birkhauser-science.com by the registered company Springer Nature Switzerland AG
The registered company address is: Gewerbestrasse 11, 6330 Cham, Switzerland

Paper in this product is recyclable.

Preface

This book is the second volume in the recently established Birkhäuser series Research Perspectives Ghent Analysis and PDE Center. This is a book series devoted to the publication of extended abstracts of seminars, conferences, workshops, and other scientific events related to the Ghent Analysis and PDE Center (GAPC). Volumes in this subseries include a collection of revised written versions of the communications or short research announcements or summaries, grouped by events or by topics.

GAPC is a research group in mathematics that focuses on different areas of analysis, as well as the study of partial differential equations (PDEs) and their applications. The main topics are functional analysis, Fourier analysis, noncommutative analysis, geometric analysis, partial differential equations of different types, harmonic analysis, functional inequalities, pseudo-differential operators, fractional derivatives, special functions, microlocal analysis, inverse problems, and imaging. The target audience of this book is any researcher working in the above fields. More details about the research activities of the center can be found on the group website:

https://analysis-pde.org

In addition to the research activities, the Ghent Analysis and PDE Center is also involved in training the next generation of mathematicians. GAPC organizes seminars, workshops, and conferences to share their research findings and collaborate with other researchers in the field. The center brings together researchers from different disciplines to collaborate on projects related to analysis and partial differential equations, and provides a platform for sharing knowledge and expertise.

The activities of the center have been supported by grants from the Research Foundation Flanders (FWO) as well as by the Methusalem program of Ghent University Special Research Fund.

This book offers an overview of some of the research results presented by the group members and guests of GAPC during the weekly Friday seminar in 2021–2022. It is an informal event of GAPC and associated researchers, where everyone can present their work or relevant literature for about 20–30 minutes. The seminar aims to exchange ideas and foster effective learning and collaboration.

In this book, group members and guests summarize their results presented during the seminar and provide outlooks for future work. In this way, the book also provides an overview of the recent developments at GAPC.

Speakers in the seminar in 2021–2022 have been invited to contribute to this book. This resulted in a total of 34 contributions, which we have subdivided into 3 parts.

1. Part I: contributions related to analysis
2. Part II: contributions related to partial differential equations
3. Part III: contributions related to mathematical modelling

In this way, the division of the book reflects the core topics of GAPC.

We are grateful to all the authors who have contributed to this volume.

Ghent, Belgium
June 2023

Michael Ruzhansky
Karel Van Bockstal

Contents

Part I
Analysis

Chapter 1
A Note on a Capelli Operator and its Resonance

Roberto Bramati

Abstract We study in parallel the resonances of the standard Laplacian on $\mathbb{R}^3$ and of a Capelli operator on $\mathbb{R}^2$ by exploiting their invariance under two appropriate group actions. In particular, we recover the well-known fact that the Laplacian on $\mathbb{R}^3$ has no resonances and we conclude that the Capelli operator on $\mathbb{R}^2$ has a single resonance at $z = 0$. The full proof of this fact is contained in a joint paper with A. Pasquale and T. Przebinda (J Lie Theory 33(1):93–132, 2023).

1.1 Introduction

This note is a short summary of a talk given at the Gent Analysis&PDE's Centre Friday seminar. The results mentioned here were obtained jointly with Angela Pasquale and Tomasz Przebinda and are contained in the paper [1] to which the reader should refer for details, proofs and precise statements.

In this note, we will give a quick introduction to the abstract resonances approach introduced in [1], proposing a parallel between the analysis of the resonances of the positive Laplacian $\Delta = -(\partial^2_{xx} + \partial^2_{yy} + \partial^2_{zz})$ on $\mathbb{R}^3$ and of the Capelli operator $C = -(x\partial_x + y\partial_y + 1)^2$ on $\mathbb{R}^2$.

The operators Δ and C are densely defined unbounded self-adjoint operators acting respectively on the Hilbert spaces $L^2(\mathbb{R}^3)$ and $L^2(\mathbb{R}^2)$. They both have a continuous spectrum $[0, +\infty)$. We consider a quadratic modification of their resolvent operators

$$(\Delta - z^2)^{-1} \qquad \text{and} \qquad (C - z^2)^{-1},$$

R. Bramati (✉)
Department of Mathematics: Analysis, Logic and Discrete Mathematics, Ghent University, Ghent, Belgium
e-mail: roberto.bramati@ugent.be

M. Ruzhansky, K. Van Bockstal (eds.), *Extended Abstracts 2021/2022*,
Research Perspectives Ghent Analysis and PDE Center 2,
https://doi.org/10.1007/978-3-031-42539-4_1

with $z \in \mathbb{C}$. The choice of z^2 instead of z will be useful due to the quadratic nature of the two operators. Spectral theory tells us that for z such that $\Im z > 0$, so that $z^2 \in \mathbb{C} \setminus [0, +\infty)$, i.e., z^2 outside of the spectrum, the resolvent operators are bounded on the respective Hilbert spaces. Moreover, they depend holomorphically on z such that $\Im z > 0$. Then one is led to wonder if, restricting the set of functions on which the resolvent operators act to a dense subspace of the Hilbert space, it is possible to extend meromorphically the resolvents across the spectrum. If the answer is positive, the resonances are the poles of such meromorphic extensions. In order to make use of Paley-Wiener type results which are fundamental when computing the meromorphic extension, it is convenient to choose as dense subspace the space of smooth functions with compact support. The resolvent operators can be seen as distribution-valued operators acting on smooth functions with compact support.

The study of resonances originated almost a hundred years ago in Quantum Mechanics and was for many decades restricted to the case of Schrödinger operators on Euclidean spaces. More recently their study was extended to other geometric settings. The most studied cases are noncompact Riemannian manifolds with bounded geometry, such as hyperbolic manifolds or symmetric spaces. The investigation of resonances in these settings, in addition to being of its own interest, is motivated by applications to geometric scattering, spectral theory, PDE's and dynamical systems. See [11] for more information about resonances and their applications.

The approach to finding resonances used in this note and in the paper [1] stems from ideas coming from the works on symmetric spaces of the noncompact type by Hilgert and Pasquale [3] and Hilgert et al. [4–6]. Their technique aims at exploiting the invariance of the operator under a group action, bringing in the powerful tools that come with invariance: harmonic analysis and representation theory. The approach of [3] was recently adapted to treat other cases, like resonances for the Laplacian on homogeneous vector bundles on symmetric spaces of real rank one [9, 10] or for the Laplacian on pseudo-Riemannian symmetric spaces [2].

The note is organized as follows. In Sect. 1.2 we will highlight which are the relevant groups and group actions for our operators Δ and C. In Sect. 1.3 we will introduce two useful function decompositions that are closely connected to the group actions. In Sect. 1.4 we will give in detail the argument for computing the meromorphic (or rather holomorphic) extension of the Laplacian. The argument for C is similar, but more involved and is treated in great detail in [1]. Finally in Sect. 1.5 we will give some general comments on the technique and briefly hint to the role of representation theory in this research area.

1.2 The Group Actions

The standard positive Laplacian Δ on $\mathbb{R}^3$, densely defined on $L^2(\mathbb{R}^3)$, is invariant under the action of the group of translations $G \cong \mathbb{R}^3$. The action is defined as

$$T(w)f(x) = f(x+w),$$

where $x, w \in \mathbb{R}^3$, and it is straightforward to see that

$$(\Delta f)(x+w) = (\Delta f(\cdot + w))(x),$$

i.e. that $T(w)\Delta = \Delta T(w)$.

The second group we will be interested in is $O_{1,1}$. We realize it as the group generated by the 2×2 matrices

$$h_a = \begin{pmatrix} a & 0 \\ 0 & a^{-1} \end{pmatrix},$$

with $a \in \mathbb{R}^*$, and the matrix

$$s = \begin{pmatrix} 0 & 1 \\ 1 & 0 \end{pmatrix}.$$

For convenience, we shall identify $h_a \equiv a$. We consider the following action ω of $O_{1,1}$ on $L^2(\mathbb{R}^2)$. For the matrices h_a the action is defined as an L^2-norm preserving rescaling by

$$\omega(a)v(x) = |a|^{-1}v(a^{-1}x).$$

For the matrix s the action is defined by

$$\omega(s)v(x') = \int_{\mathbb{R}^2} e^{-2\pi i x' J x^t} v(x)dx,$$

where

$$J = \begin{pmatrix} 0 & 1 \\ -1 & 0 \end{pmatrix}.$$

Notice that, denoting by $\mathcal{F}$ the usual Fourier transform and by $R(J)$ the right multiplication by J of the variable, we have

$$\omega(s) = R(J)\mathcal{F}, \tag{1.1}$$

so that the action of $\omega(s)$ is a skewed Fourier transform.

Now, consider the Capelli operator $C = -(x\partial_x + y\partial_y + 1)^2$. This operator is invariant under the action ω. For example, it is easy to see that $\omega(a)C = C\omega(a)$. Indeed, for $v \in C_c^\infty(\mathbb{R}^2)$, denoting by v_x and v_y the partial derivatives of v with respect to the first and the second variable,

$$\begin{aligned}\omega(a)((x\partial_x + y\partial_y + 1)v)(x, y) &= |a|^{-1}(a^{-1}xv_x(a^{-1}x, a^{-1}y)\\ &\quad + a^{-1}yv_y(a^{-1}x, a^{-1}y) + v(a^{-1}x, v^{-1}y))\\ &= (x\partial_x + y\partial_y + 1)(|a|^{-1}v(a^{-1}\cdot, a^{-1}\cdot))(x, y)\\ &= (x\partial_x + y\partial_y + 1)\omega(a)v(x, y).\end{aligned}$$

The invariance under $\omega(s)$ is less trivial to check but can be easily derived by abstract considerations. See the paper [1] for the details and Section 1.5 for further comments on how the operator C arises.

1.3 Two Useful Decompositions

We can use the Fourier transform to decompose a function in $C_c^\infty(\mathbb{R}^3)$ into its Fourier components. By the Fourier inversion formula, we have

$$f(x) = \int_{\mathbb{R}^3} e^{i\lambda x}\widehat{f}(\lambda)d\lambda = \int_{\mathbb{R}^3} e^{i\lambda\cdot} * f(x)d\lambda, \tag{1.2}$$

where $*$ denotes the usual Euclidean convolution. Each piece in the decomposition lies in a space

$$D_\lambda = \{e^{i\lambda\cdot} * f, \quad f \in C_c^\infty(\mathbb{R}^3)\}, \tag{1.3}$$

for $\lambda \in \mathbb{R}^3$, which is invariant under the action T. Indeed, consider $F_\lambda \in D_\lambda$. Then $F_\lambda(x) = e^{i\lambda x}\widehat{f}(\lambda)$ for some $f \in C_c^\infty(\mathbb{R}^3)$. It is easy to see that $T(w)F_\lambda \in D_\lambda$, since

$$\begin{aligned}T(w)F_\lambda(x) = e^{i\lambda(x+w)}\widehat{f}(\lambda) &= e^{i\lambda x}\int_{\mathbb{R}^3} f(y)e^{-i\lambda(y-w)}dy\\ &= e^{i\lambda x}\int_{\mathbb{R}^3} f(y+w)e^{-i\lambda y}dy = e^{i\lambda x}\widehat{T(w)f}(\lambda) = e^{i\lambda x}\widehat{g}(\lambda),\end{aligned}$$

where $g = T(w)f \in C_c^\infty(\mathbb{R}^3)$. Moreover the Laplacian Δ acts diagonally on the decomposition (1.2), in the sense that for each piece $F_\lambda \in D_\lambda$ we have

$$\Delta F_\lambda(x) = \Delta e^{i\lambda x}\widehat{f}(\lambda) = |\lambda|^2 e^{i\lambda x}\widehat{f}(\lambda) = |\lambda|^2 F_\lambda(x).$$

This allows us to get the following convenient formula for the resolvent of Δ:

$$(\Delta - z^2)^{-1} f(x) = \int_{\mathbb{R}^3} \frac{1}{|\lambda|^2 - z^2} e^{ix\lambda} \widehat{f}(\lambda) d\lambda. \tag{1.4}$$

We would like to have a similar decomposition also for our second case. To decompose functions in $C_c^\infty(\mathbb{R}^2)$ we use the Mellin transform. We define

$$v_\lambda(x) = \int_0^\infty a^{-1-i\lambda} v(a^{-1}x) \frac{da}{a},$$

for $\lambda \in \mathbb{C}$. Note that v_λ is a homogenous function of degree $-1 - i\lambda$. The Mellin inversion formula states that one can reconstruct the initial function through its homogeneous components by

$$v(x) = \frac{1}{2\pi} \int_{\mathbb{R}} v_\lambda(x) d\lambda. \tag{1.5}$$

We choose this decomposition because it behaves well under the action ω, as the Fourier transform behaves well under the action of T. We further split v_λ into its even and odd components as

$$v_\lambda(x) = \frac{v_\lambda(x) + v_\lambda(-x)}{2} + \frac{v_\lambda(x) - v_\lambda(-x)}{2} =: v_{0,\lambda}(x) + v_{1,\lambda}(x).$$

We see that, for $a \in \mathbb{R}^*$,

$$\begin{aligned} \omega(a) v_{0,\lambda}(x) &= |a|^{-1} v_{0,\lambda}(a^{-1}x) \\ &= |a|^{-1} \int_0^\infty b^{-1-i\lambda} v_{0,\lambda}(b^{-1}a^{-1}x) \frac{db}{b} \\ &= |a|^{i\lambda} \int_0^\infty c^{-1-i\lambda} v_{0,\lambda}\left(\frac{a}{|a|} c^{-1} x\right) \frac{dc}{c}, \end{aligned}$$

where we set $c = b|a|$. For both $a > 0$ and $a < 0$, by the parity of $v_{0,\lambda}$, we get that

$$\omega(a) v_{0,\lambda}(x) = |a|^{i\lambda} v_{0,\lambda}(x).$$

The same computation and the oddity of $v_{1,\lambda}$ show that

$$\omega(a) v_{1,\lambda}(x) = \frac{a}{|a|} |a|^{i\lambda} v_{1,\lambda}(x).$$

As for the action of $\omega(s)$, we can make use of formula (1.1) and observe that parity is preserved under the action of both the Fourier transform and right multiplication by J. Homogeneity is preserved by $R(J)$, while it is easy to see that the Fourier transform maps v_λ to a $(-1+i\lambda)$-homogeneous function. Indeed, for $t>0$ and $y\in\mathbb{R}^2$, and by the homogeneity of v_λ,

$$\begin{aligned}\mathcal{F}v_\lambda(ty) &= \int_{\mathbb{R}^2} e^{-i\lambda tyx} v_\lambda(x)dx \\ &= t^{-2}\int_{\mathbb{R}^2} e^{-i\lambda yx} v_\lambda(t^{-1}x)dx = t^{-1+i\lambda}\mathcal{F}v_\lambda.\end{aligned}$$

So the Mellin decomposition (1.5) behaves well under the action ω in the sense that thinking, say, the piece $v_{0,\lambda}$ as an element of the set $M_{0,\lambda}$ of even functions that are homogeneous of degree $(-1\pm i\lambda)$, then the image of $v_{0,\lambda}$ is still in this set. This resembles what happened for the sets D_λ defined in (1.3) under the action of translations in the Fourier transform case. The precise description of the invariant sets for the action ω requires to introduce quite some notation and this falls out of the scope of this note. All the details can be found in [1].

The Capelli operator C acts diagonally on the decomposition (1.5). Indeed

$$\begin{aligned}(x\partial_x+y\partial_y+1)v_\lambda(x,y) &= \int_0^\infty a^{-1-i\lambda}(x\partial_x+y\partial_y)v(a^{-1}x,a^{-1}y)\frac{da}{a}+v_\lambda \\ &= \int_0^\infty a^{-1-i\lambda}(a^{-1}xv_x(a^{-1}x,a^{-1}y)+a^{-1}yv_y(a^{-1}x,a^{-1}y))\frac{da}{a}+v_\lambda \\ &= -\int_0^\infty a^{-1-i\lambda}(a\partial_a)v(a^{-1}x,a^{-1}y)da+v_\lambda \\ &= \int_0^\infty(-1-i\lambda)a^{-1-i\lambda}v(a^{-1}x,a^{-1}y)\frac{da}{a}+v_\lambda = -i\lambda v_\lambda.\end{aligned}$$

Therefore $Cv_\lambda=\lambda^2 v_\lambda$. Hence, also in this case, we get a convenient formula for the resolvent of C:

$$(C-z^2)^{-1}v(x) = \int_{\mathbb{R}}(\lambda^2-z^2)^{-1}v_\lambda(x)d\lambda. \tag{1.6}$$

1.4 Resonances

We are finally ready to prove the following theorem.

Theorem 1.1 *The resolvent of the Laplacian* $(\Delta-z^2)^{-1}: C_c^\infty(\mathbb{R}^3)\to\mathcal{D}(\mathbb{R}^3)$ *as a function of* z *admits an holomorphic extension from* $\Im z>0$ *to* $\mathbb{C}$. *Therefore the Laplacian on* $\mathbb{R}^3$ *has no resonances.*

Proof We work on the explicit expression for the resolvent (1.4). Passing to polar coordinates we can rewrite, for $f \in C_c^\infty(\mathbb{R}^3)$,

$$\begin{aligned}(\Delta - z^2)^{-1} f(x) &= \int_{\mathbb{R}^3} \frac{1}{|\lambda|^2 - z^2} e^{ix\lambda} \widehat{f}(\lambda) d\lambda \\ &= \int_0^\infty \frac{1}{r^2 - z^2} \left[\left(\int_{\mathbb{S}^2} e^{ixrw} \widehat{f}(rw) dw \right) r \right] r dr \\ &=: \int_0^\infty \frac{1}{r^2 - z^2} F(r) r dr.\end{aligned}$$

We note that by the Paley-Wiener theorem, $\widehat{f}$ is entire of exponential type and rapidly decreasing. Since the integral

$$\int_{\mathbb{S}^2} e^{ixrw} \widehat{f}(rw) dw$$

is invariant under reflection of the angular variable, F is an odd function of r. Moreover, F is holomorphic as a function of $r \in \mathbb{C}$ and rapidly decreasing, due to the Paley-Wiener properties of $\widehat{f}$. Then, since

$$\frac{2r}{r^2 - z^2} = \frac{1}{r - z} + \frac{1}{r + z},$$

and by the oddity of F we get (apart from a constant factor), for all $c > 0$,

$$\begin{aligned}(\Delta - z^2)^{-1} f(x) &= \int_0^\infty \frac{F(r)}{r - z} dr + \int_0^\infty \frac{F(r)}{r + z} dr \\ &= \int_{-\infty}^\infty \frac{F(r)}{r - z} dr = \int_{-ic-\infty}^{-ic+\infty} \frac{F(r)}{r - z} dr.\end{aligned}$$

In the last equality, we changed the contour of integration, using Cauchy's Theorem and the rapid decrease of F. The last expression, as $c > 0$ varies, defines a holomorphic function on the whole complex plane $\mathbb{C}$ and therefore provides the sought holomorphic extension of the resolvent. □

Performing similar computations on formula (1.6) one can obtain the meromorphic extension of the Capelli operator C, finding the following result.

Theorem 1.2 *The resolvent of the Capelli operator* $(C - z^2)^{-1} : C_c^\infty(\mathbb{R}^2 \setminus \{0\}) \to \mathcal{D}(\mathbb{R}^2 \setminus \{0\})$ *as a function of* z *admits a meromorphic extension from* $\Im z > 0$ *to* $\mathbb{C}$ *with a single simple pole at* $z = 0$. *Therefore the Capelli operator* C *has a single resonance.*

Proof The restriction to $\mathbb{R}^2 \setminus \{0\}$ is technical but necessary. See [1] for the details of the proof. □

1.5 Further remarks

In collaboration with Pasquale and Przebinda, we recently introduced in [1] an abstract representation theoretic approach to study resonances. We built from the observation that the main tool when studying resonances is a direct integral decomposition of a unitary representation (π, V) of a Lie group G, of the form $V = \int_A^{\oplus} V_\alpha \, d\mu(\alpha)$, where V_α are suitable G-invariant spaces and A is a suitable parameter space. An invariant differential operator obtained through the derived representation of π from a G-invariant element of the center of the universal enveloping algebra of the Lie algebra of G, such as the Casimir element, acts diagonally on the pieces of the decomposition. The meromorphic extension of the resolvent of an operator acting on V can be obtained by a suitable meromorphic continuation of the spectral measure $d\mu(\alpha)$. The resonances then correspond to discrete parts of the extended spectral measure.

In this note we showed explicitly that the Fourier transform on $\mathbb{R}^3$ provides exactly such direct integral decomposition for the left regular representation of the group of translations on $L^2(\mathbb{R}^3)$. The Laplacian is the image of the Casimir element of the universal enveloping algebra of $\mathbb{R}^3$ through the derived representation of the left regular representation. Similarly, the Mellin transform on $\mathbb{R}^2$ provides the required direct integral decomposition for the representation ω of the group $O_{1,1}$ on $L^2(\mathbb{R}^2)$. The Capelli operator arises from the Casimir element of the universal enveloping algebra of $\mathfrak{o}_{1,1}$ through the derived representation of ω.

When the action of a Lie group G is present and the differential operator of interest is invariant under this action, one can study its resonances not only analytically, but also from a representation theoretic point of view. Indeed to each resonance z_0 we can associate the residue operator $\mathrm{Res}_{z_0} R$ defined as

$$\mathrm{Res}_{z_0} R(f)(y) := \mathrm{Res}_{z=z_0} \left([R(z)f](y) \right),$$

where R stands for the resolvent operator, and f is in a suitable dense subspace of V.

In our two cases, the groups act on the image $\mathrm{Res}_{z_0} R(C_c^\infty)$ of the residue operator at a resonance and thus we have representations of our groups. A fundamental aim in this research area since the work of Pasquale and Przebinda [3], is to study the representations arising in this way, determining, in particular, whether they are irreducible, finite or infinite dimensional, and if they are unitary.

For what concerns the examples studied in this note, in the case of the Laplacian of $\mathbb{R}^3$ there is nothing to say since there are no resonances. For the resonance $z = 0$ of the Capelli operator C, in the paper [1] the resonance representation was identified and described, also thanks to the fact that the unitary dual of $O_{1,1}$ is known. In the paper [1] we also study other Capelli operators and their resonances and associated representations, exploiting the rich theory of Howe's duality which offers a correspondence of Capelli operators and representations among the members of a reductive dual pair (see [7, 8]).

Acknowledgments The author is supported by the FWO Odysseus 1 grant G.0H94.18N: Analysis and Partial Differential Equations and by the Methusalem programme of the Ghent University Special Research Fund (BOF) (Grant number 01M01021).

References

1. Bramati, R., Pasquale, A., Przebinda, T.: The resonances of the Capelli operators for small split orthosymplectic dual pairs. J. Lie Theory **33**(1), 93–132 (2023)
2. Frahm, J., Spilioti, P.: Resonances and residue operators for pseudo-riemannian hyperbolic spaces. J. Math. Pures Appl. **177**, 178–197 (2023)
3. Hilgert, J., Pasquale, A.: Resonances and residue operators for symmetric spaces of rank one. J. Math. Pures Appl. **91**(5), 495–507 (2009)
4. Hilgert, J., Pasquale, A., Przebinda, T.: Resonances for the Laplacian: the cases BC_2 and C_2 (except $\mathrm{SO}_0(p, 2)$ with $p > 2$ odd). In: Geometric Methods in Physics. Trends in Mathematics, pp. 159–182. Birkhäuser, Basel (2016)
5. Hilgert, J., Pasquale, A., Przebinda, T.: Resonances for the Laplacian on products of two rank one Riemannian symmetric spaces. J. Funct. Anal. **272**(4), 1477–1523 (2017)
6. Hilgert, J., Pasquale, A., Przebinda, T.: Resonances for the Laplacian on Riemannian symmetric spaces: the case of $SL(3, \mathbb{R})/\mathbb{SO}(\mathbb{F})$. Represent. Theory **21**, 416–457 (2017)
7. Howe, R.: Remarks on classical invariant theory. Trans. Am. Math. Soc. **313**(2), 539–570 (1989)
8. Howe, R.: Transcending classical invariant theory. J. Am. Math. Soc. **2**(3), 535–552 (1989)
9. Roby, S.: Resonances of the Laplace operator on homogeneous vector bundles on symmetric spaces of real rank-one. Adv. Math. **408**(part A), 108555 (2022)
10. Roby, S.: Residue representations - the rank one case. J. Geom. Phys. **189**, 104822 (2023)
11. Zworski, M.: Mathematical study of scattering resonances. Bull. Math. Sci. **7**(1), 1–85 (2017)

Chapter 2
Schatten-von Neumann Classes $\mathscr{S}_p$ on the Torus for $0 < p \leq 2$

Duván Cardona

Abstract We provide a sharp sufficient condition on the symbol of a pseudo-differential operator on the torus $\mathbb{T}^n$ in order to guarantee its membership in the Schatten-von Neumann ideals $\mathscr{S}_p(L^2(\mathbb{T}^n))$ for lower orders of p, namely when $0 < p \leq 2$.

2.1 Introduction

Let $\mathbb{T}^n = \mathbb{R}^n/\mathbb{Z}^n$ be the n-torus. The purpose of this note is to present a sharp condition for the membership of pseudo-differential operators to the Schatten classes $\mathscr{S}_p(L^2(\mathbb{T}^n))$ of compact linear operators on $L^2(\mathbb{T}^n)$ in terms of symbol criteria in the case where $0 < p \leq 2$.

In terms of the Fourier transform on the n-torus

$$\widehat{f}(\xi) := \int_{\mathbb{T}^n} e^{-2\pi i x\cdot\xi} f(x)dx, \ \ \xi \in \mathbb{Z}^n, \ f \in L^1(\mathbb{Z}^n),$$

any bounded linear operator A on $L^2(\mathbb{Z}^n)$ admits the pseudo-differential representation

$$Af(x) = \sum_{\xi\in\mathbb{Z}^n} e^{2\pi i x\cdot\xi} a(x,\xi)\widehat{f}(\xi), \ f \in L^2(\mathbb{T}^n).$$

D. Cardona (✉)
Department of Mathematics: Analysis, Logic and Discrete Mathematics, Ghent University, Ghent, Belgium
e-mail: duvan.cardonasanchez@ugent.be

M. Ruzhansky, K. Van Bockstal (eds.), *Extended Abstracts 2021/2022*, Research Perspectives Ghent Analysis and PDE Center 2,
https://doi.org/10.1007/978-3-031-42539-4_2

By following the standard terminology we say that A is the pseudo-differential operator associated to the symbol $a := a(x,\xi)$. Moreover, the symbol can be obtained from values of the operator by the following identity (see [22, Chapter 4])

$$a(x,\xi) = e_\xi^{-1} A e_\xi, \quad x \in \mathbb{T}^n, \ \xi \in \mathbb{Z}^n,$$

where we have used the notation $e_\xi(x) := e^{i2\pi x\cdot\xi}$ for the standard trigonometric polynomials of the torus.

The aim of this paper is to provide a sharp sufficient condition for the membership of pseudo-differential operators in the Schatten classes on $L^2(\mathbb{T}^n)$, for lower-order exponents. More precisely, we consider the case of the spectral index $0 < p \leq 2$.[1]

We recall that for any $p > 0$, a compact operator $T : L^2(\mathbb{T}^n) \to L^2(\mathbb{T}^n)$ belongs to the Schatten von-Neumann ideal $\mathscr{S}_p(L^2(\mathbb{T}^n))$, if the sequence of its singular values $\{s_n(T)\}_{n\in\mathbb{N}}$ (formed by the eigenvalues of the operator $\sqrt{T^*T}$) belongs to $\ell^p(\mathbb{N})$, that is, if $\sum_{n=1}^{\infty} s_n(T)^p < \infty$.

The problem of finding necessary and sufficient conditions for the membership of pseudo-differential operators on manifolds M (on in the Euclidean case) has a long tradition. Indeed, as it was pointed out in [4] if one considers the problem of classifying pseudo-differential in the Schatten von-Neumann classes $\mathscr{S}_r(L^2(\mathbb{R}^n))$, whose symbols belong to the Hörmander classes $S^m_{\rho,\delta}(\mathbb{R}^n)$, the Beals-Fefferman classes $S^{M_1,M_2}_{\Phi,\phi}(\mathbb{R}^n)$, or the Hörmander classes $S(m,g)$ the subject becomes more classical, see Sect. 2.3 for details.

2.2 Schatten Classes on the Torus for Small Exponents

Here, we consider the case of the torus $M = \mathbb{T}^n$, motivated by the recent work [4] where the authors have investigated necessary and sufficient conditions for the membership of pseudo-differential operators on compact Lie groups, even in the particular case of the torus $M = \mathbb{T}^n$ where several atypical properties and exotic examples for Schatten properties were presented, see Subsection 1.3 of [4] for details.

Although in general non-commutative compact Lie groups the characterisation in terms of a symbol criterion of the membership of non-invariant pseudo-differential operators in Schatten von-Neuman ideals is still an open problem,[2] with the more recent advances discussed in [4] (with necessary and sufficient conditions in the

[1] When $p = \infty$, one usually writes $\mathscr{S}_\infty(L^2(\mathbb{T}^n)) = \mathscr{B}(L^2(\mathbb{T}^n))$ for the algebra of bounded linear operators on $L^2(\mathbb{T}^n)$.

[2] with a complete characterisation under the ellipticity condition, when the operators are Fourier multipliers, or when the index p of the Schatten class is an integer. See [4] for the state of the art of this problem in the setting on compact manifolds and for some open problems on the subject.

setting of the global Hörmander classes) and the references therein (see e.g. [5, 7–19]), here we consider the problem when $0 < p \leq 2$ in the case of the torus $\mathbb{T}^n$. The following is our main theorem.

Theorem 2.1 *Let $A : L^2(\mathbb{T}^n) \to L^2(\mathbb{T}^n)$ be a bounded pseudo-differential operator with symbol $a := a(x, \xi)$. Then, for any $0 < p \leq 2$, if*

$$\|a\|^p_{\ell^p(\mathbb{Z}^n, L^2(\mathbb{T}^n))} := \sum_{\xi \in \mathbb{Z}^n} \|a(\cdot, \xi)\|^p_{L^2(\mathbb{T}^n)} < \infty, \tag{2.1}$$

then the operator A belongs to the Schatten class $\mathcal{S}_p(L^2(\mathbb{T}^n))$ and the inequality

$$\|A\|_{\mathcal{S}_p(L^2(\mathbb{T}^n))} \leq \|a\|_{\ell^p(\mathbb{Z}^n, L^2(\mathbb{T}^n))}$$

holds.

Our proof of Theorem 2.1 will be presented in an upcoming work as well as an analysis for this problem when $p > 2$. Below, we discuss some historical aspects of the subject. We also note that the case $p = 2$ in Theorem 2.1 is a very well-known consequence of the Plancherel theorem. We observe that in the case where A is a Fourier multiplier, that is, its symbol is independent of $x \in \mathbb{T}^n$, our theorem is sharp in the sense that (2.1) is a necessary and sufficient condition in order to deduce that $A \in \mathcal{S}_p$. However, in the non-invariant case, one can apply Proposition 23 of [1] to prove that there exists a positive operator $A \in \mathcal{S}_p(L^2(\mathbb{T}^n))$, with $0 < p \leq 2$, such that $a \notin \ell^p(\mathbb{Z}^n, L^2(\mathbb{T}^n))$.

2.3 State-of-the-Art

We finish this note by making a brief overview of the *state-of-the-art* about the membership of pseudo-differential operators on Schatten classes. The problem of finding sufficient conditions for Schatten properties of pseudo-differential operators can be traced back to Hörmander [20], where he observed that the distribution of the eigenvalues of an elliptic pseudo-differential operator $A = \mathrm{Op}^w(a)$, and then its Schatten properties, is encoded in terms of the level sets of the symbol a. Indeed, he showed that the spectral formula

$$N(\lambda) \sim \int_{\{(x,\xi) \in T^*M : a(x,\xi) < \lambda\}} dx\, d\xi,$$

holds for any $\lambda > 0$. Here, we have denoted by $N(\lambda)$ the Weyl eigenvalue counting function of the operator A. To our knowledge, the first results of this type can be traced back to H. Weyl who considered the case of second-order differential operators, and to Courant [20, Page 297]. In [20, Theorem 3.9], Hörmander proved the following sufficient condition

$$m \in L^1(\mathbb{R}^{2n}),\ a \in S(m,g),\ \text{implies that } \mathrm{Op}^w(a) \in \mathcal{S}_1(L^2(\mathbb{R}^n)),$$

with the metric g and the weight function m satisfying suitable conditions. In [21], Hörmander also characterised the L^2 continuity of Weyl operators with the symbols in $S(m,g)$ as follows

$$\{\mathrm{Op}^w(a) : a \in S(m,g)\} \subseteq \mathcal{S}_\infty(L^2(\mathbb{R}^n)) \ \text{ if and only if } \ m \in L^\infty. \tag{2.2}$$

As it was pointed out by J. Toft in [4], by adding some additional conditions on the metric m and on the weight g, Buzano and Nicola in [2], extended (2.2) into

$$\{\mathrm{Op}^w(a) : a \in S(m,g)\} \subseteq \mathcal{S}_p(L^2(\mathbb{R}^n)) \text{ if and only if } m \in L^p,$$

for every $p \in [1,\infty]$. In [23], Toft showed that (2.2) still holds true without the additional assumptions on m and g in [2].

In [3] the Schatten characterization

$$\mathrm{Op}^w(a) \in \mathcal{S}_p(L^2(\mathbb{R}^n)) \ \text{ if and only if } \ a \in L^p,$$

provided $a \in S(m,g)$ and $h_g^N m \in L^p$ for some $N \geq 0$. Here, $h_g \leq 1$ is the Planck's function. For further Schatten properties of pseudo-differential operators on $\mathbb{R}^n$, see e.g. [24, 25] and for Schatten properties on compact manifolds, we refer the reader to the works [5, 7–19].

Acknowledgments The author was supported by the FWO Odysseus 1 grant G.0H94.18N: Analysis and Partial Differential Equations, by the Methusalem programme of the Ghent University Special Research Fund (BOF) (Grant number 01M01021).

The author thanks Michael Ruzhansky, Joachim Toft, Karel Van Bockstal, and Julio Delgado for several discussions about some projects related to this note.

References

1. Bingyang, H., Khoi, L.H., Zhu, K.: Frames and operators in Schatten classes. Houston J. Math. **41**(4), 1191–1219 (2015)
2. Buzano, E., Nicola, N.: Pseudo-differential operators and schatten-von Neumann classes. In: Boggiatto, P., Ashino, R., Wong, M.W. (eds.) Advances in Pseudo-Differential Operators, Proceedings of the Fourth ISAAC Congress, Operator Theory: Advances and Applications. Birkhäuser Verlag, Basel (2004)
3. Buzano, E., Toft, J.: Schatten-von Neumann properties in the Weyl calculus. J. Funct. Anal. **259**(12), 3080–3114 (2010)
4. Cardona, D., Chatzakou, M., Ruzhansky, M., Toft, J.: Schatten-von Neumann properties for Hörmander classes on compact Lie groups (2023). arXiv:2301.04044

5. Cardona, D., Del Corral, C.: The Dixmier trace and the noncommutative residue for multipliers on compact manifolds. In: Advances in Harmonic Analysis and Partial Differential Equations. Trends in Mathematics, pp. 121–163. Birkhäuser, Cham (2020)
6. Cardona, D., Ruzhansky, M.: Subelliptic pseudo-differential operators and Fourier integral operators on compact Lie groups. MSJ Memoir. Math. Soc. Japan (2020). arXiv:2008.09651
7. Cardona, D., Delgado, J., Ruzhansky, M.: A note on the local Weyl formula on compact Lie groups. J. Lie Theory (2022). arXiv:2210.00311
8. Chatzakou, M., Delgado, J., Ruzhansky, M.: On a class of anharmonic oscillators. J. Math. Pures Appl. **153**(9), 1–29 (2021)
9. Chatzakou, M., Delgado, J., Ruzhansky, M.: On a class of anharmonic oscillators II. General case. Bull. Sci. Math. **180**, 103196, 22pp. (2022)
10. Delgado, J.: Trace formulas for nuclear operators in spaces of Bochner integrable functions. Monatsh. Math. **172**(3–4), 259–275 (2013)
11. Delgado, J.: On the r-nuclearity of some integral operators on Lebesgue spaces. Tohoku Math. J. **67**(1), 125–135 (2015)
12. Delgado, J., Ruzhansky, M.: Schatten classes on compact manifolds: kernel conditions. J. Funct. Anal. **267**(3), 772–798 (2014)
13. Delgado, J., Ruzhansky, M.: Kernel and symbol criteria for Schatten classes and r-nuclearity on compact manifolds. C. R. Math. Acad. Sci. Paris **352**(10), 779–784 (2014)
14. Delgado, J., Ruzhansky, M.: L^p-nuclearity, traces, and Grothendieck-Lidskii formula on compact Lie groups. J. Math. Pures Appl. **102**(1), 153–172 (2014)
15. Delgado, J., Ruzhansky, M.: Schatten classes and traces on compact groups. Math. Res. Lett. **24**(4), 979–1003 (2017)
16. Delgado, J., Ruzhansky, M.: The bounded approximation property of variable Lebesgue spaces and nuclearity. Math. Scand. **122**(2), 299–319 (2018)
17. Delgado, J., Ruzhansky, M.: Fourier multipliers, symbols, and nuclearity on compact manifolds. J. Anal. Math. **135**(2), 757–800 (2018)
18. Delgado, J., Ruzhansky, M.: Schatten-von Neumann classes of integral operators. J. Math. Pures Appl. **154**(9), 1–29 (2021)
19. Delgado, J., Ruzhansky, M., Wang, B.: Approximation property and nuclearity on mixed-norm Lp, modulation and Wiener amalgam spaces. J. Lond. Math. Soc. **94**(2), 391–408 (2016)
20. Hörmander, L.: On the asymptotic distribution of the eigenvalues of pseudodifferential operators in $\mathbb{R}^n$. Ark. Mat. **17**, 297–313 (1979)
21. Hörmander, L.: The Weyl calculus of pseudo-differential operators. Commun. Pure Appl. Math. **32**, 359–443 (1979)
22. Ruzhansky, M., Turunen, V.: Pseudo-Differential Operators and Symmetries: Background Analysis and Advanced Topics. Birkhäuser-Verlag, Basel (2010)
23. Toft, J.: Schatten-von Neumann properties in the Weyl calculus, and calculus of metrics on symplectic vector spaces. Ann. Global Anal. Geom. **30**(2), 169–209 (2006)
24. Toft, J.: Continuity and Schatten properties for pseudo-differential operators on modulation spaces. Modern Trends in Pseudo-Differential Operators. Operator Theory: Advances and Applications, vol. 172, pp. 173–206. Birkhäuser, Basel (2007)
25. Toft, J.: Schatten properties, nuclearity and minimality of phase shift invariant spaces. Appl. Comput. Harmon. Anal. **46**(1), 154–176 (2019)

Chapter 3
Log-Sobolev and Nash Inequalities on Graded Groups

Marianna Chatzakou

Abstract In this work, we give a brief exposition of the results on important functional inequalities in the setting of graded Lie groups that appeared in the joint work Chatzakou et al. (Logarithmic Sobolev inequalities on Lie groups (2021). arXiv:2106.15652) with A. Kassymov and M. Ruzhansky. These inequalities include the logarithmic Sobolev inequality on graded groups, its semi-probabilistic equivalent on stratified groups, referred to as the "semi-Gaussian" logarithmic Sobolev inequality, and the Nash inequality on graded groups.

3.1 Introduction

It is well-known that in Euclidean spaces $\mathbb{R}^n$ the Sobolev inequality

$$\|u\|_{L^{p^*}(\mathbb{R}^n)} \leq C\|\nabla u\|_{L^p(\mathbb{R}^n)},$$

where $1 < p < n$ and $p^* = \frac{np}{n-p}$, is a very important tool in PDEs and variational problems, while the study of the best constant $C(n, p)$ plays an important role, see [11] and [1]. Later on Folland and Stein [7] extended the aforementioned inequality to any stratified group $\mathbb{G}$, with the latter taking the from:

$$\|u\|_{L^{p^*}(\Omega)} \leq C\left(\int_\Omega |\nabla_H u|^p dx\right)^{\frac{1}{p}}, \ 1 < p < Q, \ p^* = \frac{Qp}{Q-p}, \quad u \in C_0^\infty(\mathbb{G}),$$

for any $\Omega \subset \mathbb{G}$ open subset in $\mathbb{G}$, where ∇_H stands for the horizontal gradient and Q for the homogeneous dimension of the group $\mathbb{G}$.

M. Chatzakou (✉)
Department of Mathematics: Analysis, Logic and Discrete Mathematics, Ghent University, Ghent, Belgium
e-mail: marianna.chatzakou@ugent.be

M. Ruzhansky, K. Van Bockstal (eds.), *Extended Abstracts 2021/2022*, Research Perspectives Ghent Analysis and PDE Center 2,
https://doi.org/10.1007/978-3-031-42539-4_3

On the other hand, still in the Euclidean setting, we have the logarithmic Sobolev inequality of the following form:

$$\int_{\mathbb{R}^n} \frac{|u|^p}{\|u\|_{L^p(\mathbb{R}^n)}^p} \log\left(\frac{|u|^p}{\|u\|_{L^p(\mathbb{R}^n)}^p}\right) dx \leq \frac{n}{p} \log\left(C \frac{\|\nabla u\|_{L^p(\mathbb{R}^n)}^p}{\|u\|_{L^p(\mathbb{R}^n)}^p}\right). \tag{3.1}$$

We refer to [13] for the case $p = 2$ and to [5] for a general overview of the previous inequality.

The analogue of the logarithmic Sobolev inequality for the case $p = 2$ with respect to the Gaussian measure $d\mu(x) = (2\pi)^{-n/2}e^{-|x|^2/2}dx$ reads as follows:

$$\int_{\mathbb{R}^n} |u(x)|^2 \log\left(\frac{|u(x)|}{\|u\|_{L^2(\mu)}}\right) d\mu(x) \leq \int_{\mathbb{R}^n} |\nabla u(x)|^2 d\mu(x), \tag{3.2}$$

for any suitable u, and was shown by Gross in [8]. The importance of the inequality (3.2) is that the normalization constant $(2\pi)^{-n/2}$ is uniformly bounded for any dimension n which allows it to pass to infinite dimensional spaces. Inequality (3.2) is also known as the "Gaussian inequality".

For the purposes of the current article, it is important to note that the logarithmic Sobolev inequality (3.1) for the case $p = 2$ is equivalent to the Nash inequality which can be stated as

$$\|u\|_{L^2(\mathbb{R}^n)}^{1+n/2} \leq C(n)\|u\|_{L^1(\mathbb{R}^n)}\|\nabla u\|_{L^2(\mathbb{R}^n)}^{n/2}, \tag{3.3}$$

as well as to the Gross inequality given in (3.2). In this work, we show that this is true in more generality. To be more precise let us summarize the results presented in the current work.

1. The L^2-version of the logarithmic Sobolev inequality in the setting of graded groups implies the Nash in the setting;
2. The L^2-version of the logarithmic Sobolev inequality in the setting of stratified groups is equivalent to the "semi-Gaussian" logarithmic Sobolev inequality in the stratified setting.

Of course, proving the above results requires proving the logarithmic Sobolev inequality in the setting of graded groups (and consequently also in the setting of stratified groups). Let us point out that we use the term "semi-Gaussian" for the counterpart of the Gaussian inequality in our setting, to indicate that the appearing measure is not a probability measure on the whole manifold.

We refer to the work [4] for the detailed proof of the above results, as well as for the corresponding weighted versions, and their non-refined formulation for general Lie groups. Additionally, in [2] one can find the limiting case for $p = 1$ of the logarithmic Sobolev inequality, known as "Shannon inequality". Regarding other functional inequalities in the setting of general Lie groups, we refer to the work [3] where the authors prove logarithmic Hardy and Rellich inequalities in the setting, including their refined version for the setting of graded Lie groups.

3.2 Preliminary Notions on Graded Lie Groups

In this section, we introduce the necessary notions and definitions that are related to the present exposition of results. We note that the analysis of such groups started with the monograph of Folland and Stein [7]. Another comprehensive analysis of such groups with additional incorporated notions was done by Fischer and Ruzhansky in [6], and is the one that we follow here. For more details, we refer the interested reader to [6].

Let $\mathbb{G}$ be a nilpotent Lie group that is connected and simply connected and let $\mathfrak{g}$ be its Lie algebra. If $\mathfrak{g}$ admits a vector space decomposition of the form

$$\mathfrak{g} = \oplus_{j=1}^{\infty} V_j, \quad \text{s.t. } [V_i, V_j] \subset V_{i+j}, \tag{3.4}$$

where all but finitely many of the V_j's are 0, then $\mathfrak{g}$ as well as $\mathbb{G}$ are call graded. The vector space V_j shall be called the jth statum of the Lie algebra.

Trivially, the vector fields $X_1, \cdots, X_N$ viewed as elements of the vector space $\mathfrak{g}$ with the binary operation $[\cdot,\cdot] : \mathfrak{g} \times \mathfrak{g} \to \mathfrak{g}$ defined as $[X_i, X_j] := X_i X_j - X_j X_1$, span the vector space $\mathfrak{g}$; i.e., we can write

$$\mathfrak{g} = \text{span}\{X_1, \cdots, X_N\},$$

The graded Lie algebras whose first stratum $V_1 = \{X_1, \cdots, X_{N_1}\}$ generates after iterative commutators of $X_1, \cdots, X_{N_1}$ the whole of $\mathfrak{g}$ are called stratified. Consequently, the corresponding Lie groups are also called stratified.

A graded Lie group $\mathbb{G}$ is naturally a homogeneous Lie group when, assuming that $\mathbb{G}$ has topological dimension N, $\mathbb{G}$ is equipped with the natural dilations $\delta_r : \mathbb{R}^N \to \mathbb{R}^N$ of the form

$$\delta_r(x) := (r^{\nu_1} x_1, \ldots, r^{\nu_N} x_N), \ \nu_1, \ldots, \nu_n > 0.$$

A mapping δ_r for $r > 0$ is an automorphism of the Lie group $\mathbb{G}$ and, of course also of the Lie algebra $\mathfrak{g}$. The powers ν_j are called weights and their sum gives the homogeneous dimension of $\mathbb{G}$; i.e., if Q denotes the homogeneous dimension of $\mathbb{G}$, then we have

$$Q = \nu_1 + \cdots + \nu_N.$$

A homogeneous Lie group $\mathbb{G}$ can be equipped with a homogeneous quasi-norm $|\cdot|: \mathbb{G} \to [0,\infty)$ that is a continuous non-negative function satisfying the properties:

1. $|x| = |x^{-1}|$ for all $x \in \mathbb{G}$;
2. $|\lambda x| = \lambda |x|$ for all $x \in \mathbb{G}$;
3. $|x| = 0$ if and only if $x = 0$,

and does not necessarily satisfy the triangle inequality with a constant equal to 1.

Homogeneous Lie groups are nilpotent Lie groups that are characterised as being the nilpotent Lie groups that admit a Rockland operator; see Proposition 4.1.3 in [FR16]. A Rockland operator, usually denoted by $\mathcal{R}$, is roughly speaking a generalisation of the sub-Laplacian $\mathcal{L} = X_1^2 + \cdots + X_{N_1}^2$ operator on stratified groups, or even simpler, of the usual Laplace operator on $\mathbb{R}^n$ when the derivatives can be of higher order. An example of such a generalisation of the sub-Laplacian $\mathcal{L}$ in the setting of a graded Lie group is the Rockland operator

$$\mathcal{R} = (-1)^{\frac{\nu_0}{\nu_j}} \sum_{j=1}^{N} c_j X_j^{2\frac{\nu_0}{\nu_j}},$$

where the involved vector fields X_j appear in different orders that are determined by the dilation structure of $\mathbb{G}$. Thus, Rockland operators are differential operators that are homogeneous (with respect to the dilations) of positive degree, left-invariant and hypo-elliptic.

Finally, to introduce the Sobolev spaces in our setting, let us fix a graded Lie group $\mathbb{G}$ and a Rockland operator $\mathcal{R}$ of degree ν. Extending the Euclidean concept of Sobolev spaces, the operator $\mathcal{R}$ gives rise to the Sobolev spaces in $\mathbb{G}$. Particularly, the homogeneous and in-homogeneous Sobolev spaces are the Banach spaces endowed with the norms

$$\|f\|_{\dot{L}_a^p(\mathbb{G})} := \|\mathcal{R}^{\frac{a}{\nu}} f\|_{L^p(\mathbb{G})},$$

and

$$\|f\|_{L_a^p(\mathbb{G})} := (\|f\|_{L^p(\mathbb{G})} + \|\mathcal{R}^{\frac{a}{\nu}} f\|_{L^p(\mathbb{G})})^{\frac{1}{p}},$$

respectively, for some suitable f that makes the above norms finite.

3.3 Auxiliary and Main Results

As mentioned earlier, the below results that are given here without proof are particular cases of the ones presented in [4]. Before starting the exposition of the selected results, let us mention the two key results that consist of the basic ideas for the development of the corresponding proofs.

3.3.1 Auxiliary Results

The first auxiliary result is the logarithmic Hölder inequality on general measure spaces and its proof makes use of the Hölder inequality on general measure spaces. The rest of the results in the present section are stated for a graded Lie group $\mathbb{G}$ of homogeneous dimension Q and for positive Rockland operators $\mathcal{R}_1$ and $\mathcal{R}_2$ of homogeneous dimensions ν_1 and ν_2, respectively that are assumed to be fixed in the sequel.

Lemma 3.1 *Let X be a measure space and let $f \in L^p(X) \cap L^q(X) \setminus \{0\}$ with $1 < p < q < \infty$. The next inequality holds true:*

$$\int_X \frac{|f|^p}{\|f\|^p_{L^p(X)}} \log\left(\frac{|f|^p}{\|f\|^p_{L^p(X)}}\right) dx \leq \frac{q}{q-p} \log\left(\frac{\|f\|^p_{L^q(X)}}{\|f\|^p_{L^p(X)}}\right).$$

The next result is known as the Gagliardo-Nirenberg inequality and can be found in [10].

Theorem 3.1 *For $a_1 > a_2 \geq 0$, $p \in \left(1, \frac{Q}{a_1}\right)$ and $\frac{Qp}{Q-a_2p} < q < \frac{Qp}{Q-a_1p}$, we have*

$$\int_{\mathbb{G}} |f(x)|^q dx$$
$$\leq C\left(\int_{\mathbb{G}} |\mathcal{R}_1^{\frac{a_1}{\nu_1}} f(x)|^p dx\right)^{\frac{Q(q-p)-a_2pq}{(a_1-a_2)p^2}} \left(\int_{\mathbb{G}} |\mathcal{R}_2^{\frac{a_2}{\nu_2}} f(x)|^p dx\right)^{\frac{a_1pq-Q(q-p)}{(a_1-a_2)p^2}}, \tag{3.5}$$

for any $f \in \dot{L}^p_{a_1,a_2}(\mathbb{G})$, where the optimal constant $C = C(\mathcal{R}_1, \mathcal{R}_2, a_1, a_2, p, q)$ depends on the indicated parameters and is explicitly given in [10], see also [4].

Before moving on to present the two main results of this work, let us state the (unweighted) logarithmic Sobolev inequality in the setting of a graded Lie group, with the latter inequality giving rise to the L^2-version of the logarithmic Sobolev inequality in the stratified case. For the proofs of the consecutive results, we refer to [4].

Theorem 3.2 (Unweighted Logarithmic Sobolev Inequality) *Let $1 < p < \infty$ and $0 < a < \frac{Q}{p}$. Then for any $f \neq 0$ we have*

$$\int_{\mathbb{G}} \frac{|f|^p}{\|f\|^p_{L^p(\mathbb{G})}} \log\left(\frac{|f|^p}{\|f\|^p_{L^p(\mathbb{G})}}\right) dx \leq \frac{Q}{ap} \log\left(A \frac{\|f\|^p_{\dot{L}^p_a(\mathbb{G})}}{\|f\|^p_{L^p(\mathbb{G})}}\right), \tag{3.6}$$

with $\|f\|_{\dot{L}^p_a(\mathbb{G})} \equiv \|\mathcal{R}_1^{\frac{a}{\nu_1}} f\|_{L^p(\mathbb{G})}$, *for*

$$A = \inf_{q:\, p<q<\frac{Qp}{Q-ap}} C^{\frac{ap^2}{Q(q-p)}}, \tag{3.7}$$

where C *is the best constant given in* (3.5) *for* $a_1 = a, a_2 = 0$. *The next result is the* L^2*-version of the logarithmic Sobolev inequality in the stratified setting, which is a special case of Theorem 3.2.*

Corollary 3.1 *Assume that* $\mathbb{G}$ *is a stratified group. For any* f *such that* $\|f\|_{L^2(\mathbb{G})} = 1$, *the following* L^2*-logarithmic Sobolev inequality is satisfied*

$$\int_{\mathbb{G}} |f|^2 \log|f|\, dx \leq \frac{Q}{4} \log\left(A \int_{\mathbb{G}} |\nabla_H f|^2\, dx \right), \tag{3.8}$$

where

$$A = \inf_{2<q<\frac{2Q}{Q-2}} C^{\frac{4}{Q(q-2)}},$$

for C *being the constant given in* (3.5) *for* $a_1 = 1, a_2 = 0$ *and* $p = 2$ *when* $\mathcal{R}_1$ *is chosen to be the sub-Laplacian* $\mathcal{L}$ *on* $\mathbb{G}$.

3.3.2 Main Results

Theorem 3.3 (Nash Inequality) *Let* $\mathbb{G}$ *be a graded Lie group of homogeneous dimension* Q. *Then for* $0 < a < \frac{Q}{2}$ *Nash's inequality reads as*

$$\|f\|_{L^2(\mathbb{G})}^{2+\frac{4a}{Q}} \leq A_2 \|f\|_{L^1(\mathbb{G})}^{\frac{4a}{Q}} \|f\|_{\dot{L}^2_a(\mathbb{G})}^2, \tag{3.9}$$

where A_2 *is the given by* (3.7) *for* $p = 2$. *In particular, if* $\mathbb{G}$ *is a stratified group, then Nash's inequality becomes*

$$\|u\|_{L^2(\mathbb{G})}^{2+\frac{4}{Q}} \leq A_2 \|f\|_{L^1(\mathbb{G})}^{\frac{4}{Q}} \|\nabla_H f\|_{L^2(\mathbb{G})}^2, \tag{3.10}$$

where ∇_H *stands for the horizontal gradient on* $\mathbb{G}$.

Let us point out, that in the case where $\mathbb{G}$ is a stratified group, Nash's inequality (3.10) generalises the one shown by Varopoulos, Saloff-Coste and Coulhon in [12] which is exactly (3.10) for $a = 1$. Inequality (3.10) with the addition of parameter a already extends the original Nash inequality [9] in the Euclidean setting $\mathbb{R}^n$.

At this point, let us introduce a decomposition of an element x in the stratified group $\mathbb{G} \cong \mathbb{R}^N$: we shall write $x = (x', x'') \in \mathbb{R}^{N_1} \times \mathbb{R}^{N-N_1}$ for the decomposition of the element x as two elements, with x' belonging in the sub-manifold $\mathbb{R}^{N_1}$ of topological dimension the dimension of the first stratum of $\mathfrak{g}$ and x'' lying in the remaining manifold.

Theorem 3.4 *Let $\mathbb{G} \cong \mathbb{R}^N$ be a stratified group with homogeneous dimension Q. Then the following "semi-Gaussian" log-Sobolev inequality is satisfied*

$$\int_{\mathbb{G}} |g|^2 \log |g| \, d\mu \leq \int_{\mathbb{G}} |\nabla_H g|^2 \, d\mu \,, \tag{3.11}$$

for any g such that $\|g\|_{L^2(\mu)} = 1$, where $\mu = \mu_1 \otimes \mu_2$, for μ_2 being the Lebesgue measure dx'' on $\mathbb{R}^{N-N_1}$ and μ_1 being the Gaussian measure on $\mathbb{R}^{N_1}$ given by $d\mu_1 = \gamma e^{-\frac{|x'|^2}{2}} dx'$, for $x' \in \mathbb{R}^{N_1}$, where $|x'|$ is the Euclidean norm of x', and

$$\gamma := (4^{-1} Q e^{\frac{2N_1}{Q}-1} A)^{Q/2} \,, \tag{3.12}$$

for A as in (3.1). *Moreover, inequality* (3.11) *imply inequality* (3.8) *and vice versa.*

Finally, we want to point out that in the Euclidean setting inequality (3.11) boils down to the classical Gaussian inequality (3.2); indeed, see Section 7 in [4], in this case, the first stratum covers the whole manifold, while the normalisation constant can be explicitly computed as $\gamma = (2\pi)^{-\frac{n}{2}}$ as desired.

Acknowledgments Marianna Chatzakou is supported by the FWO Odysseus 1 grant G.0H94.18N: Analysis and Partial Differential Equations, and by the Methusalem programme of the Ghent University Special Research Fund (BOF) (Grant number 01M01021), and is a postdoctoral fellow of the Research Foundation—Flanders (FWO) under the postdoctoral grant No 12B1223N.

References

1. Aubin, T.: Problèmes isoperimétriques et espaces de Sobolev. J. Differ. Geom. **11**, 573–598 (1976)
2. Chatzakou, M., Kassymov, A., Ruzhansky, M.: Anisotropic Shannon inequality (2021). arXiv:2106.14182
3. Chatzakou, M., Kassymov, A., Ruzhansky, M.: Logarithmic Hardy inequalities on Lie groups (2021). arXiv:2107.04874
4. Chatzakou, M., Kassymov, A., Ruzhansky, M.: Logarithmic Sobolev inequalities on Lie groups (2021). arXiv:2106.15652
5. Del Pino, M., Dolbeault, J.: The optimal Euclidean L^p-Sobolev logarithmic inequality. J. Funct. Anal. **197**(1), 151–161 (2003)
6. Fischer, V., Ruzhansky, M.: Quantization on Nilpotent Lie Groups. Progress in Mathematics, vol. 314. Birkhäuser, Basel (2016)

7. Folland, G., Stein, E.M.: Hardy Spaces on Homogeneous Groups. Mathematical Notes, vol. 28. Princeton University Press, Princeton (1982)
8. Gross, L.: Logarithmic Sobolev inequalities. Am. J. Math. **97**(4), 1061–1083 (1975)
9. Nash, J.: Continuity of solutions of parabolic and elliptic equations. Am. J. Math. **80**(4), 931–954 (1958)
10. Ruzhansky, M., Tokmagambetov, N., Yessirkegenov, N.: Best constants in Sobolev and Gagliardo-Nirenberg inequalities on graded groups and ground states for higher order nonlinear subelliptic equations. Calc. Var. Partial Differ. Equ. **59**(175), 23pp. (2020)
11. Talenti, G.: Best constant in Sobolev inequality. Ann. di Mat. Pura ed Appl. **110**(1), 353–372 (1976)
12. Varopoulos, N.T., Saloff-Coste, L., Coulhon, T.: Analysis and Geometry on Groups. Cambridge University Press, Cambridge (1993)
13. Weissler, F.B.: Logarithmic Sobolev inequalities for the heat-diffusion semigroup. Trans. Am. Math. Soc. **237**, 255–269 (1978)

Chapter 4
One-Sided Hardy-Littlewood Maximal Function on Generalised Lorentz Spaces

Abhishek Ghosh and Parasar Mohanty

Abstract We provide a proof of the equivalence between the weak- and strong-type boundedness of the one-sided maximal function on $\mathbb{R}$ on weighted Lorentz spaces. As a consequence, we obtain a characterisation of the weights.

4.1 Introduction and Preliminaries

For a locally integrable function f on $\mathbb{R}$, the one-sided Hardy-Littlewood maximal function $M^+ f$ is defined by

$$M^+ f(x) = \sup_{h>0} \frac{1}{h} \int\limits_{x}^{x+h} |f(y)|\, dy.$$

Weighted inequalities for one-sided Hardy-Littlewood maximal functions were pioneered by E. Sawyer in [9]. We also refer to the influential works [5, 7] and references therein for historical developments. This article deals with the one-sided Hardy-Littlewood maximal function on generalised Lorentz spaces. One-sided Hardy-Littlewood maximal function maps the Lorentz space $L^{p,q}(\mathbb{R})$ to itself for $1 < p < \infty, 1 \leq q < \infty$ or $p = q = \infty$. This is a consequence of the classical Hardy inequality.

The theory of generalised Lorentz spaces and maximal functions was systematically developed in [3] by M. Carro et al., we also refer to the articles [2, 4, 8] for early pioneering works. Recently, in [1], E. Agora et al. provided an elegant proof of

A. Ghosh (✉)
Tata Institute of Fundamental Research, Centre for Applicable Mathematics, Bangalore, India
e-mail: abhi21@tifrbng.res.in

P. Mohanty
Department of Mathematics and Statistics, Indian Institute of Technology Kanpur, Kanpur, India
e-mail: parasar@iitk.ac.in

M. Ruzhansky, K. Van Bockstal (eds.), *Extended Abstracts 2021/2022*,
Research Perspectives Ghent Analysis and PDE Center 2,
https://doi.org/10.1007/978-3-031-42539-4_4

the equivalence of the weak- and strong-type boundedness of the Hardy-Littlewood maximal function in this setting. This leads us to investigate the boundedness of the one-sided Hardy-Littlewood maximal function on generalised Lorentz spaces. Referring [8], we say $u \in A^+(p,q)$ if there is a constant $C > 0$ such that

$$\|\chi_{(x,y)}\|_{L^{p,q}(u)}\|\chi_{(y,z)}u^{-1}\|_{L^{p',q'}(u)} \leq C(z-x), \tag{4.1}$$

holds for all $x, y, z \in \mathbb{R}$ with $x < y < z$. When $p = q$, we simply recover A_p^+ weights. It was proved in [8] that $\|M^+ f\|_{L^{p,q}(u)} \leq C\|f\|_{L^{p,q}(u)}$ if and only if $u \in A^+(p,q) \iff u \in A_p^+$ for $1 < p < \infty, 1 < q \leq \infty$. As the $A^+(p,q)$ condition is necessary for M^+ to be bounded from $L^{p,q}(u)$ to $L^{p,\infty}(u)$, the above theorem produces an equivalence between the weak and strong type boundedness of M^+, i.e., M^+ is bounded from $L^{p,q}(u)$ to $L^{p,\infty}(u)$ if and only if M^+ is bounded from $L^{p,q}(u)$ to itself for $1 < p < \infty$ and $1 < q \leq \infty$. However, the case $q = 1$ was neither considered in [8], nor could we find it in any other literature, so we record it here as a consequence of our result.

Theorem 4.1 *Let $1 < p < \infty$ and $1 \leq q < \infty$ and u be a weight on $\mathbb{R}$. Then M^+ is bounded from $L^{p,1}(u)$ to $L^{p,q}(u)$ if and only if $u \in A_p^+$.*

To state our results, we need the following definitions and preliminaries. Let w be a weight on $\mathbb{R}^+$ and $W(t) := \int_0^t w(s)ds$. For a weight u on $\mathbb{R}$, we denote $\lambda_f^u(t) := u(\{x : |f(x)| > t\})$. For $0 < p, r < \infty$, the generalised Lorentz spaces are defined as follows

$$\Lambda_u^{p,r}(w) := \left\{f : \|f\|_{\Lambda_u^{p,r}(w)} = \left(p\int_0^\infty t^{r-1}W^{\frac{r}{p}}(\lambda_f^u(t))dt\right)^{\frac{1}{r}} < \infty\right\},$$

$$\Lambda_u^{p,\infty}(w) := \{f : \|f\|_{\Lambda^{p,\infty}(w)} = \sup_{t>0} tW(\lambda_f^u(t))^{\frac{1}{p}} < \infty\}.$$

We simply denote $\Lambda_u^{p,p}(w)$ as $\Lambda_u^p(w)$. Classical $L^{p,r}(u)$ spaces can be recovered as $\Lambda_u^{p,r}(1)$ or $\Lambda_u^r(t^{\frac{r}{p}-1})$. We summarise a few facts about these spaces. For more details on these spaces, we refer to [3] and references therein. We say that a weight w on $\mathbb{R}^+$ satisfies the condition Δ_2 if $W(2r) \lesssim W(r)$. Throughout this article, we assume $w \in \Delta_2$ and, under this hypothesis, $\Lambda_u^{p,r}(w)$ is a quasi-normed space (see Lemma 2.2.10 in [3]). Also, $\|.\|_{\Lambda_u^{p,r}(w)}$ decreases as r increases and $\Lambda_u^{p,q_0}(w) \subset \Lambda_u^{p,q_1}(w)$ continuously for $0 < q_0 \leq q_1 \leq \infty$. $\Lambda_u^\infty(w)$ is identified with $L^\infty(\mathbb{R})$. Now, we are in a position to state our main result, which extends the results in [1] to the one-sided setting.

Theorem 4.2 *Let $1 < p < \infty$ and $1 < q < \infty$ and u be a weight on $\mathbb{R}$. Then the following are equivalent*

(i) M^+ *is bounded from* $\Lambda_u^{p,q}(w)$ *to* $\Lambda_u^{p,q}(w)$,
(ii) M^+ *is bounded from* $\Lambda_u^{p,q}(w)$ *to* $\Lambda_u^{p,\infty}(w)$.

We are only going to prove (ii) implies (i) as the other direction is obvious. In the next result, we concentrate on the characterisation of weights for which M^+ is bounded on generalised Lorentz spaces.

Definition 4.1 Let w be a weight on $\mathbb{R}^+$, u be a weight on $\mathbb{R}$, and $W(t) := \int_0^t w(s)ds$. For $0 < p < \infty$, we say that the pair (w, u) is in the class $\mathcal{Z}_p^+$ if there exist $r \in (0, p)$ and a constant $C > 0$ such that for every finite family of numbers $x_j < y_j < z_j$, and measurable sets $\{F_j\}_{j=1}^M$ with $F_j \subseteq (y_j, z_j)$ for $j = 1, \dots, M$, we have

$$\frac{W(u(\bigcup_j (x_j, y_j)))^{\frac{1}{r}}}{W(u(\bigcup_j F_j))^{\frac{1}{r}}} \leq C \max_j \left(\frac{(z_j - x_j)}{|F_j|} \right). \tag{4.2}$$

Theorem 4.3 *Let $0 < p, q < \infty$. M^+ is bounded from $\Lambda_u^{p,q}(w)$ to $\Lambda_u^{p,q}(w)$ if and only if $(w, u) \in \mathcal{Z}_p^+$.*

As a consequence of Theorems 4.2 and 4.3 we also obtain that for $1 < p < \infty$, M^+ is bounded from $\Lambda_u^p(w)$ to $\Lambda_u^{p,\infty}(w)$ if and only if $(w, u) \in \mathcal{Z}_p^+$.

4.2 Main Results

In this section, we will prove Theorem 4.1 and the equivalence of weak- and strong-type boundedness of the one-sided Hardy-Littlewood maximal function on generalised Lorentz spaces; our approach is motivated by Hunt and Kurtz [6], see also [3].

Theorem 4.4 *Let $0 < p < \infty$ and $0 < s < \infty$. If there is a constant $C > 0$ such that for all measurable sets F the following*

$$\int_0^1 \lambda^{s-1} W(u(\{x : M^+\chi_F(x) > \lambda\}))^{\frac{s}{p}} d\lambda \leq C W(u(F))^{\frac{s}{p}} \tag{4.3}$$

holds, then there exists some $r \in (0, p)$ such that M^+ maps $\Lambda_u^r(w)$ to itself boundedly.

We need the following lemma to prove Theorem 4.4.

Lemma 4.1 *For any measurable set F and $0 < b < 1$, let us denote $F_b := \{x : M^+\chi_F(x) > b\}$. Then for $0 < a < 1$,*

$$\{x : M^+\chi_F(x) > a\, b\} \subseteq \{x : M^+\chi_{F_b}(x) > a\}.$$

Proof Observe the fact that for a.e. $x \in F_b$ we have $M^+\chi_{F_b}(x) = 1 > a$. The proof shall be completed if we can prove that for $x \notin F_b$, $M^+(\chi_F)(x) \le bM^+(\chi_{F_b})(x)$. The remainder of the proof is devoted to proving the claim.

Let $x \notin F_b$ and $h > 0$ be such that $[x, x+h) \cap F \neq \emptyset$. Then $[x, x+h) \cap F_b$ can be written as disjoint union of open intervals $\{(x_j, y_j)\}_j$ with the property that $\frac{1}{y_j - x_j}\int_{x_j}^{y_j} \chi_F \le M^+\chi_F(x_j) \le b$ for all j. This implies that

$$
\begin{aligned}
|[x, x+h) \cap F_b| = \sum_j (y_j - x_j) \ge \frac{1}{b}\sum_j \int_{x_j}^{y_j} \chi_F = \frac{1}{b}\int_{\{z \in [x, x+h): M^+\chi_F(z) > b\}} \chi_F \\
\ge \frac{1}{b}|[x, x+h) \cap F|.
\end{aligned}
$$

Hence for $x \notin F_b$, $M^+(\chi_F)(x) \le bM^+(\chi_{F_b})(x)$ and we are done. □

Now we are in a position to prove Theorem 4.4.

Proof We start with the following claim.
Claim: There exist a $\delta > 0$ and a constant $C > 0$ such that

$$
\|M^+\chi_F\|_{\Lambda_u^{\tilde{p},\tilde{s}}(w)} \le C\|\chi_F\|_{\Lambda_u^{\tilde{p}}(w)}, \tag{4.4}
$$

holds for all measurable sets F, where $\tilde{p} = \frac{p(s-\delta)}{s}$ and $\tilde{s} = s - \delta$.

Recall that for a measurable set F and $t > 0$, $F_t := \{x : M^+(\chi_F)(x) > t\}$. Now, let us compute the norm $\|M^+\chi_F\|_{\Lambda_u^{\tilde{p},\tilde{s}}(w)}$.

$$
\begin{aligned}
\|M^+\chi_F\|^{\tilde{s}}_{\Lambda_u^{\tilde{p},\tilde{s}}(w)} &= \int_0^1 t^{\tilde{s}-1} W^{\frac{\tilde{s}}{\tilde{p}}}(u(F_t))\, dt = \int_0^1 t^{s-\delta-1} W^{\frac{s}{p}}(u(F_t))\, dt \\
&= \sum_{k=0}^{\infty}\int_0^1 t^{s-1} W^{\frac{s}{p}}(u(F_t)) \frac{1}{k!}\left(\delta \log \frac{1}{t}\right)^k dt.
\end{aligned} \tag{4.5}
$$

The claim will be proved if we can show that there is a constant $C > 0$ such that for all measurable sets F and for all $k = 0, 1, \ldots,$ the following holds:

$$
\int_0^1 t^{s-1} W^{\frac{s}{p}}(u(F_t)) \frac{1}{k!}\left(\log \frac{1}{t}\right)^k dt \le C(C)^k W^{\frac{s}{p}}(u(F)). \tag{4.6}
$$

Once we prove (4.6) then multiplying both sides of (4.6) by β^k for some sufficiently small $\beta \in (0, 1)$ and summing it over $k \geq 0$, we get the following for $\delta = \frac{\beta}{C}$.

$$
\begin{aligned}
\|M^+ \chi_F\|^{\tilde{s}}_{\Lambda_u^{\tilde{p},\tilde{s}}(w)} &= \sum_{k=0}^{\infty} \int_0^1 t^{s-1} W^{\frac{s}{p}}(u(F_t)) \frac{1}{k!} \left(\frac{\beta}{C} \log \frac{1}{t} \right)^k dt \\
&\leq C W^{\frac{s}{p}}(u(F)) \sum_{k=0}^{\infty} \beta^k \\
&\lesssim C W^{\frac{s}{p}}(u(F)).
\end{aligned}
$$

Thus, the claim is proved assuming (4.6).

The rest of the proof is devoted to prove (4.6). We use induction. By the hypothesis, the inequality (4.6) is true for all measurable sets F when $k = 0$. Let us assume that (4.6) is true for k and all subsets F and we will show that it is true for $k + 1$. We apply (4.6) for the set F_a where $a \in (0, 1)$, then

$$
I := \int_0^1 t^{s-1} W^{\frac{s}{p}}(u((F_a)_t)) \frac{1}{k!} (\log \frac{1}{t})^k \, dt \leq C(C)^k W^{\frac{s}{p}}(u(F_a)). \tag{4.7}
$$

Lemma 4.1 implies that $(F_a)_t \supseteq F_{at}$, thus we obtain

$$
\begin{aligned}
I &\geq \int_0^1 t^{s-1} W^{\frac{s}{p}}(u(F_{at})) \frac{1}{k!} \left(\log \frac{1}{t} \right)^k dt \\
&= a^{-s} \int_0^a x^{s-1} W^{\frac{s}{p}}(u(F_x)) \frac{1}{k!} \left(\log \frac{a}{x} \right)^k dx.
\end{aligned}
$$

This estimate together with (4.7) implies that

$$
\frac{1}{a} \int_0^a x^{s-1} W^{\frac{s}{p}}(u(F_x)) \frac{1}{k!} \left(\log \frac{a}{x} \right)^k dx \leq C(C)^k a^{s-1} W^{\frac{s}{p}}(u(F_a)) \tag{4.8}
$$

holds for all $a \in (0, 1)$. Now integrating (4.8) with respect to a and using induction hypothesis we get

$$
II := \int_0^1 \frac{1}{a} \int_0^a x^{s-1} W^{\frac{s}{p}}(u(F_x)) \frac{1}{k!} \left(\log \frac{a}{x} \right)^k dx \, da \leq C(C)^{k+1} W^{\frac{s}{p}}(u(F)).
$$

Changing the order of integration in II, we obtain

$$
II = \int_0^1 x^{s-1} W^{\frac{s}{p}}(u(F_x)) \frac{1}{(k+1)!} \left(\log \frac{1}{x} \right)^{k+1} dx \leq C(C)^{k+1} W^{\frac{s}{p}}(u(F)).
$$

Thus the inequality (4.6) is true for $k+1$ and hence by induction (4.6) is true for all $k = 0, 1, 2, \ldots$ and for all measurable sets F.

Hence, the proof of the claim is complete. Since $\tilde{p} < p$, invoking the interpolation Theorem 2.6.10 in [3] with indices $p_0 = q_0 = \tilde{p}$ and $p_1 = q_1 = \infty$ concludes that there exists $r \in (0, p)$ such that M^+ maps $\Lambda_u^r(w)$ to itself boundedly. □

As an application of Theorem 4.4 we conclude the proof of Theorem 4.1.

Proof of Theorem 4.1 It is natural to see that $L^{p,q}(u)$ can be identified with $\Lambda_u^{p,q}(1)$ spaces for $0 < p, q < \infty$. For $u \in A_p^+$, there is an $\epsilon > 0$ such that $u \in A_{p-\epsilon}^+$ (see [9]). Hence $M^+ : L^{p-\epsilon}(u) \to L^{p-\epsilon,\infty}(u)$. Since M^+ is of strong type (∞, ∞), using off-diagonal Marcinkiewicz interpolation, M^+ is bounded from $L^{p,1}(u)$ to $L^{p,q}(u)$.

For the necessary part, assume that M^+ is bounded from $L^{p,1}(u)$ to $L^{p,q}(u)$ which means that M^+ is bounded from $\Lambda_u^{p,1}(1)$ to $\Lambda_u^{p,q}(1)$. Theorem 4.4 implies that there exists $1 < r < p$ such that $\|M^+ f\|_{\Lambda_u^r(1)} \lesssim \|f\|_{\Lambda_u^r(1)}$, i.e., $\|M^+ f\|_{L^r(u)} \lesssim \|f\|_{L^r(u)}$. Now the Marcinkiewicz interpolation theorem implies that $M^+ : L^p(u) \to L^p(u)$ and this, in turn, implies that $u \in A_p^+$. □

To prove Theorem 4.2 we need the following crucial lemma which is motivated by [1].

Lemma 4.2 *For* $0 < \lambda < 1$ *and any measurable set* F *in* $\mathbb{R}$,

$$\chi_{\{M^+\chi_F(x)>\lambda\}}(x) \lesssim \frac{1}{\lambda(1-\log\lambda)} M^+(\chi_{\{M^+\chi_F(x)>\lambda\}} M^+\chi_F)(x) \quad a.e\ x \in \mathbb{R}. \tag{4.9}$$

Proof Fix $\lambda \in (0, 1)$. Let F be a measurable set and $x \in \{M^+\chi_F > \lambda\}$. Then there exists $h > 0$ such that $\frac{|F\cap I|}{h} > \lambda$, where $I = (x, x+h)$. Since $t(1-\log t)$ is an increasing function in $(0, 1)$, and using the equivalence $(M^+ f)^* \simeq f^{**}$, we obtain

$$\begin{aligned}
\lambda(1-\log\lambda) &\le \frac{1}{|I|}\int_0^{|I|} \frac{min(t, |F\cap I|)}{t} dt \\
&\le \frac{1}{|I|}\int_0^{|I|} \left(\frac{1}{t}\int_0^t \chi_{F\cap I}^*(s)\, ds\right) dt \lesssim \frac{1}{|I|}\int_0^{|I|} (M^+\chi_{F\cap I})^*(t) dt \\
&\le \frac{1}{|I|} \sup_{|A|=|I|} \Bigg(\int_{A\cap(-\infty,x)} M^+(\chi_{F\cap(x,x+h)})(y)\, dy \\
&\quad + \int_{A\cap(x,x+h)} M^+(\chi_{F\cap(x,x+h)})(y)\, dy \Bigg).
\end{aligned} \tag{4.10}$$

For $y \in (-\infty, x)$, $M^+(\chi_{F\cap(x,x+h)})(y) = \sup_{r>0} \frac{1}{r}\int_y^{y+r} \chi_{F\cap(x,x+h)}$. Observe that if $r > 0$ is such that $y + r > x + h$, then $\frac{1}{r}\int_y^{y+r} \chi_{F\cap(x,x+h)} \leq \frac{1}{h}\int_x^{x+h} \chi_F$. Again, for $r > 0$ with $x < y + r < x + h$, we have $\frac{1}{r}\int_y^{y+r} \chi_{F\cap(x,x+h)} \leq \frac{1}{r}\int_x^{x+r} \chi_F$. Hence, for $y \in (-\infty, x)$, $M^+(\chi_{F\cap(x,x+h)})(y) \leq M^+(\chi_F)(x)$. Thus from (4.10) we get

$$\begin{aligned}
\lambda(1 - \log\lambda) &\leq \frac{1}{|I|}\left(|I|M^+(\chi_F)(x) + \int_x^{x+h} M^+\chi_F(y)dy\right) \\
&\leq M^+(\chi_F)(x) + M^+(M^+\chi_F)(x) \lesssim M^+(M^+\chi_F)(x) \\
&\lesssim M^+(\chi_{\{M^+\chi_F>\lambda\}}M^+\chi_F)(x) + M^+(\chi_{\{M^+\chi_F\leq\lambda\}}M^+\chi_F)(x) \\
&\lesssim M^+(\chi_{\{M^+\chi_F>\lambda\}}M^+\chi_F)(x),
\end{aligned}$$

where in the last inequality we have used the fact that $M^+(\chi_{\{M^+\chi_F\leq\lambda\}}M^+\chi_F)(x) \leq \lambda$ and by Lebesgue differentiation theorem $\lambda \leq M^+(\chi_{\{M^+\chi_F>\lambda\}}M^+\chi_F)(x)$. Hence, the proof is complete. □

Now, we present the main result of this section.

Proof of Theorem 4.2 As mentioned earlier, we are only going to prove that (ii) implies (i). Fix $1 < p, q < \infty$, and assume that M^+ is bounded from $\Lambda_u^{p,q}(w)$ to $\Lambda_u^{p,\infty}(w)$. We start with the strategy of the proof. It is enough to find some $s \in (1, \infty)$ and a constant $C > 0$ such that

$$\int_0^1 \lambda^{s-1} W(u(\{x : M^+\chi_F(x) > \lambda\}))^{s/p}\, d\lambda \leq CW(u(F))^{s/p} \tag{4.11}$$

holds for all measurable set F. Now Theorem 4.4 implies that there exists $r \in (0, p)$ such that $\|M^+ f\|_{\Lambda_u^r(w)} \leq C\|f\|_{\Lambda_u^r(w)}$. Since $0 < r < p$, invoking the interpolation Theorem 2.6.10 in [3] with indices $p_0 = q_0 = r$ and $p_1 = q_1 = \infty$ concludes that M^+ is bounded from $\Lambda_u^{p,q}(w)$ to itself.

The remainder of the proof is devoted to the proof of (4.11). Using Lemma 4.2, we obtain that the integral in (4.11) is bounded by the following

$$\int_0^1 \lambda^{s-1} W(u(\{x : M^+(\chi_{\{M^+\chi_F(x)>\lambda\}}M^+\chi_F)(x) > c\lambda(1 - \log\lambda)\}))^{s/p} d\lambda.$$

Since M^+ is bounded from $\Lambda_u^{p,q}(w)$ to $\Lambda_u^{p,\infty}(w)$, the above integral is dominated by

$$\int_0^1 \lambda^{s-1}\left(\frac{1}{\lambda^p(1 - \log\lambda)^p}\|\chi_{\{M^+\chi_F(x)>\lambda\}}M^+\chi_F\|^p_{\Lambda_u^{p,q}(w)}\right)^{s/p} d\lambda. \tag{4.12}$$

At this point we need to estimate the norm $\|\chi_{\{M^+\chi_F(x)>\lambda\}} M^+\chi_F\|^q_{\Lambda_u^{p,q}(w)}$ and it is the sum of the following two terms

$$p\int_0^\lambda t^{q-1} W(u(\{x : \chi_{\{M^+\chi_F>\lambda\}}(x) M^+\chi_F(x) > t\}))^{q/p}\, dt,$$

$$\text{and } p\int_\lambda^1 t^{q-1} W(u(\{x : \chi_{\{M^+\chi_F>\lambda\}}(x) M^+\chi_F(x) > t\}))^{q/p}\, dt.$$

Since for the first term $t \le \lambda$, we have $\{x : \chi_{\{M^+\chi_F>\lambda\}}(x) M^+\chi_F(x) > t\} \subseteq \{x : M^+\chi_F(x) > \lambda\}$. This implies that the first term is dominated by

$$C\,\lambda^q W(u(\{x : M^+\chi_F(x) > \lambda\}))^{q/p} \le C\|M^+(\chi_F)\|^q_{\Lambda_u^{p,\infty}(w)} \le C\|\chi_F\|^q_{\Lambda_u^{p,q}(w)}.$$

Similarly, the second term is bounded by

$$\begin{aligned} C\int_\lambda^1 t^q W(u(\{x : M^+\chi_F(x) > t\}))^{q/p} \frac{dt}{t} &\le C\int_\lambda^1 \|\chi_F\|^q_{\Lambda_u^p(w)} \frac{dt}{t} \\ &\le C\left(\log\frac{1}{\lambda}\right)\|\chi_F\|^q_{\Lambda_u^p(w)}, \end{aligned}$$

where in the last but one inequality we have used the boundedness of M^+ from $\Lambda_u^{p,q}(w)$ to $\Lambda_u^{p,\infty}(w)$. Combining the above estimates, we obtain the following: There is a constant $C > 0$ such that

$$\|\chi_{\{M^+\chi_F>\lambda\}} M^+\chi_F\|^q_{\Lambda_u^{p,q}(w)} \le C(1 - \log\lambda)\|\chi_F\|^q_{\Lambda_u^p(w)}\ \forall\, 0 < \lambda < 1, \tag{4.13}$$

and for any measurable set F. The estimate (4.13) implies that the integral in (4.12) is dominated by

$$C\|\chi_F\|^s_{\Lambda_u^{p,q}(w)} \int_0^1 \frac{1}{\lambda}\frac{1}{(1-\log\lambda)^{s/q'}}\, d\lambda \le C\|\chi_F\|^s_{\Lambda_u^{p,q}(w)},$$

if we choose s such that $s > q'$. This completes the proof of claim (4.11) and Theorem 4.2. □

4.2.1 Characterisation of Weights and Concluding Remarks

We start with the following preparatory lemma.

Lemma 4.3 *Let $0 < r < \infty$. Then there exists a constant $C > 0$ such that*

$$\|M^+\chi_F\|_{\Lambda_u^{r,\infty}(w)} \leq C\|\chi_F\|_{\Lambda_u^r(w)}, \tag{4.14}$$

for all measurable sets $F \subset \mathbb{R}$ if and only if there is a constant $C > 0$ such that for every finite family of numbers $x_j < y_j < z_j$ and measurable sets $\{F_j\}_{j=1}^M$ with $F_j \subseteq (y_j, z_j)$ for $j = 1, \ldots, M$, we have

$$\frac{W(u(\bigcup_j (x_j, y_j)))^{\frac{1}{r}}}{W(u(\bigcup_j F_j))^{\frac{1}{r}}} \leq C \max_j \left(\frac{(z_j - x_j)}{|F_j|}\right). \tag{4.15}$$

Proof First, we prove the necessary part. Fix $x_j < y_j < z_j$ and the measurable sets $\{F_j\}_{j=1}^M$ with $F_j \subseteq (y_j, z_j)$ for $j = 1, \ldots, M$. Let $F = \bigcup_j F_j$ and $t > 0$ be such that $\frac{1}{t} > \max_j \frac{(z_j - x_j)}{|F_j|}$. Observe that $(x_j, y_j) \subseteq \{x : M^+\chi_F(x) > t\}$ for all $j = 1, \ldots, M$. Now (4.14) implies that

$$tW\big(u\big(\cup_j (x_j, y_j)\big)\big)^{\frac{1}{r}} \leq tW(u(\{x : M^+\chi_F(x) > t\}))^{\frac{1}{r}} \leq CW\big(u\big(\bigcup_j F_j\big)\big)^{\frac{1}{r}}.$$

Since $t > 0$ is arbitrary, we get the desired condition (4.15).

Conversely, let F be any measurable set and recall that for any $t > 0$, $F_t = \{x : M^+\chi_F(x) > t\}$. Using Riesz's rising sun lemma (see Lemma 2.1 in [9]), we decompose F_t as union of pairwise disjoint bounded open intervals $\{(a_j, b_j)\}_j$ with the following property

$$\frac{1}{b_j - x}\int_x^{b_j} \chi_F \geq t \text{ for all } x \in [a_j, b_j). \tag{4.16}$$

Now we decompose again the intervals $(a_j, b_j) = \bigcup_{l\geq 1}(x_j^{l-1}, x_j^l]$ where $\{x_j^l\}$ is an increasing sequence in (a_j, b_j) defined as: $x_j^0 = a_j$ and $\int_{x_j^{l-1}}^{x_j^l} \chi_F = \int_{x_j^l}^{b_j} \chi_F$ for all $l \geq 1$. Construction of the sequence ensures that $\int_{x_j^{l-1}}^{b_j} \chi_F = 4\int_{x_j^l}^{x_j^{l+1}} \chi_F$. This together with (4.16) implies that $\frac{4|F\cap[x_j^l, x_j^{l+1}]|}{b_j - x_j^{l-1}} \geq t$. Denote $F_j^l := F \cap (x_j^l, x_j^{l+1}]$ for all j and $l \geq 1$. Now condition (4.15) implies the following

$$t\ W(u(F_t))^{\frac{1}{r}} = t\ W\big(u\big(\cup_j \cup_l (x_j^{l-1}, x_j^l)\big)\big)^{\frac{1}{r}} \leq C\frac{W(u(\cup_j\cup_l(x_j^{l-1}, x_j^l)))^{\frac{1}{r}}}{\sup\frac{(x_j^{l+1} - x_j^{l-1})}{|F_j^l|}}$$

$$\leq C\ W(u(\cup_{j,l}F_j^l))^{\frac{1}{r}} \leq CW(u(F))^{\frac{1}{r}}.$$

Taking supremum over $t > 0$ completes the proof. □

Proof of Theorem 4.3 Assume M^+ is bounded from $\Lambda_u^{p,q}(w)$ to itself. Theorem 4.4 implies that there exist $r \in (0, p)$ and a constant $C > 0$ such that

$$\|M^+\chi_F\|_{\Lambda_u^{r,\infty}(w)} \leq C\|\chi_F\|_{\Lambda_u^r(w)} \tag{4.17}$$

holds for all measurable sets F. Then the desired condition follows from the necessary part of Theorem 4.3.

For the sufficiency, let $(u, w) \in \mathcal{Z}_p^+$. Then there exists a $r \in (0, p)$ such that (4.2) holds. Then the lemma 4.3 implies that there exists a constant $C > 0$ such that

$$\|M^+\chi_F\|_{\Lambda_u^{r,\infty}(w)} \leq C\|\chi_F\|_{\Lambda_u^r(w)}$$

holds for all measurable sets F in $\mathbb{R}$. Since $0 < r < p$ and M^+ map L^∞ to L^∞, the proof is complete once we apply interpolation Theorem 2.6.10 in [3]. □

Remark 4.1 As an application of Theorem 4.2 and the Lebesgue differentiation theorem, we can conclude that there is no weight u such that M^+ maps $L^{1,q}(u)$ to $L^{1,\infty}(u)$ boundedly for $1 < q < \infty$.

We conclude with the following example.

Example Take $w = \chi_{(0,1)}$ and $u(x) = e^x$, then $(w, u) \in \mathcal{Z}_p^+$ for all $1 < p < \infty$. To see this one can use the fact that the pair (w, u) satisfies

$$\frac{W(u(x-h, x))}{W(u(F))} \leq C\frac{|I|}{|F|}, \tag{4.18}$$

where $I = (x - h, x + h)$ and $F \subseteq [x, x + h)$. Now the condition (4.2) holds with $r = 1$ from (4.18). Moreover, it is easy to see that u cannot satisfy the following condition

$$\frac{W(u(I))}{|I|^p} \lesssim \frac{W(u(F))}{|F|^p} \quad \text{with } F \subset I$$

for any $p > 0$ and therefore the Hardy-Littlewood maximal function M is not bounded on $\Lambda_u^p(w)$ for any $p > 0$. □

References

1. Agora, E., Antezana, J., Carro, M.J.: Weak-type boundedness of the Hardy-Littlewood maximal operator on weighted Lorentz spaces. J. Fourier Anal. Appl. **22**(6), 1431–1439 (2016)
2. Ariño, M.A., Muckenhoupt, B.: Maximal functions on classical Lorentz spaces and Hardy's inequality with weights for nonincreasing functions. Trans. Am. Math. Soc. **320**(2), 727–735 (1990)
3. Carro, M.J., Raposo, J.A., Soria, J.: Recent developments in the theory of Lorentz spaces and weighted inequalities. Mem. Am. Math. Soc. **187**(877), xii+128pp. (2007)

4. Chung, M.H., Hunt, R.A., Kurtz, D.S.: The Hardy-Littlewood maximal function on L(p,q) spaces with weights. Ind. Univ. Math. J. **31**(1), 109–120 (1982)
5. Forzani, L., Martín-Reyes, F.J., Ombrosi, S.: Weighted inequalities for the two-dimensional one-sided Hardy-Littlewood maximal function. Trans. Am. Math. Soc. **363**(4), 1699–1719 (2011)
6. Hunt, R.A., Kurtz, D.S.: The Hardy-Littlewood maximal function on L(p,1). Ind. Univ. Math. J. **32**(1), 155–158 (1983)
7. Martín-Reyes, F.J.: New proofs of weighted inequalities for the one-sided Hardy-Littlewood maximal functions. Proc. Am. Math. Soc. **117**(3), 691–698 (1993)
8. Ortega Salvador, P.: Weighted Lorentz norm inequalities for the one-sided Hardy-Littlewood maximal functions and for the maximal ergodic operator. Can. J. Math. **46**(5), 1057–1072 (1994)
9. Sawyer, E.: Weighted inequalities for the one-sided Hardy-Littlewood maximal functions. Trans. Am. Math. Soc. **297**(4), 367–374 (1986)

Chapter 5
Remarks on Gradient Yamabe Solitons

Brian Grajales, Enrique López, and Matheus Hudson

Abstract This note studies gradient Yamabe solitons realised as warped metrics. First, we prove that the potential function and the scalar curvature depend solely on the product on the base. Building on this, we establish some results by analysing the warped function.

5.1 Introduction

The Yamabe flow was introduced by Hamilton in [4] to construct Yamabe metrics on compact Riemannian manifolds without boundary. This flow, named after Hidehiko Yamabe, provides a powerful tool for studying and understanding the geometric properties of manifolds. One of the main motivations behind the study of the Yamabe flow is its connection to the famous Yamabe problem, which seeks to find a metric on a given Riemannian manifold that has constant scalar curvature within its conformal class. The Yamabe flow naturally arises as a gradient flow associated with this problem, aiming to deform the metric towards a solution with constant scalar curvature. The Yamabe solitons are important in the literature because they are self-similar solutions of the Yamabe flow. More precisely, a complete Riemannian metric g on an n-dimensional smooth manifold M^n is a gradient Yamabe soliton if there

B. Grajales (✉)
IMECC-Departamento de Matemática, Universidade Estadual de Campinas, Rua Sérgio Buarque de Holanda 651, Campinas, SP, Brazil
e-mail: grajales@ime.unicamp.br

E. López
Departamento de Matemáticas, Escuela Superior Politécnica del Litoral, Guayaquil, Ecuador
e-mail: enfelope@espol.edu.ec

M. Hudson
Departamento de Matemática, Universidade Federal do Amazonas, Manaus, Brazil
e-mail: matheushudson@ufam.edu.br

M. Ruzhansky, K. Van Bockstal (eds.), *Extended Abstracts 2021/2022*,
Research Perspectives Ghent Analysis and PDE Center 2,
https://doi.org/10.1007/978-3-031-42539-4_5

exists a smooth vector field X and a constant $\lambda \in \mathbb{R}$ such that

$$Sg + \mathscr{L}_X g = \lambda g, \tag{5.1}$$

where S is the scalar curvature of g, and $\mathscr{L}_X$ the Lie derivative with respect to X. For $\lambda > 0$, $\lambda = 0$, or $\lambda < 0$, (M^n, g, X) is called shrinking, steady, or expanding, respectively. If $X = \nabla\varphi$, the equation above can also be written as

$$Sg + \text{Hess}\,\varphi = \lambda g,$$

where Hess φ denotes the Hessian of φ. The function φ is called the potential function of a gradient Yamabe soliton (M^n, g, φ). The soliton is trivial whenever $\nabla\varphi$ is parallel.

The concept of *warped product* was introduced by Bishop and O'Neil in [2]. Given two Riemannian manifolds (B^n, g_B) and (F^m, g_F), and a positive smooth warping function f defined on B^n, let us consider on the product manifold $B^n \times F^m$, the warped metric $g = \pi_1^* g_B + (f \circ \pi_1)^2 \pi_2^* g_F$, where π_1 and π_2 are the natural projections on B^n and F^m, respectively. Under these conditions, the product manifold is called the warped product of B and F. If f is a constant, then $(B^n \times F^m, g)$ is called a standard Riemannian product. When a gradient Yamabe soliton is realised as a warped metric g on $B^n \times F^m$, for simplicity we say that $B^n \times F^m$ is a gradient Yamabe soliton warped product.

5.2 Gradient Yamabe Warped Product

Following Tokura et al. [1], the below theorem does not need the assumption that the function of fibre is constant to guarantee which scalar curvature of fibre is also constant and reciprocally, actually, this assumption is always true in the gradient Yamabe warped product.

Theorem 5.1 *Let $M = B^n \times_f F^m$ be a warped product with metric $g = g_B + f^2 g_F$, where $f > 0$ is a smooth function on B. If (M, g) is a gradient Yamabe soliton, then the potential function depends only on the base, and the scalar curvature of the fibre S_F is constant.*

Proof Suppose that g is a Yamabe soliton warped metric with potential function η. Then

$$\nabla_g^2 \eta = (-S_g + \lambda) g, \tag{5.2}$$

for some constant $\lambda \in \mathbb{R}$.

Using the known formula for the Hessian of a smooth function η on the warped product, see Bishop and O'Neill [2], we obtain

$$0 = \left(\nabla_g^2 \eta\right)(U, V) = U(V(\eta)) - (\nabla_U V)(\eta) = U(V(\eta)) - \frac{U(f)}{f} V(\eta),$$

for all U, V horizontal and vertical vector fields, respectively. Next, we compute

$$U\left(V\left(\frac{\eta}{f}\right)\right) = U\left(\frac{V(\eta)}{f}\right) = \frac{1}{f}\left[U(V(\eta)) - \frac{U(f)}{f} V(\eta)\right] = 0.$$

Therefore, the function $V\left(\frac{\eta}{f}\right)$ depends only on the fibre F^m. Thus, without loss of generality, we can write that $\eta = \varphi + fh$, for some functions φ on B^n and h on F^m. Next, we observe that, for horizontal vector fields, the Eq. (5.2) becomes

$$\nabla_B^2 \varphi(X, Y) + h \nabla_B^2 f(X, Y) = (\lambda - S_g) g_B(X, Y), \tag{5.3}$$

which follows from Lemma 7.4 in [2] or, alternatively, [6, Corollary 43]. This lemma also implies

$$S_g = S_B - \frac{2m}{f} \Delta_B f + \frac{S_F}{f^2} - m(m-1) \frac{|\nabla_B f|^2}{f^2}. \tag{5.4}$$

From Eqs. (5.3) and (5.4), we get

$$V(h) \nabla_B^2 f(X, Y) = -V\left(\frac{S_F}{f^2}\right) g_B(X, Y). \tag{5.5}$$

Note that h is constant if and only if S_F is constant. Now, we are going to proceed by contradiction; suppose that h is not constant. We claim that

$$\nabla_B^2 f = \frac{a}{f^2} g_B, \tag{5.6}$$

for some constant $a \neq 0$.

Indeed, take a horizontal vector field X such that $g_B(X, \nabla f) = 0$. From (5.5), we get

$$V(h) \nabla_B^2 f(\nabla_B f, X) = 0.$$

Since h is nonconstant, there exists a vertical vector field V such that $V(h) \neq 0$, and then $\nabla_B^2 f(\nabla_B f, X) = 0$. Therefore, there exists a smooth function ψ on B^n so that $\nabla_B^2 f(\nabla_B f) = \psi \nabla_B f$. From (5.5), one has

$$\left(V(h)\psi + \frac{V(S_F)}{f^2}\right)|\nabla_B f|^2 = 0. \tag{5.7}$$

Consider the closed set $A = \{x \in B^n; |\nabla_B f|(x) = 0\}$. Note that the set $int(A)$ is empty or, equivalently, A^C is a dense set in B^n. In fact, if there exists $x_0 \in int(A)$, then $\nabla_B^2 f(x_0) = 0$ implies that S_F is constant from (5.5), which is a contradiction. From (5.7) and by continuity, we have that

$$V(h)\psi + \frac{V(S_F)}{f^2} = 0$$

on $B^n \times F^m$. So,

$$V(h)X(\psi f^2) = 0.$$

As $V(h) \neq 0$ it follows that $\psi f^2 = a$, for some constant $a \neq 0$, and then

$$aV(h) = -V(S_F). \tag{5.8}$$

Again from (5.5)

$$V(h)\left(\nabla_B^2 f - \frac{a}{f^2} g_B\right) = 0,$$

we conclude the proof of (5.6). Follow up right away the identity

$$|\nabla f|^2 + \frac{2a}{f} = b, \tag{5.9}$$

for some constant b. Since $\eta = \varphi + fh$, and by using the Yamabe soliton equation on F^m (see [6, Corollary 43]), we obtain

$$f\nabla_F^2 h + f(\nabla_B f(\varphi) + h|\nabla_B f|^2)g_F = (\lambda - S_g)f^2 g_F. \tag{5.10}$$

Taking the divergence of this previous equation with respect to g_F and (5.4), we get

$$f div_F \nabla_F^2 h + f|\nabla_B f|^2 \nabla_F h = -\nabla_F S_F. \tag{5.11}$$

Taking the covariant derivative $\nabla_B f$, we have

$$|\nabla_B f|^2 div_F \nabla_F^2 h + (|\nabla_B f|^4 + 2f\nabla_B^2 f(\nabla_B f, \nabla_B f))\nabla_F h = 0.$$

Thus

$$div_F \nabla_F^2 h + \left(|\nabla_B f|^2 + \frac{2a}{f} \right) \nabla h = 0,$$

by (5.9) we have

$$div_F \nabla_F^2 h = -b \nabla_F h.$$

So, by (5.11), (5.8) and (5.9) we have

$$\frac{a}{f} \nabla_F h = 0.$$

Since $a \neq 0$, we have a contradiction. Therefore, h and S_F are constant. □

5.3 Applications

In view of Theorem 5.1, we shall assume that $\eta = \varphi$ for some $\varphi \in C^{\infty}(B^n)$ and that S_F is constant. The following lemma will be used to prove a weaker version of [1, Theorem 1.3], which states that a gradient Yamabe soliton with a compact Riemannian base is trivial.

Lemma 5.1 *Let (M, φ, g) be a gradient Yamabe warped product with $\varphi \in C^{\infty}(B^n)$. Then*

$$\Delta_{n \ln f} \varphi = 0.$$

Proof Suppose g is a Yamabe soliton warped metric with potential function φ. Then on the basis, we get

$$\nabla_B^2 \varphi = (-S_g + \lambda) g_B, \tag{5.12}$$

for some constant $\lambda \in \mathbb{R}$. Tracing the above equation we have

$$\Delta_B \varphi = n(-S_g + \lambda), \tag{5.13}$$

On the other hand, on the fibre, we know that $f(\lambda - S_g) = \nabla_B f(\varphi)$. Thus

$$\frac{1}{n} \Delta_B \varphi = \frac{\nabla_B f(\varphi)}{f},$$

Hence,

$$\frac{1}{n}\Delta_{(n\ln f)}\varphi = 0.$$

□

Proposition 5.1 *Let $M = B^n \times_f F^m$ be a gradient Yamabe soliton warped product with potential function φ. If φ reaches its maximum or minimum, then φ is constant.*

Proof By Lemma 5.1, we have that

$$\Delta_{(n\ln f)}\varphi = 0$$

so the result follows from the strong maximum principle. □

Corollary 5.1 ([1]) *Let $M = B^n \times_f F^m$ be a gradient Yamabe soliton warped product with potential function φ. If B^n is compact, then φ is constant and M is trivial.*

We end this section with the following proposition, which gives us a criterion for the potential function to be constant.

Proposition 5.2 *Let $M = B^n \times_f F^m$ be a gradient Yamabe soliton warped product with potential function φ and assume that B^n is complete and noncompact. If $\varphi \in L^p(M, f^{-n}dV)$ for $1 < p < \infty$, then φ is constant.*

Acknowledgments Brian Grajales is supported by the grant 2023/04083-0 (São Paulo Research Foundation FAPESP). Enrique López is supported by Escuela Superior Politécnica del Litoral. Matheus Hudson is partially supported by grant 062.00931/2013 (Fundação de Apoio à Pesquisa do Estado do Amazonas) and grant 001 (Coordenação de Aperfeiçoamento de Pessoal de Nível Superior).

References

1. Adriano, L., Barboza, M., Pina, R., Tokura, W.: On warped product gradient Yamabe solitons. J. Math. Anal. Appl. **473**(1), 201–214 (2019)
2. Bishop, R.L., O'Neill, B.: Manifolds of negative curvature. Trans. Am. Math. Soc. **145**, 1–49 (1969)
3. Chu, Y., Wang, X.: On the scalar curvature estimates for gradient yamabe solitons. Kodai Math. J. **36**, 246–257 (2013)
4. Hamilton, R.S.: The Ricci flow on surfaces, Mathematics and general relativity. Contemp. Math. **71**, 237–262 (1988)
5. Huang, G., Li, H.: On a classification of the quasi Yamabe gradient solitons. Methods Appl. Anal. **21**, 379–390 (2014)
6. O'Neill, B.: Semi-Riemannian Geometry with Applications to Relativity. Academic Press, London (1983)
7. Pigola, S., Rimoldi, M., Setti, A.G.: Remarks on non-compact gradient Ricci solitons. Math. Z. **268**, 777–790 (2011)

Chapter 6
Boundedness of Fourier Multipliers on Fundamental Domains of Lattices

Arne Hendrickx

Abstract This short paper is an overview of my research about L^p-L^q boundedness of Fourier multipliers on fundamental domains of lattices in $\mathbb{R}^d$.We start with an outline of Fourier analysis on fundamental domains, then use interpolation techniques to obtain some inequalities and conclude with a Hörmander-type boundedness theorem. At the end, I present some current research perspectives.

6.1 Introduction

Ever since the publication of Hörmander's paper [4] mathematicians have tried to generalise his results about the L^p-L^q boundedness of Fourier multipliers to other settings. Especially several settings within harmonic analysis are perfectly suited for this kind of investigation. Recently, there has been growing interest in crystals and quasicrystals within the mathematical community, the theory of which is sometimes named "aperiodic order".

Hörmander proved in the Euclidean case [4, Theorem 1.11] that a Fourier multiplier with symbol $\sigma(\xi)$ has a bounded L^p-L^q extension provided that $1 < b < \infty$, $1 < p \leq 2 \leq q < \infty$ with $\frac{1}{p} + \frac{1}{q} = \frac{1}{b}$ and $\sigma(\xi)$ is a measurable function satisfying

$$m\{\xi \in \mathbb{R}^d : |\sigma(\xi)| \geq s\} \leq \frac{C}{s^b} \quad \text{for all} \quad s > 0,$$

where $C > 0$ is some constant. The goal of my paper [3] was to investigate this theorem in the setting of fundamental domains of lattices in $\mathbb{R}^d$.

A. Hendrickx (✉)
Department of Mathematics: Analysis, Logic and Discrete Mathematics, Ghent University, Ghent, Belgium
e-mail: arnhendr.Hendrickx@UGent.be

M. Ruzhansky, K. Van Bockstal (eds.), *Extended Abstracts 2021/2022*,
Research Perspectives Ghent Analysis and PDE Center 2,
https://doi.org/10.1007/978-3-031-42539-4_6

6.2 Main Results

Let us first introduce the necessary concepts and notation. A fundamental domain of a lattice $L = A\mathbb{Z}^d$, where the generator matrix A is an invertible matrix, is a measurable set $\Omega \subseteq \mathbb{R}^d$ such that $\Omega + L = \mathbb{R}^d$ as a direct sum. It follows easily that Ω is homeomorphic to $A\mathbb{T}^d$. Fuglede's theorem [2, Sect. 6] ensures that $\{e^{2\pi i\kappa\cdot x} \mid \kappa \in L^{\perp}\}$ is an orthonormal basis for $L^2(\Omega)$, which we endow with the normalised measure $\mathrm{d}x / |\Omega|$, where the dual lattice is defined as $L^{\perp} := \left(A^{\mathrm{T}}\right)^{-1} \mathbb{Z}^d$. In this way, one can define the Fourier transform $\mathcal{F}_\Omega : L^2(\Omega) \to \ell^2(L^{\perp})$ by

$$\widehat{f}(\kappa) := \mathcal{F}_\Omega f(\kappa) := \frac{1}{|\Omega|} \int_\Omega f(x)\, e^{-2\pi i\kappa\cdot x}\, \mathrm{d}x$$

with inverse transform

$$\check{f}(x) := \mathcal{F}_\Omega^{-1} f(x) := \sum_{\kappa \in L^{\perp}} f(\kappa)\, e^{2\pi i\kappa\cdot x}.$$

A more detailed summary of the necessary concepts can be found in [3].

Our strategy for proving the Hörmander-type L^p-L^q boundedness theorem for Fourier multipliers on fundamental domains of lattices follows the approach that Hörmander outlined in [4]. We begin by remarking that we have a Plancherel formula, which leads to the Hausdorff-Young inequality via Riesz-Thorin interpolation. Then, interpolating between the Plancherel formula and the Hausdorff-Young inequality by means of the Marcinkiewicz interpolation, we obtain Paley's inequality. Finally, we get the Hausdorff-Young-Paley inequality via Stein-Weiss interpolation in the following form:

Theorem 6.1 (Hausdorff-Young-Paley Inequality) *Let $1 < p \leq 2$, and let $1 < p \leq b \leq p' < \infty$ with $\frac{1}{p} + \frac{1}{p'} = 1$. Suppose that $\varphi(\kappa)$ is a positive function on $L^{\perp}$ satisfying*

$$M_\varphi := \sup_{s>0} s \sum_{\substack{\kappa \in L^{\perp} \\ \varphi(\kappa) \geq s}} 1 < \infty.$$

Then for every $f \in L^p(\Omega)$ we have

$$\left(\sum_{\kappa \in L^{\perp}} \left| \widehat{f}(\kappa)\, \varphi(\kappa)^{\frac{1}{b} - \frac{1}{p'}} \right|^b \right)^{\frac{1}{b}} \lesssim M_\varphi^{\frac{1}{b} - \frac{1}{p'}} \, \|f\|_{L^p(\Omega)} .$$

This Hausdorff-Young-Paley inequality is an essential ingredient for our proof of L^p-L^q boundedness of Fourier multipliers on fundamental domains.

Let $C^\infty_{\text{per}}(\Omega)$ be the space of all smooth L-periodic functions on Ω, where the L-periodicity means that $f(x+\lambda) = f(x)$ for all $x \in \Omega$ and $\lambda \in L$. An operator $A : C^\infty_{\text{per}}(\Omega) \to C^\infty_{\text{per}}(\Omega)$ is called a *Fourier multiplier* with symbol $\sigma : L^\perp \to \mathbb{C}$ if for all $f \in C^\infty_{\text{per}}(\Omega)$ and $\kappa \in L^\perp$ it holds that

$$\widehat{Af}(\kappa) = \sigma(\kappa)\,\widehat{f}(\kappa).$$

We are ready to state the main results about L^p-L^q boundedness of Fourier multipliers on fundamental domains of lattices. As a special case, we first mention the special case of L^2-L^2 boundedness, which is better treated separately for technical reasons.

Theorem 6.2 *Let $A : C^\infty_{per}(\Omega) \to C^\infty_{per}(\Omega)$ be a Fourier multiplier with bounded symbol $\sigma \in \ell^\infty(L^\perp)$. Then we have*

$$\|A\|_{\mathcal{L}(L^2(\Omega),L^2(\Omega))} := \sup_{f\neq 0} \frac{\|Af\|_{L^2(\Omega)}}{\|f\|_{L^2(\Omega)}} = \|\sigma\|_{\ell^\infty(L^\perp)}\,.$$

In particular, A can be extended to a bounded linear operator from $L^2(\Omega)$ to $L^2(\Omega)$.

Indeed, we can also obtain an analogue of Hörmander's theorem [4, Th. 1.11] in our setting, which reads as follows.

Theorem 6.3 *Let $1 < p \leq 2 \leq q < \infty$ with p and q not both equal to 2, and let $A : C^\infty_{per}(\Omega) \to C^\infty_{per}(\Omega)$ be a Fourier multiplier with symbol σ satisfying*

$$\sup_{s>0} s \left(\sum_{\substack{\kappa \in L^\perp \\ |\sigma(\kappa)|\geq s}} 1 \right)^{\frac{1}{p}-\frac{1}{q}} < \infty. \tag{6.1}$$

Then

$$\|A\|_{\mathcal{L}(L^p(\Omega),L^q(\Omega))} := \sup_{f\neq 0} \frac{\|Af\|_{L^q(\Omega)}}{\|f\|_{L^p(\Omega)}} \lesssim \sup_{s>0} s \left(\sum_{\substack{\kappa \in L^\perp \\ |\sigma(\kappa)|\geq s}} 1 \right)^{\frac{1}{p}-\frac{1}{q}}.$$

In particular, A can be extended to a bounded linear operator from $L^p(\Omega)$ to $L^q(\Omega)$.

Actually, we can use the fact that Ω is compact and thus has a finite Lebesgue measure to show L^p-L^q boundedness for all $1 < p, q < \infty$ under a suitable growth condition for the symbol of the Fourier multiplier. As an application, we can get several Sobolev space embeddings.

6.3 Research Perspectives

Although [3] dealt with the question about L^p-L^q boundedness of Fourier multipliers on fundamental domains of lattices under a Hörmander-type condition, we can also wonder about the boundedness of Fourier multipliers between different spaces. Currently, I am working on Titchmarsh theorems for Fourier transforms of functions in additive and multiplicative Hölder-Lipschitz spaces and a generalisation via moduli of continuity. This may allow us to deduce boundedness results for Fourier multipliers between general Lipschitz-type spaces, most notably including Hölder-Lipschitz spaces and Dini-Lipschitz spaces.

Once this question has been dealt with, I am planning to move on to the study of pseudo-differential operators on fundamental domains of lattices. It would also be interesting to make a connection to pseudo-differential operators on lattices as in [1]. This theory can then be seen as a stepping stone to the setting of more complex crystals and quasicrystals, where several questions and problems can be investigated.

Acknowledgments This research has been partially supported by Fonds Wetenschappelijk Onderzoek—Vlaanderen (FWO) with the PhD Fellowship Fundamental Research grant 1187323N.

References

1. Botchway, L.N.A., Kibiti, P.G., Ruzhansky, M.: Difference equations and pseudo-differential operators on $\mathbb{Z}^n$. J. Funct. Anal. **278**(11), 108473 (2020)
2. Fuglede, B.: Commuting self-adjoint partial differential operators and a group theoretic problem. J. Funct. Anal. **16**(1), 101–121 (1974)
3. Hendrickx, A.: L^p-L^q boundedness of Fourier multipliers on fundamental domains of lattices in $\mathbb{R}^d$. J. Fourier Anal. Appl. **28**(60), 1–30 (2022)
4. Hörmander, L.: Estimates for translation invariant operators in L^p spaces. Acta Math. **104**(1–2), 93–140 (1960)

Chapter 7
Pointwise Domination and Weak L^1 Boundedness of Littlewood-Paley Operators via Sparse Operators

Mahdi Hormozi

Abstract This contribution is based on joint work with Hormozi et al. (J. Fourier Anal. Appl. 29:49, 2023). We simplify and shorten the proofs of the main results of Shi et al. (J. Math. Pures Appl. 101:394–413, 2014) significantly. In particular, the new proof for Shi et al. (J. Math. Pures Appl. 101:394–413, 2014; Theorem 1.1) is quite short and, unlike the original proof, does not rely on the properties of the "Marcinkiewicz function". This allows us to get a precise linear dependence on Dini constants with a subsequent application to Littlewood–Paley operators by the well-known techniques.

7.1 Introduction

Write $\Gamma_\alpha(x)$ for the cone in $\mathbb{R}^{n+1}_+$ of aperture $\alpha > 1$ centered at x, that is

$$\Gamma_\alpha(x) = \{(y,t) \in \mathbb{R}^{n+1}_+ : |x-y| < \alpha t\}.$$

Let $S_{\alpha,\phi}$ be the square function defined by means of a standard kernel ϕ as follows:

$$S_{\alpha,\phi} f(x) = \Big(\iint_{\Gamma_\alpha(x)} |f \star \phi_t(y)|^2 \frac{dydt}{t^{n+1}} \Big)^{\frac{1}{2}}, \tag{7.1}$$

where $\phi_t(x) = t^{-n}\phi(x/t)$ and $\star$ refers to the convolution operation of two functions.

The study on the linear/multilinear square functions has important applications in PDEs and other fields of mathematics. For further details on the theory of linear multilinear square functions and their applications, we refer to [8, 9, 11] and the references therein. In [9], Lerner proved sharp weighted norm inequalities for $S_{\alpha,\phi} f$

M. Hormozi (✉)
School of Mathematics, Institute for Research in Fundamental Sciences (IPM), Tehran, Iran

M. Ruzhansky, K. Van Bockstal (eds.), *Extended Abstracts 2021/2022*, Research Perspectives Ghent Analysis and PDE Center 2,
https://doi.org/10.1007/978-3-031-42539-4_7

by applying *intrinsic square functions* introduced in [12]. Later on, Lerner himself improved the result –he obtained sharp dependence on α– in [8] by using the local mean oscillation formula.

Recall that a modulus of continuity is an increasing concave function defined on $(0, 1)$. Let $\varphi : [0, 1] \to [0, \infty)$ be a modulus of continuity which satisfies the Dini condition. That is, $\varphi : [0, 1] \to [0, \infty)$ is an increasing function such that $[\varphi]_{\text{Dini}} := \int_0^1 \frac{\varphi(t)}{t} dt + \varphi(1) < \infty$.

Let us recall the definition of multilinear square functions considered in this paper. Let $\psi(x, \mathbf{y}) := \psi(x, y_1, \ldots, y_m)$ be a locally integrable function defined away from the diagonal $x = y_1 = \cdots = y_m$ in $(\mathbb{R}^n)^{m+1}$. Let φ and w be moduli of continuity.

Assume that there is a positive constant A so that the following conditions hold:

- Size condition:

$$|\psi(x, \mathbf{y})| \leq A \left(1 + \sum_{i=1}^{m} |x - y_i|\right)^{-nm} w\left(\frac{1}{1 + \sum_{i=1}^{m} |x - y_i|}\right).$$

- Smoothness condition:

$$|\psi(x, \mathbf{y}) - \psi(x', \mathbf{y})|$$

$$\leq A \left(1 + \sum_{i=1}^{m} |x - y_i|\right)^{-nm} w\left(\frac{1}{1 + \sum_{i=1}^{m} |x - y_i|}\right) \varphi\left(\frac{|x - x'|}{1 + \sum_{i=1}^{m} |x - y_i|}\right),$$

whenever $|x - x'| < \frac{1}{2} \max_j |x - y_j|$, and

$$\left|\psi(x, \mathbf{y}) - \psi(x, y_1, \ldots, y_i', \ldots, y_m)\right|$$

$$\leq A \left(1 + \sum_{i=1}^{m} |x - y_i|\right)^{-nm} w\left(\frac{1}{1 + \sum_{i=1}^{m} |x - y_i|}\right) \varphi\left(\frac{|y_i - y_i'|}{1 + \sum_{i=1}^{m} |x - y_i|}\right),$$

whenever $|y_i - y_i'| < \frac{1}{2} \max_j |x - y_j|$ for $i = 1, 2, \ldots, m$.

Let

$$L := t + |x + z - y_1| + |x + z - y_2|.$$

For $t > 0$, define the operator ψ_t by

$$\psi_t \mathbf{f}(x) := \frac{1}{t^{nm}} \int_{(\mathbb{R}^n)^m} \psi\left(\frac{x}{t}, \frac{y_1}{t}, \ldots, \frac{y_m}{t}\right) \prod_{j=1}^{m} f_j(y_j) dy_j,$$

for all $\mathbf{f} = (f_1, \ldots, f_m) \in S(\mathbb{R}^n) \times \cdots \times S(\mathbb{R}^n)$.

Given $\alpha > 0$ and $\lambda > 2m$, the multilinear square functions $S_{\alpha,\psi}$ and g_λ^* are defined by

$$S_{\alpha,\psi}\mathbf{f}(x) := \left(\iint_{\Gamma_\alpha(x)} |\psi_t \mathbf{f}(y)|^2 \frac{dydt}{t^{n+1}}\right)^{\frac{1}{2}}$$

and

$$g_\lambda^* \mathbf{f}(x) := \left(\iint_{\mathbb{R}^{n+1}_+} \left(\frac{t}{t+|x-y|}\right)^{n\lambda} |\psi_t \mathbf{f}(y)|^2 \frac{dydt}{t^{n+1}}\right)^{\frac{1}{2}}.$$

We recall some of the basic properties of the multilinear Littlewood–Paley functions.

- Let $\alpha \geq 1$. Then $\|S_{\alpha,\psi} f\|_{L^2} = \alpha^{n/2} \|S_{1,\psi} f\|_{L^2}$ for all $f \in L^2(\mathbb{R}^n)$.
- $$\|S_{\alpha,\psi}\mathbf{f}\|_{L^{1/m,\infty}} \lesssim_m \alpha^{nm} \|S_{1,\psi}\mathbf{f}\|_{L^{1/m,\infty}}$$
  for all $\mathbf{f} \in L^1(\mathbb{R}^n)^m$.
- $S_{\alpha,\psi} \lesssim g^*_{\lambda,\psi}$.
- Let $x \in \mathbb{R}^n$. Then
  $$g^*_{\lambda,\psi}\mathbf{f}(x) \lesssim \sum_{k=0}^{\infty} 2^{-\frac{k\lambda n}{2}} S_{2^{k+1},\psi}\mathbf{f}(x)$$
  for all $\mathbf{f} \in L^1(\mathbb{R}^n)^m$.

Example Let $m = 1$. We exhibit some examples of ψ and moduli of continuity w and φ for which the size and smoothness conditions hold and $S_{\alpha,\psi}$ is L^2-bounded. For the sake of simplicity, we consider the operator $S_{1,\psi}$. Let $2 < \kappa \leq 4$. Define

$$\psi(x) = \psi(x_1, x_2, \ldots, x_n) = \frac{\sin x_1}{(1+|x|^2)^{\frac{n}{2}} \log^\kappa(2+|x|^2)} \quad (x \in \mathbb{R}^n).$$

Let $l \in \mathbb{N}$. Then

$$|\mathcal{F}\psi(\xi)| \lesssim \frac{1}{1+|\xi|^l} \log^{1-\kappa}\left(2 + \frac{1}{|\xi|}\right) \quad (\xi \in \mathbb{R}^n).$$

Thus $\mathcal{F}\psi$ decays rapidly at ∞ and 0 for any n if $\kappa > 1$. We consider the integral operator

$$S_{1,\psi} f(x) = \left(\iint_{\Gamma_1(x)} |f \star \psi_t(y)|^2 \frac{dydt}{t^{n+1}}\right)^{\frac{1}{2}},$$

which was defined in (7.1) with ϕ replaced by ψ. We can estimate the L^2-norm of $S_{1,\psi}$ with ease by the use of the Fourier transform:

$$\|S_{1,\psi} f\|_{L^2} \lesssim \|f\|_{L^2} \sqrt{\int_0^\infty \min\left(\frac{1}{t^2}, \log^{2-2\kappa}\left(2+\frac{1}{t}\right)\right)\frac{dt}{t}} \sim \|f\|_{L^2}.$$

Let us verify that size condition and smoothness condition hold for

$$w(t) = \varphi(t) := \log^{-\frac{\kappa}{2}}\left(2 + \frac{1}{\min(1,t)}\right).$$

It is noteworthy that w fails the log-Dini condition. It is easy to check the size condition since $|\sin x_1| \leq 1$ for all $x_1 \in \mathbb{R}$ and $\kappa > 2$. On the other hand, since

$$|\nabla \psi(x+h)| \lesssim \frac{1}{(1+|x|^2)^{\frac{n}{2}} \log^{\kappa}(2+|x|^2)}$$

and

$$\frac{|h|}{\log(2+|x|)} \lesssim \frac{1}{\log\left(2+\frac{1+|x|}{|h|}\right)}$$

if $|h| < \frac{|x|}{2}$, the smoothness condition is satisfied. □

In many papers (e.g. [1, 2, 11]), a priori boundedness of g_λ^* is assumed. That is, for some $1 < q_1, \cdots, q_m < \infty$ with $\frac{1}{q} = \frac{1}{q_1} + \cdots \frac{1}{q_m}$, there exist a constant $C > 0$ such that

$$\left\| g_\lambda^*(\mathbf{f}) \right\|_{L^q} \leq C \prod_{i=1}^m \|f_i\|_{L^{q_i}}.$$

Let $0 < \delta, \gamma \leq 1$ and $\omega(t) = t^\delta$ and $\varphi(t) = t^\gamma$. By exploiting the well-known boundedness of Marcinkiewicz function and more explicit decomposition of the operator, the following theorem is proved in [11].

Theorem 7.1 (Weak End-Point Estimates) *Suppose that $\lambda > 2m$, $0 < \gamma < \min\{n(\lambda - 2m)/2, \delta\}$ and $f_1, \cdots, f_m \in L^1$. Then for any $\mu > 0$, there exists a constant $C > 0$ such that*

$$\left|\left\{x \in \mathbb{R}^n : |g_\lambda^*(\mathbf{f})(x) > \mu\right\}\right| \leq \frac{C}{\mu^{\frac{1}{m}}} \prod_{i=1}^m \|f_i\|_{L^1}^{\frac{1}{m}}.$$

In the next theorem, notwithstanding the generalisation, we simplified and shortened the proof of Theorem 7.1 significantly. In particular, unlike the proof given in [11], does not rely on the properties of the "Marcinkiewicz function". Also, using the basic properties of the multilinear Littlewood–Paley functions, a priori boundedness of $S_{1,\psi}$ is assumed solely. It is noteworthy not to assume the log-Dini condition on w and φ, that is, we do not have to assume that

$$\int_0^1 w(t)\log\frac{1}{t}\cdot\frac{dt}{t}<\infty$$

or that

$$\int_0^1 \varphi(t)\log\frac{1}{t}\cdot\frac{dt}{t}<\infty.$$

Theorem 7.2 (Theorem 3.1 in [4]) *Let $\lambda > 2m$. Let $\sqrt[m]{\varphi}$, $\sqrt[m]{w}$ be moduli of continuity satisfying the Dini condition. Then for any $\rho > 0$ and $\alpha \geq 1$,*

$$\begin{aligned}&\|S_{\alpha,\psi}\|_{L^1\times L^1\times\cdots\times L^1\to L^{1/m,\infty}}\\ &\qquad\lesssim \alpha^{nm}(A[\sqrt[m]{w}]^m_{Dini}(1+[\sqrt[m]{\varphi}]^m_{Dini})+\|S_{1,\psi}\|_{L^{p_1}\times L^{p_2}\times\cdots\times L^{p_m}\to L^p})\end{aligned}$$

and

$$\begin{aligned}&\|g^*_{\lambda,\psi}\|_{L^1\times L^1\times\cdots\times L^1\to L^{1/m,\infty}}\\ &\qquad\lesssim A[\sqrt[m]{w}]^m_{Dini}(1+[\sqrt[m]{\varphi}]^m_{Dini})+\|S_{1,\psi}\|_{L^{p_1}\times L^{p_2}\times\cdots\times L^{p_m}\to L^p}.\end{aligned}$$

We remark that the Dini-condition of $\varphi^{\frac{1}{m}}$ and $w^{\frac{1}{m}}$ is needed. Thus we need a stronger type of Dini condition, that is, the one of $\varphi^{\frac{1}{m}}$ and $w^{\frac{1}{m}}$ when we consider the weak endpoint estimate via the Calderón–Zymgund decomposition.

Consider m weights $w_1,\ldots,w_m$ and denote $\vec{w}=(w_1,\ldots,w_m)$. Also let $1 < p_1,\ldots,p_m < \infty$ and p be numbers such that $\frac{1}{p}=\frac{1}{p_1}+\cdots+\frac{1}{p_m}$ and denote $\vec{p}=(p_1,\ldots,p_m)$. Set

$$\nu_{\mathbf{w}}:=\prod_{i=1}^m w_i^{\frac{p}{p_i}}.$$

We say that $\mathbf{w}$ satisfies the $A_{\mathbf{p}}$-condition if

$$[\mathbf{w}]_{A_{\mathbf{p}}}:=\sup_Q\left(\fint_Q \nu_{\mathbf{w}}\right)\prod_{j=1}^m\left(\fint_Q w_j^{1-p_j'}\right)^{\frac{p}{p_j'}}<\infty.$$

The class $A_{\mathbf{p}}$ collects all $\mathbf{w}$ for which $[\mathbf{w}]_{A_{\mathbf{p}}}$ is finite. When $p_j = 1$, $\left(⨍_Q w_j^{1-p_j'} \right)^{\frac{p}{p_j'}}$ is understood as $(\inf_Q w_j)^{-p}$.

The following theorem is proved in [11].

Theorem 7.3 (Weighted Estimates) *Let* $\lambda > 2m$, $\mathbf{w} \in A_{\mathbf{p}}$ $0 < \gamma < \min\{n(\lambda - 2m)/2, \delta\}$ *and* $f_1, \cdots, f_m \in L^1$. *Then for any* $\mu > 0$, *and* $1 < p_1, \cdots, p_m < \infty$ *with* $\frac{1}{p} = \frac{1}{p_1} + \cdots \frac{1}{p_m}$, *there exist a constant* $C > 0$ *such that*

$$\left\| g_\lambda^*(\mathbf{f}) \right\|_{L^p(\nu_{\mathbf{w}})} \leq C \prod_{i=1}^{m} \| f_i \|_{L^{p_i}(w_i)}.$$

In the last years, there have been several advances in the fruitful area of weighted inequalities concerning the precise determination of the optimal bounds of the weighted operator norm of linear and bilinear Calderón-Zygmund operators with a Dini continuous kernel in terms of the $A_{\mathbf{p}}$ constant of weights (see e.g. [7, 10] and the references therein).

The algorithm to obtain sparse domination is formulated in [10] in general and can be used to study both standard Calderón-Zygmund operators and square functions. However, in order to obtain estimates for kernels satisfying the Dini condition, the main obstacle is the endpoint estimate and its bound.

The thrust of relaxing the log-Dini condition to the Dini condition comes from the works [5, 6]. In fact, in these papers, the authors obtained the sharp estimates for singular integral operators whose kernels satisfy the Dini condition. Thus, it is natural to ask ourselves whether a counterpart to the Littlewood–Paley operators is available.

In [4], we got a precise linear dependence on Dini constants. By means of the product, we relax the log-Dini condition in the pointwise bound to the classical Dini condition. This solves an open problem (see e.g. [3, pp. 37–38]).

In order the present the main results of [4], we recall we the following notation:

- For $\kappa, x > 0$, write $\log^\kappa x = (\log x)^\kappa$.
- By a cube, we mean a compact or right-open cube whose edges are parallel to coordinate axes. For $k = 1, 2, 3$, let $\mathcal{D}^{(k)}(\mathbb{R})$ be the minimal family satisfying the following conditions:
  - $\{[3j + k - 1, 3j + k)\}_{j \in \mathbb{Z}} \subset \mathcal{D}^{(k)}(\mathbb{R})$.
  - If $I_1 \in \mathcal{D}^{(k)}(\mathbb{R})$ satisfies $\ell(I_1) = 2\ell(I_2)$ or $\ell(I_2) = 2\ell(I_1)$ and $\sharp(\overline{I_1} \cap \overline{I_2}) = 1$, then $I_2 \in \mathcal{D}^{(k)}(\mathbb{R})$.

For $\mathbf{k} = (k_1, k_2, \ldots, k_n) \in \{1, 2, 3\}^n$, we define

$$\mathcal{D}^{(\mathbf{k})}(\mathbb{R}^n) = \{I_1 \times I_2 \times \cdots \times I_n \ : \ I_j \in \mathcal{D}^{(k_j)}(\mathbb{R}), \ell(I_1) = \ell(I_2) = \cdots = \ell(I_n)\}.$$

- Let Q be a cube. For $f \in L^1(Q)$, we use the symbol $\langle f \rangle_{1,Q}$ to denote the integral average over a cube Q.
- Let $0 < \eta < 1$. We say that a family $\mathcal{S}$ is said to be η-sparse if

$$\left| \bigcup_{R \in \mathcal{S}, R \subsetneq Q} R \right| \leq (1-\eta)|Q|.$$

The next theorem shows the pointwise control by sparse operators (with precise constants), of multilinear Littlewood–Paley functions with Dini-continuous kernels.

Theorem 7.4 (Theorem 2.2 in [4]) *For any $\alpha \geq 1$ and for all compactly supported $f \in L^1(\mathbb{R}^n)$, there exist sparse families $\mathcal{S}^{(\mathbf{k})} \subset \mathcal{D}^{(\mathbf{k})}$ (depending on f) for each $\mathbf{k} = (k_1, k_2, \ldots, k_n) \in \{1,2,3\}^n$ such that*

$$S_{\alpha,\psi} f \cdot 1_{Q_0}$$

$$\lesssim \alpha^n \log^{\frac{3}{2}}(2+\alpha)([\varphi]_{Dini}[w]_{Dini} + \|S_{1,\psi}\|_{L^2 \to L^2}) \sum_{\mathbf{k} \in \{1,2,3\}^n} \Big[\sum_{P \in \mathcal{S}^{(\mathbf{k})}} \langle f \rangle_{1,P}^2 1_P \Big]^{\frac{1}{2}}.$$

Here, the implicit constant in $\lesssim$ is independent of α, $[w]_{Dini}$ and $[\varphi]_{Dini}$.

Applying Theorem 7.4, one can obtain the weighted bounds for the multilinear Littlewood–Paley functions with Dini-continuous kernels.

Theorem 7.5 (Theorem 3.3 in [4]) *Let $\sqrt[m]{\varphi}$, $\sqrt[m]{w}$ be functions satisfying the Dini condition. Let $\alpha \geq 1$, $\mathbf{w} \in A_{\mathbf{p}}$ and $\frac{1}{p} = \frac{1}{p_1} + \cdots + \frac{1}{p_m}$ with $1 < p_1, \ldots, p_m < \infty$. Write*

$$K := \alpha^{nm} \log^{\frac{1}{2}+m}(2+\alpha)$$

$$\times \left(A[\sqrt[m]{w}]^m_{Dini}(1 + [\sqrt[m]{\varphi}]^m_{Dini}) + \|S_{1,\psi}\|_{L^{p_1} \times L^{p_2} \times \cdots \times L^{p_m} \to L^p}\right) [\mathbf{w}]_{A_{\mathbf{p}}}^{\max(\frac{1}{2}, \frac{p_1'}{p}, \ldots, \frac{p_m'}{p})}.$$

Then

$$\|g_\lambda^*(\mathbf{f})\|_{L^p(\nu_{\mathbf{w}})} \lesssim K \prod_{i=1}^m \|f_i\|_{L^{p_i}(w_i)}$$

for all $\mathbf{f} = \{f_i\}_{i=1}^m$ satisfying $f_i \in L^{p_i}(w_i)$ for each i, where the implicit constant is independent of α and $\mathbf{w}$.

Acknowledgments The research is financially supported by a grant from IPM.

References

1. Bui, T.A., Hormozi, M.: Weighted bounds for multilinear square functions. Potential Anal. **46**, 135–148 (2017)
2. Cao, M., Hormozi, M., Ibañez-Firnkorn, G., Rivera-Ríos, I.P., Si, Z., Yabuta, K.: Weak and strong type estimates for the multilinear Littlewood–Paley operators. J. Fourier Anal. Appl. **27**(62), 42pp. (2021)
3. Cao, M., Yabuta, K.: The multilinear Littlewood–Paley operators with minimal regularity conditions. J. Fourier Anal. Appl. **25**(3), 1203–1247 (2019)
4. Hormozi, M., Sawano, Y., Yabuta, K.: On the weak boundedness of multilinear Littlewood–Paley functions. J. Fourier Anal. Appl. **29**, 49 (2023)
5. Hytönen, T., Roncal, L., Tapiola, O.: Quantitative weighted estimates for rough homogeneous singular integrals. Isr. J. Math. **218**(1), 133–164 (2017)
6. Lacey, M.: An elementary proof of the A_2 bound. Isr. J. Math. **217**(1), 181–195 (2017)
7. Damián, W., Hormozi, M., Li, K.: New bounds for bilinear Calderón-Zygmund operators and applications. Rev. Mat. Iberoam. **34**(3), 1177–1210 (2018)
8. Lerner, A.K.: On sharp aperture-weighted estimates for square functions. J. Fourier Anal. Appl. **20**(4), 784–800 (2014)
9. Lerner, A.K.: Sharp weighted norm inequalities for Littlewood–Paley operators and singular integrals. Adv. Math. **226**, 3912–3926 (2011)
10. Lerner, A.K.: On pointwise estimates involving sparse operators. New York J. Math. **22**, 341–349 (2016)
11. Shi, S., Xue, Q., Yabuta, K.: On the boundedness of multilinear Littlewood–Paley g_λ^* function. J. Math. Pures Appl. **101**, 394–413 (2014)
12. Wilson, J.M.: The intrinsic square function. Rev. Mat. Iberoam. **23**, 771–791 (2007)

Chapter 8
$H^p \to L^p$ Boundedness of Fourier Multipliers on Graded Lie Groups

Qing Hong and Guorong Hu

Abstract In this contribution, we discuss the joint work with Hong et al. (Fourier multipliers for Hardy spaces on graded Lie groups. In: Proceedings of the Royal Society of Edinburgh Section A: Mathematics, 153(5), 1729–1750, 2022. https://doi.org/10.1017/prm.2022.71) wherein we have investigated the $H^p(G) \to L^p(G)$, $0 < p \leq 1$, boundedness of Fourier multiplier operators on an arbitrary graded Lie group G, where $H^p(G)$ is the Hardy spaces on G. Our main result extends those obtained by Fischer and Ruzhansky (Colloq Math 165:1–30, 2021), who proved the $L^1(G) \to L^{1,\infty}(G)$ and $L^p(G) \to L^p(G)$, $1 < p < \infty$, boundedness of such Fourier multiplier operators.

8.1 Introduction

A Lie group G is said to be graded if it is connected and simply connected, and its Lie algebra $\mathfrak{g}$ is endowed with a vector space decomposition $\mathfrak{g} = \oplus_{k=1}^{\infty} \mathfrak{g}_k$ (where all but finitely many of the $\mathfrak{g}_k$'s are $\{0\}$) such that $[\mathfrak{g}_k, \mathfrak{g}_{k'}] \subset \mathfrak{g}_{k+k'}$ for all $k, k' \in \mathbb{N}$. Any graded Lie group G must be nilpotent, and the exponential map $\exp : \mathfrak{g} \to G$ is a diffeomorphism. Examples of graded Lie groups include the Euclidean space $\mathbb{R}^n$, the Heisenberg group $\mathbb{H}^n$ and, more generally, all stratified Lie groups. For a complete description of the notions of graded Lie groups, we refer to [5].

The Lie algebra $\mathfrak{g}$ is equipped with a natural family of dilations $\{\delta_r\}_{r>0}$ which are linear mappings from $\mathfrak{g}$ to $\mathfrak{g}$ determined by

$$\delta_r X = r^k X \quad \text{for } X \in \mathfrak{g}_k.$$

We choose and fix a basis $\{X_1, \cdots, X_n\}$ of $\mathfrak{g}$ which is adapted to the gradation. This means that $\{X_1, \cdots, X_{n_1}\}$ (possibly $\emptyset$) is a basis of $\mathfrak{g}_1$, $\{X_{n_1+1}, \cdots, X_{n_1+n_2}\}$

Q. Hong · G. Hu (✉)
School of Mathematics and Statistics, Jiangxi Normal University, Nanchang, Jiangxi, China
e-mail: qhong@mail.bnu.edu.cn; hugr@mail.ustc.edu.cn

M. Ruzhansky, K. Van Bockstal (eds.), *Extended Abstracts 2021/2022*, Research Perspectives Ghent Analysis and PDE Center 2,
https://doi.org/10.1007/978-3-031-42539-4_8

(possibly $\emptyset$) is a basis of $\mathfrak{g}_2$, and so on. For each $j \in \{1, \cdots, n\}$, let v_j be the unique positive integer such that $X_j \in \mathfrak{g}_{v_j}$. Then we have

$$\delta_r X_j = r^{v_j} X_j, \quad j = 1, \cdots, n.$$

The integers $v_1, \cdots, v_n$ become weights of the dilations $\{\delta_t\}_{t>0}$, and we define the homogeneous dimension of G to be

$$Q = \sum_{k=1}^{\infty} k(\dim \mathfrak{g}_k) = \sum_{j=1}^{n} v_j.$$

To introduce Fourier analysis on graded Lie groups, we need to recall some basic representation theory of Lie groups. A representation π of a Lie group G on a Hilbert space $\mathcal{H}_\pi \neq \{0\}$ is a homomorphism from G into the group of bounded linear operators on $\mathcal{H}_\pi$ with bounded inverse. More precisely,

- for every $x \in G$, the linear mapping $\pi(x) : \mathcal{H}_\pi \to \mathcal{H}_\pi$ is bounded and has bounded inverse;
- for every $x, y \in G$, we have $\pi(xy) = \pi(x)\pi(y)$.

We say that a representation π of G is irreducible if it has no closed invariant subspaces. A representation π is said to be unitary if $\pi(x)$ is unitary for every $x \in G$, and is said to be strongly continuous if the mapping $\pi : G \to \mathscr{L}(\mathcal{H}_\pi)$ is continuous with respect to the strong operator topology in $\mathscr{L}(\mathcal{H}_\pi)$, where $\mathscr{L}(\mathcal{H}_\pi)$ denotes the Banach space of all bounded linear operators on $\mathcal{H}_\pi$. Two representations π_1 and π_2 are said to be equivalent if there exists a bounded linear mapping $A : \mathcal{H}_{\pi_1} \to \mathcal{H}_{\pi_2}$ between their representation spaces with a bounded inverse such that the relation $A\pi_1(x) = \pi_2(x)A$ holds for all $x \in G$. The set of all equivalence classes of strongly continuous irreducible unitary representations of G is called the unitary dual of G and is denoted by $\widehat{G}$. In what follows, we will identify one representation π with its equivalent class $[\pi]$.

For a unitary representation of G, the corresponding infinitesimal representation which acts on the universal enveloping algebra $\mathfrak{U}(\mathfrak{g})$ of the Lie algebra $\mathfrak{g}$ is still denoted by π. This is characterized by its action on $\mathfrak{g}$:

$$\pi(X) = \partial_{t=0}\pi(e^{tX}), \quad X \in \mathfrak{g}.$$

The infinitesimal action acts on the space $\mathcal{H}_\pi^\infty$ of smooth vectors, that is, the space of vectors $v \in \mathcal{H}_\pi$ such that the function $G \ni x \mapsto \pi(x)v \in \mathcal{H}_\pi$ is of class C^∞.

The Fourier coefficients or group Fourier transform of a function $f \in L^1(G)$ at $\pi \in \widehat{G}$ is defined by

$$\mathcal{F}_G f(\pi) \equiv \widehat{f}(\pi) \equiv \pi(f) := \int_G f(x)\pi(x)^* d\mu(x).$$

It is readily seen that

$$\|\widehat{f}(\pi)\|_{\mathscr{L}(\mathcal{H}_\pi)} \leq \|f\|_{L^1(G)}.$$

There exists a unique positive Borel measure $\widehat{\mu}$ on $\widehat{G}$, called the Plancherel measure, such that for any continuous function f on G with compact support, one has

$$\int_G |f(x)|^2 d\mu(x) = \int_{\widehat{G}} \|\mathcal{F}_G f(\pi)\|^2_{HS(\mathcal{H}_\pi)} d\widehat{\mu}(\pi),$$

where $\|\cdot\|_{HS(\mathcal{H}_\pi)}$ denotes the Hilbert-Schmidt norm on the space $HS(\mathcal{H}_\pi) \sim \mathcal{H}_\pi \otimes \mathcal{H}_\pi^*$ of Hilbert-Schmidt operators on the Hilbert space $\mathcal{H}_\pi$. This implies that the Fourier transform $\mathcal{F}_G$ extends to a unitary operator from $L^2(G)$ onto $L^2(\widehat{G})$.

By the general theory on locally compact unimodular groups of type I (see e.g. [4]), if T is an L^2-bounded operator on G which commutes with left-translations, then there exists a field of bounded operators $\widehat{T}(\pi)$ such that for all $f \in L^2(G)$,

$$\mathcal{F}_G(Tf)(\pi) = \widehat{T}(\pi)\widehat{f}(\pi) \quad \text{a.e. } \pi \in \widehat{G}.$$

Moreover, we have

$$\|T\|_{\mathscr{L}(L^2(G))} = \sup_{\pi \in \widehat{G}} \|\widehat{T}(\pi)\|_{\mathscr{L}(\mathcal{H}_\pi)},$$

where the supremum here is understood as the essential supremum with respect to the Plancherel measure μ. Conversely, given any $\sigma = \{\sigma(\pi), \pi \in \widehat{G}\} \in L^\infty(\widehat{G})$, there is a corresponding operator T_σ given by

$$\mathcal{F}_G(T_\sigma f)(\pi) = \sigma(\pi)\widehat{f}(\pi), \quad f \in L^2(G).$$

By the Plancherel theorem, T_σ is bounded on $L^2(G)$ with $\|T_\sigma\|_{\mathscr{L}(L^2(G))} = \|\sigma\|_{L^\infty(\widehat{G})}$. The operator T_σ is called the Fourier multiplier operator associated with σ.

A left-invariant differential operator $\mathcal{R}$ on G is called a Rockland operator if it is homogeneous of positive degree and for each unitary irreducible non-trivial representation π of G, the operator $\pi(\mathcal{R})$ is injective on $\mathcal{H}_\pi^\infty$. Rockland operators may be defined on any homogeneous group, however it turns out that the existence of a Rockland operator on a homogeneous group implies that (the Lie algebra of) the group admits a gradation. This is the reason why we and the authors in [2] consider the setting of graded Lie groups. On any graded Lie group G, the operator

$$\sum_{j=1}^{n} (-1)^{\frac{\nu_0}{\nu_j}} c_j X_j^{2\frac{\nu_0}{\nu_j}}$$

with $c_j > 0$ is a Rockland operator of homogeneous degree $2\nu_0$ if ν_0 is any common multiple of $\upsilon_1, \cdots, \upsilon_n$.

8.2 Weak (1, 1) and L^p Boundedness of T_σ

Fischer and the third-named author [2] first investigated the weak (1, 1) and L^p boundedness of the Fourier multiplier operator T_σ. To generalise the classical Mihlin condition to the setting of graded Lie groups, they introduce the notion of "difference operators" Δ^α, which act on the field $\sigma = \{\sigma(\pi), \pi \in \widehat{G}\}$ of operators and play a role of derivatives with respect to the variable ξ in the Euclidean setting. See [2] and [5] for details.

Let us recall the main result in [2]:

Theorem A ([2, Theorem 1.1]) *Let G be a graded Lie group with homogeneous dimension Q. Let $\sigma = \{\sigma(\pi), \pi \in \widehat{G}\}$ be a measurable field of operators in $L^\infty(\widehat{G})$. Assume that there exist a positive Rockland operator $\mathcal{R}$ (of homogeneous degree ν) and an integer $N > Q/2$ divisible by the dilation weights $\upsilon_1, \cdots, \upsilon_n$ such that*

$$\sup_{\pi \in \widehat{G}} \left\| \Delta^\alpha \sigma\, \pi(\mathcal{R})^{\frac{[\alpha]}{\nu}} \right\|_{\mathscr{L}(\mathcal{H}_\pi)} < \infty \tag{8.1}$$

and

$$\sup_{\pi \in \widehat{G}} \left\| \pi(\mathcal{R})^{\frac{[\alpha]}{\nu}}\, \Delta^\alpha \sigma \right\|_{\mathscr{L}(\mathcal{H}_\pi)} < \infty, \tag{8.2}$$

hold for all $\alpha \in \mathbb{N}_0^n$ with $[\alpha] \le N$. Then the Fourier multiplier operator T_σ defined via

$$\mathcal{F}_G(T_\sigma f)(\pi) = \sigma(\pi)\widehat{f}(\pi) \tag{8.3}$$

is of weak type (1, 1), and is bounded on $L^p(G)$ for all $1 < p < \infty$.

Obviously, Theorem A is a generalisation of the classical Mihlin multiplier theorem.

Examining the proof of Theorem A (given in [2]), we find that the condition (8.1) is sufficient to give the weak (1, 1) estimate of T_σ. The latter, along with the $L^2(G)$-boundedness of T_σ (which follows from the Plancherel theorem) and an interpolation argument yields the $L^p(G)$-boundedness of T_σ for $1 < p \le 2$. Note that $T_\sigma^* = T_{\sigma^*}$, where $\sigma^* = \{\sigma(\pi)^*, \pi \in \widehat{G}\}$, and that if σ satisfies (8.2) then σ^* satisfies (8.1). Hence, if σ satisfies (8.1) and (8.2), then both T_σ and T_σ^* are bounded on $L^p(G)$ for $1 < p \le 2$, which implies that T_σ is bounded on $L^p(G)$ for all $1 < p < \infty$.

8.3 $H^p(G) \to L^p(G)$ Boundedness of T_σ

Our main purpose is to extend Theorem A to the case $0 < p \leq 1$ by investigating the $H^p(G) \to L^p(G)$ boundedness of T_σ, where $H^p(G)$ is the Hardy space on G. Our main result is the following.

Theorem 8.1 *Let G be a graded Lie group with homogeneous dimension Q. Let $\sigma = \{\sigma(\pi), \pi \in \widehat{G}\}$ be a measurable field of operators in $L^\infty(\widehat{G})$. Let $0 < p \leq 1$. Assume that there exist a positive Rockland operator $\mathcal{R}$ (of homogeneous degree ν) and an integer $N > Q(1/p - 1/2)$ divisible by the dilation weights $v_1, \cdots, v_n$ such that*

$$\sup_{\pi \in \widehat{G}} \left\| \Delta^\alpha \sigma \, \pi(\mathcal{R})^{\frac{[\alpha]}{\nu}} \right\|_{\mathscr{L}(\mathcal{H}_\pi)} < \infty$$

holds for all $\alpha \in \mathbb{N}_0^n$ with $[\alpha] \leq N$, Then the Fourier multiplier operator T_σ defined by (8.3) *is bounded from $H^p(G)$ to $L^p(G)$.*

Some remarks and discussions concerning Theorem 8.1 are in order.

1. Theorem 8.1 can be viewed as a generalisation of the Calderón-Torchinsky multiplier theorem [3].
2. Taking $p = 1$ in Theorem 8.1, we have the $H^1(G) \to L^1(G)$ boundedness of T_σ, under the assumption that σ satisfies (8.1) for some integer $N > Q/2$ that is divisible by the dilation weights $v_1, \cdots, v_n$. Thus (by interpolation), our result also implies the $L^p(G)$-boundedness of T_σ stated in Theorem A under the same assumptions.
3. The optimality of the Mihlin-Hörmander condition for spectral/Fourier multipliers on Lie groups is a very deep problem. It is known that on any 2-step stratified group the sufficient and necessary condition for L^p-boundedness of a spectral multiplier $F(\mathcal{L})$ (where $\mathcal{L}$ is a sub-Laplaican) is that F satisfies a scale-invariant smoothness condition of order $s > n/2$, where n is the topological dimension of the group (see [6, 7]). It is natural to ask whether the condition $N > Q(1/p - 1/2)$ in Theorem 8.1 can be replaced by $N > n(1/p - 1/2)$.
4. It is also interesting to study $H^p(G) \to L^p(G)$ boundedness of Fourier multiplier operators on arbitrary compact Lie groups.

Acknowledgments G. Hu and Q. Hong were supported by the NNSF of China (Grant Nos. 11901256 and 12001251) and the NSF of Jiangxi Province (Grant Nos. 20192BAB211001 and 20202BAB211001).

References

1. Hong, Q., Hu, G., Ruzhansky, M.: Fourier multipliers for Hardy spaces on graded Lie groups. In: Proceedings of the Royal Society of Edinburgh Section A: Mathematics, **153**(5), 1729–1750 (2022). https://doi.org/10.1017/prm.2022.71
2. Fischer, V., Ruzhansky, M.: Fourier multipliers on graded Lie groups. Colloq. Math. **165**, 1–30 (2021)
3. Calderón, A.P., Torchinsky, A.: Parabolic maximal functions associated with a distribution. II. Adv. Math. **24**, 101–171 (1977)
4. Dixmier, J.: C^*-algebras. Translated from the French by Francis Jellett, North-Holland Mathematical Library, vol. 15, North-Holland, Amsterdam (1977)
5. Fischer, V., Ruzhansky, M.: Quantization on Nilpotent Lie Groups. Progress in Mathematics, vol. 314. Birkhäuser, Basel (2016)
6. Martini, A., Müller, D.: Spectral multiplier theorems of Euclidean type on new classes of 2-step stratified groups. Proc. Lond. Math. Soc. **109**, 1229–1263 (2014)
7. Martini, A., Müller, D.: Spectral multipliers on 2-step groups: topological versus homogeneous dimension. Geom. Funct. Anal. **26**, 680–702 (2016)

Chapter 9
On a Reverse Integral Hardy Inequality on Polarisable Metric Measure Space

Aidyn Kassymov

Abstract In this short note, we give reverse integral Hardy inequalities on polarisable metric measure space in the case $q \le p < 0$ (with two negative exponents). Also, as for applications, we present the reverse Hardy-Littlewood-Sobolev and Stein-Weiss type inequalities with two negative exponents on homogeneous Lie groups and with arbitrary quasi-norm, the result which appears to be new already in the Euclidean space.

9.1 Introduction

In the famous work [1], G.H. Hardy showed the following (direct) integral inequality:

$$\int_a^\infty \frac{1}{x^p}\left(\int_a^\infty f(t)dt\right)^p dx \le \left(\frac{p}{p-1}\right)^p \int_a^\infty f^p(x)dx,$$

where $f \ge 0$, $p > 1$, and $a > 0$.

The reverse integral Hardy inequality was proved in [2], where the authors obtained it in the following form:

$$\left(\int_a^b \left(\int_a^x f(t)dt\right)^q u(x)dx\right)^{\frac{1}{q}} \ge C\left(\int_a^b f^p(x)v(x)dx\right)^{\frac{1}{p}},$$

A. Kassymov (✉)
Institute of Mathematics and Mathematical Modeling, Almaty, Kazakhstan
e-mail: kassymov@math.kz

M. Ruzhansky, K. Van Bockstal (eds.), *Extended Abstracts 2021/2022*, Research Perspectives Ghent Analysis and PDE Center 2,
https://doi.org/10.1007/978-3-031-42539-4_9

and the conjugate of this inequality is:

$$\left(\int_a^b \left(\int_x^b f(t)dt\right)^q u(x)dx\right)^{\frac{1}{q}} \geq C\left(\int_a^b f^p(x)v(x)dx\right)^{\frac{1}{p}},$$

where $f \geq 0$, $u, v \geq 0$ and $p, q < 0$. The reverse Hardy inequalities were also studied in [3].

We recall a polarisable metric measure space as a metric measure space with the following polar decomposition. That is, we consider metric spaces $\mathbb{X}$ with a Borel measure dx allowing for the following *polar decomposition* at $a \in \mathbb{X}$: we assume that there is a locally integrable function $\lambda \in L^1_{loc}$ such that for all $f \in L^1(\mathbb{X})$ we have

$$\int_{\mathbb{X}} f(x)dx = \int_0^\infty \int_{\Sigma_r} f(r,\omega)\lambda(r,\omega)d\omega_r dr, \tag{9.1}$$

for some set $\Sigma_r = \{x \in \mathbb{X} : d(x,a) = r\} \subset \mathbb{X}$ with a measure on it denoted by $d\omega$, and $(r, \omega) \to a$ as $r \to 0$.

The condition (9.1) is rather general (see [4]) since we allow the function λ to depend on the whole variable $x = (r, \omega)$. Since $\mathbb{X}$ does not necessarily have a differentiable structure, the function $\lambda(r, \omega)$ can not generally be obtained as the Jacobian of the polar change of coordinates. However, if such a differentiable structure exists on $\mathbb{X}$, the condition (9.1) can be obtained as the standard polar decomposition formula. In particular, as several examples of $\mathbb{X}$: Euclidean space, homogeneous groups, hyperbolic spaces, and Cartan-Hadamard manifolds for which the condition (9.1) is satisfied.

In [4], the (direct) integral Hardy inequality on metric measure spaces was established with applications to homogeneous Lie groups, hyperbolic spaces, Cartan-Hadamard manifolds with negative curvature and on general Lie groups with Riemannian distance for $1 < p \leq q < \infty$.

9.2 Main Results

As the main results of this section, we show the reverse integral Hardy inequality as well as its conjugate [5].

Theorem 9.1 *Assume that $p, q < 0$ are such that $q \leq p < 0$. Let $\mathbb{X}$ be a metric measure space with a polar decomposition at $a \in \mathbb{X}$. Suppose that $u, v \geq 0$ are locally integrable functions on $\mathbb{X}$. Then the inequality*

$$\left[\int_{\mathbb{X}} \left(\int_{B(a,|x|_a)} f(y)dy\right)^q u(x)dx\right]^{\frac{1}{q}} \geq C_1(p,q)\left(\int_{\mathbb{X}} f^p(x)v(x)dx\right)^{\frac{1}{p}} \tag{9.2}$$

holds for all non-negative real-valued measurable functions f, if and only if

$$
\begin{aligned}
0 < D_1 &= \inf_{x \neq a} \mathcal{D}_1(|x|_a) \\
&= \inf_{x \neq a} \left[\left(\int_{B(a,|x|_a)} u(y)dy \right)^{\frac{1}{q}} \left(\int_{B(a,|x|_a)} v^{1-p'}(y)dy \right)^{\frac{1}{p'}} \right],
\end{aligned}
$$

and $\mathcal{D}_1(|x|_a)$ is non-decreasing. Moreover, the biggest constant $C_1(p,q)$ in (9.2) satisfies

$$
D_1 \geq C_1(p,q) \geq |p|^{\frac{1}{q}} (p')^{\frac{1}{p'}} D_1,
$$

where $\frac{1}{p} + \frac{1}{p'} = 1$.

The conjugate version of the integral Hardy inequality states as follows:

Theorem 9.2 *Assume that $p, q < 0$ such that $q \leq p < 0$. Let $\mathbb{X}$ be a metric measure space with a polar decomposition at $a \in \mathbb{X}$. Suppose that $u, v \geq 0$ are locally integrable functions on $\mathbb{X}$. Then the inequality*

$$
\left[\int_{\mathbb{X}} \left(\int_{\mathbb{X} \setminus B(a,|x|_a)} f(y)dy \right)^q u(x)dx \right]^{\frac{1}{q}} \geq C_2(p,q) \left(\int_{\mathbb{X}} f^p(x)v(x)dx \right)^{\frac{1}{p}}
$$

holds for all non-negative real-valued measurable functions f, if and only if

$$
\begin{aligned}
0 < D_2 &= \inf_{x \neq a} \mathcal{D}_2(|x|_a) \\
&= \inf_{x \neq a} \left[\left(\int_{\mathbb{X} \setminus B(a,|x|_a)} u(y)dy \right)^{\frac{1}{q}} \left(\int_{\mathbb{X} \setminus B(a,|x|_a)} v^{1-p'}(y)dy \right)^{\frac{1}{p'}} \right],
\end{aligned}
$$

and $\mathcal{D}_2(|x|_a)$ is non-increasing. Moreover, the biggest constant C_2 satisfies

$$
D_2 \geq C_2(p,q) \geq |p|^{\frac{1}{q}} (p')^{\frac{1}{p'}} D_2,
$$

where $\frac{1}{p} + \frac{1}{p'} = 1$.

9.3 Applications

Let us present one of the main results of this note.

Theorem 9.3 (The Reverse Hardy-Littlewood-Sobolev Inequality) *Let $\mathbb{G}$ be a homogeneous Lie group of homogeneous dimension $Q \geq 1$ with arbitrary quasi-*

norm $|\cdot|$. *Assume that* $q < p < 0$, $\lambda < 0$ *such that* $\frac{1}{p'} + \frac{1}{q} + \frac{\lambda}{Q} = 0$, *where* $\frac{1}{p} + \frac{1}{p'} = 1$ *and* $\frac{1}{q} + \frac{1}{q'} = 1$. *Then for all non-negative functions* $f \in L^{q'}(\mathbb{G})$ *and* $0 < \int_{\mathbb{G}} h^p(x)dx < \infty$, *we get*

$$\int_{\mathbb{G}}\int_{\mathbb{G}} f(x)|y^{-1}x|^{\lambda}h(y)dxdy \geq C\left(\int_{\mathbb{G}} f^{q'}(x)dx\right)^{\frac{1}{q'}}\left(\int_{\mathbb{G}} h^p(x)dx\right)^{\frac{1}{p}}, \tag{9.3}$$

where C *is a positive constant independent of* f *and* h.

Remark 9.1 Inequality (9.3) seems to be new even in the Abelian (Euclidean) case $\mathbb{G} = (\mathbb{R}^n, +)$, $Q = n$ and $|\cdot| = |\cdot|_E$, where $|\cdot|_E$ is the Euclidean distance.

Let us now show the reverse Stein-Weiss type inequality. Usually, in the case $1 < p \leq q < \infty$, from Stein-Weiss inequality, we can obtain the Hardy-Littlewood-Sobolev inequality, but in the case $q < p < 0$, from the following reverse Stein-Weiss type inequality, we can not obtain reverse Hardy-Littlewood-Sobolev inequality

Theorem 9.4 *Let* $\mathbb{G}$ *be a homogeneous Lie group of homogeneous dimension* $Q \geq 1$ *with arbitrary quasi-norm* $|\cdot|$. *Assume that* $q \leq p < 0$, $\lambda < 0$, $\beta > -\frac{Q}{p'}$, $\alpha > -\frac{Q}{q}$ *and* $\frac{1}{p'} + \frac{1}{q} + \frac{\alpha+\beta+\lambda}{Q} = 0$, *where* $\frac{1}{p} + \frac{1}{p'} = 1$ *and* $\frac{1}{q} + \frac{1}{q'} = 1$. *Then for all non-negative functions* $f \in L^{q'}(\mathbb{G})$ *and* $0 < \int_{\mathbb{G}} h^p(x)dx < \infty$, *we get*

$$\int_{\mathbb{G}}\int_{\mathbb{G}} |x|^{\alpha} f(x)|y^{-1}x|^{\lambda}h(y)|y|^{\beta}dxdy$$
$$\geq C\left(\int_{\mathbb{G}} f^{q'}(x)dx\right)^{\frac{1}{q'}}\left(\int_{\mathbb{G}} h^p(x)dx\right)^{\frac{1}{p}}, \tag{9.4}$$

where C *is a positive constant independent of* f *and* h.

Remark 9.2 Inequality (9.4) seems to be new even in the Abelian (Euclidean) case $\mathbb{G} = (\mathbb{R}^n, +)$, $Q = n$ and $|\cdot| = |\cdot|_E$.

Remark 9.3 Particularly, from $\alpha > -\frac{Q}{q} > 0$, we cannot obtain the reverse Hardy-Littlewood-Sobolev in (9.4).

Let us now present an improved reverse Stein-Weiss-type inequality on homogeneous Lie groups. We called the following improved version because we need only one condition for the parameters α and β.

Theorem 9.5 *Let* $\mathbb{G}$ *be a homogeneous group of homogeneous dimension* $Q \geq 1$ *and let* $|\cdot|$ *be an arbitrary homogeneous quasi-norm on* $\mathbb{G}$. *Assume that* $q \leq p < 0$, $\lambda < 0$, *and* $\frac{1}{p'} + \frac{1}{q} + \frac{\alpha+\beta+\lambda}{Q} = 0$, *where* $\frac{1}{p} + \frac{1}{p'} = 1$ *and* $\frac{1}{q} + \frac{1}{q'} = 1$. *Then for all non-negative functions* $f \in L^{q'}(\mathbb{G})$ *and* $0 < \int_{\mathbb{G}} h^p(x)dx < \infty$, (9.4) *holds, that is,*

$$\int_{\mathbb{G}}\int_{\mathbb{G}} |x|^{\alpha} f(x)|y^{-1}x|^{\lambda} h(y)|y|^{\beta} dxdy \geq C\left(\int_{\mathbb{G}} f^{q'}(x)dx\right)^{\frac{1}{q'}}\left(\int_{\mathbb{G}} h^{p}(x)dx\right)^{\frac{1}{p}},$$

if one of the following conditions is satisfied:

(a) $\beta > -\frac{Q}{p'}$;

(b) $\alpha > -\frac{Q}{q}$.

Acknowledgments The first and second authors were supported in parts by the FWO Odysseus 1 grant no. G.0H94.18N: Analysis and Partial Differential Equations, by the Methusalem programme of the Ghent University Special Research Fund (BOF) (grant no. 01M01021) and by the EPSRC (grant no. EP/R003025/2). Also, this research has been funded by the Science Committee of the Ministry of Science and Higher Education of the Republic of Kazakhstan (grant no. AP19676031) and partially supported by the collaborative research programme "Qualitative analysis for nonlocal and fractional models" from Nazarbayev University. This research also was supported by Ministry of Higher Education and Science of Republic of Kazakhstan (Grant No.AP23484106).

References

1. Hardy, G.H.: Note on a theorem of Hilbert. Math. Z. **6**(3–4), 314–317 (1920)
2. Beesack, P.R., Heinig, H.P.: Hardy's inequalities with indices less than 1. Proc. Am. Math. Soc. **83**(3), 532–536 (1981)
3. Kufner, A., Kuliev, K., Kulieva, G.: The Hardy inequality with one negative parameter. Banach J. Math. Anal. **2**(2), 76–84 (2008)
4. Ruzhansky, M., Verma, D.: Hardy inequalities on metric measure spaces. Proc. R. Soc. A. **475**(2223), 20180310 (2019)
5. Kassymov, A., Ruzhansky, M., Suragan, D.: Hardy inequalities on metric measure spaces, III: the case $q \leq p \leq 0$ and applications. Proc. R. Soc. A. **479**(2269), 20220307 (2023)

Chapter 10
Logarithmic Sobolev Inequalities of Fractional Order on Noncommutative Tori

Gihyun Lee

Abstract In this paper, we prove a version of the logarithmic Sobolev inequality of fractional order on noncommutative n-tori for any dimension $n \geq 2$.

10.1 Introduction

Among all noncommutative spaces studied in Alain Connes' noncommutative geometry program [11] noncommutative tori are the most extensively studied ones. One of the reasons behind the extensive research conducted on noncommutative tori is the presence of counterparts to various mathematical tools employed in the study of analysis and geometry. For example, counterparts to the notions such as vector bundles, Fourier series, spaces of smooth functions, Sobolev spaces and pseudodifferential calculus are established and available in the setting of noncommutative tori (see, e.g., [8, 10, 17, 18, 24, 34] and the references therein). Moreover, noncommutative tori are utilized in the mathematical modeling of physical phenomena, such as the quantum Hall effect [2], topological insulators [4, 25] and string theory [12, 29].

Let us now briefly review the classical Sobolev inequalities on Euclidean spaces. The classical Sobolev inequality states that if a function f defined on $\mathbb{R}^n$, along with its first derivatives, belongs to $L_p(\mathbb{R}^n)$ and if $q = (\frac{1}{p} - \frac{1}{n})^{-1}$ is finite, then f is in $L_q(\mathbb{R}^n)$. This inequality has broad applications in various fields of analysis, including the study of partial differential equations. However, the classical Sobolev inequality has the following limitations. As evident from the value of q we see that, as the dimension n increases, the difference between q and p narrows. Consequently, as n grows, the improvement in summability obtained from the differentiability decreases. For this reason, in the infinite-dimensional setting such

G. Lee (✉)
Department of Mathematics: Analysis, Logic and Discrete Mathematics, Ghent University, Ghent, Belgium
e-mail: Gihyun.Lee@ugent.be

M. Ruzhansky, K. Van Bockstal (eds.), *Extended Abstracts 2021/2022*, Research Perspectives Ghent Analysis and PDE Center 2,
https://doi.org/10.1007/978-3-031-42539-4_10

as quantum field theory, we cannot expect to obtain an inequality that precisely corresponds to the classical Sobolev inequality. Motivated by this observation, Gross [15] introduced and proved the following logarithmic Sobolev inequality on Euclidean spaces.

$$\int_{\mathbb{R}^n} |f(x)|^2 \log |f(x)| \, d\nu(x) \leq \int_{\mathbb{R}^n} |\nabla f(x)|^2 \, d\nu(x) + \|f\|_2^2 \log \|f\|_2.$$

Here ν denotes the Gaussian measure on $\mathbb{R}^n$ and $\|\cdot\|_2$ denotes the L_2-norm associated with ν. As mentioned in [15] this inequality can be utilized in the infinite-dimensional setting, because the coefficients in the inequality do not depend on the dimension n.

Since the introduction of Gross' logarithmic Sobolev inequality, various methods have been employed to establish logarithmic Sobolev inequalities in different settings. Given the vast number of papers on this topic, it is not feasible to encompass all the references here. However, for a partial overview, let us list a few results which can be found in the literature. Rosen presented and proved a logarithmic Sobolev inequality on weighted $\mathbb{R}^n$ [27]. Weissler proved a logarithmic Sobolev inequality on the circle [33]. Gross [16] and Chatzakou et al. [7] investigated logarithmic Sobolev inequalities on Lie groups. Stroock-Zegarlinski [31] and Bodineau-Helffer [3] obtained logarithmic Sobolev inequalities for spin systems. Brannan-Gao-Junge derived a version of the logarithmic Sobolev inequality by utilizing the lower bound of the Ricci curvature of a compact Riemannian manifold [5].

In the case of noncommutative tori, as mentioned above, there exist counterparts to the tools used in Fourier theory for ordinary tori (see [8]). Therefore, the arguments employed on ordinary tori can often be adapted to the setting of noncommutative tori. By using this harmonic analysis technique of noncommutative tori and the theory of operator algebras Xiong-Xu-Yin provided a detailed account on the construction of Sobolev, Besov and Triebel-Lizorkin spaces on noncommutative tori ([34]; see also [18, 28, 30] for Sobolev spaces on noncommutative tori). The embedding theorems between these spaces are also proved in [34]. In addition, in [20, 21] McDonald-Ponge proved versions of Sobolev inequalities on noncommutative tori as consequences of the Lieb-Thirring inequalities on (curved) noncommutative tori.

However, to the best of the author's knowledge, logarithmic Sobolev inequalities in the setting of noncommutative tori have not been studied much in the literature. The logarithmic Sobolev inequality on noncommutative 2-tori by Khalkhali-Sadeghi [19] is the only existing result on this topic. They attempted to establish logarithmic Sobolev inequality on noncommutative 2-tori by adopting Weissler's proof of the logarithmic Sobolev inequality on the circle [33]. However, due to technical issues arising from the noncommutativity, they were only able to obtain the following form of logarithmic Sobolev inequality for strictly positive elements of the form $x = \sum_{k\in\mathbb{Z}} x_k U_1^k U_2^{kl}$, where $0 \neq l \in \mathbb{Z}$:

$$\tau\left(x^2 \log x\right) \leq \sum_{k\in\mathbb{Z}} (1 + |l|)\, |k|\, |x_k|^2 + \|x\|_{L_2}^2 \log \|x\|_{L_2}.$$

We refer to Sect. 10.2 for notations and background material on noncommutative tori.

In this paper, we prove the following logarithmic Sobolev inequalities of fractional order on noncommutative tori for any dimension $n \geq 2$.

Theorem 10.1 *Let* $0 < x \in C^\infty(\mathbb{T}_\theta^n)$, $a > 0$ *and* $0 < s < \frac{n}{2}$. *Then there is a constant* $C(n, s, a) > 0$ *depends only on* n, s *and* a *such that*

$$\tau\left[x^2 \log\left(\frac{x^2}{\|x\|_{L_2}^2}\right)\right] \leq C(n, s, a)\, \|x\|_{W_2^s}^2 - \frac{n}{s}(\log a + 1)\, \|x\|_{L_2}^2. \tag{10.1}$$

Our proof of this theorem is based on the short proof of fractional order logarithmic Sobolev inequality on $\mathbb{R}^n$ presented in a recent article by Chatzakou-Ruzhansky [6]. In the setting of noncommutative tori, there are tools correspond to the key tools used in the proof by Chatzakou-Ruzhansky [6], which enables us to apply their approach immediately to noncommutative tori. Instead of Jensen's inequality for concave functions and probability measures used in the proof by Chatzakou-Ruzhansky, we utilize Jensen's operator inequality known in the operator algebraic setting ([9, 14]; see also [23]) and the embedding theorem between Sobolev spaces proved in [34] to prove Theorem 10.1.

Although the main result of this paper, Theorem 10.1, is the first result on logarithmic Sobolev inequality of *fractional order* on noncommutative tori, it still has certain limitations and room for improvement. First, the Sobolev norm used on the right-hand side of (10.1) needs to be replaced by the homogeneous Sobolev norm,

$$\|x\|_{\dot{W}_2^s(\mathbb{T}_\theta^n)}^2 := \sum_{k\in\mathbb{Z}^n} |k|^{2s}\, |x_k|^2, \qquad s > 0.$$

However, an inequality between homogeneous Sobolev norms on noncommutative tori which can be applied to the arguments in this paper is missing in the literature. Although McDonald-Ponge's Sobolev inequalities [20, 21] deal with homogeneous Sobolev norms, these results cannot be directly applied to the arguments of this paper. In [20, 21] Sobolev inequalities are proven only for zero mean value elements, i.e., elements x with $\tau(x) = 0$. But in order for the logarithm defined by holomorphic functional calculus appearing in (10.1) to make sense, our focus should be restricted to strictly positive elements, and strictly positive elements cannot have a zero mean. Another issue is the lack of a result on the sharpness of Sobolev inequalities on noncommutative tori. In [6] Chatzakou-Ruzhansky utilized the sharp constant of the Sobolev inequality on $\mathbb{R}^n$ obtained in [13] to get an explicit expression of the constant appearing in their logarithmic Sobolev inequality and study its behavior. Similarly, if we could determine the sharpness of Sobolev inequalities on noncommutative tori, we could expect to improve the constant on the right-hand side of (10.1) or get an explicit expression of it. In particular, the sharpness of Sobolev inequalities would enable us to verify the conjecture on the logarithmic Sobolev inequality on noncommutative 2-tori stated in [19] by setting $a = \frac{1}{e}$ in (10.1).

This paper is organized as follows. In Sect. 10.2, we gather some background material on noncommutative tori, Sobolev spaces and Jensen's operator inequality used in this paper. In Sect. 10.3, we prove the logarithmic Sobolev inequalities of fractional order on noncommutative tori, the main result of this paper (Theorem 10.1).

10.2 Preliminaries

In this section, we gather background material on noncommutative tori, Sobolev spaces and Jensen's operator inequality used in the proof of the main theorem in Sect. 10.3.

10.2.1 Noncommutative Tori

We refer to [11, 17, 26] for a more detailed account on noncommutative tori.

Let θ be a skew-symmetric $n \times n$ matrix over $\mathbb{R}$. The noncommutative torus associated with θ, denoted by $\mathbb{T}^n_\theta$, is a noncommutative space in the sense of Alain Connes' noncommutative geometry [11]. The C^*-algebra $C(\mathbb{T}^n_\theta)$ and the von Neumann algebra $L_\infty(\mathbb{T}^n_\theta)$ of $\mathbb{T}^n_\theta$ are generated by the unitaries $U_1, \ldots, U_n$ subject to the relations,

$$U_k U_j = e^{2\pi i \theta_{jk}} U_j U_k, \qquad 1 \leq j, k \leq n.$$

These unitaries $U_1, \ldots, U_n$ can be concretely realized as unitary operators on $L_2(\mathbb{T}^n)$ (see, e.g., [17]), and hence both $C(\mathbb{T}^n_\theta)$ and $L_\infty(\mathbb{T}^n_\theta)$ are $*$-subalgebras of $\mathscr{L}(L_2(\mathbb{T}^n))$, the C^*-algebra of bounded linear operators on $L_2(\mathbb{T}^n)$. If $\theta = 0$, then we recover the spaces of continuous functions $C(\mathbb{T}^n)$ and essentially bounded functions $L_\infty(\mathbb{T}^n)$ on the ordinary n-torus $\mathbb{T}^n = \mathbb{R}^n/(2\pi\mathbb{Z})^n$.

In what follows, for any $k = (k_1, \ldots, k_n) \in \mathbb{Z}^n$, we shall denote $U_1^{k_1} \cdots U_n^{k_n}$ by U^k. Given any $T \in \mathscr{L}(L_2(\mathbb{T}^n))$, we define

$$\tau(T) = \langle T1|1\rangle_{L_2(\mathbb{T}^n)} = (2\pi)^{-n} \int_{\mathbb{T}^n} (T1)(x)\, dx.$$

This defines a tracial state on both $C(\mathbb{T}^n_\theta)$ and $L_\infty(\mathbb{T}^n_\theta)$. By a direct calculation, it can be shown that $\tau(U^k) = 0$ for $0 \neq k \in \mathbb{Z}^n$ and $\tau(1) = 1$. For $x, y \in C(\mathbb{T}^n_\theta)$ we define

$$\langle x|y\rangle_{L_2(\mathbb{T}^n_\theta)} = \tau(xy^*).$$

Let us denote by $L_2(\mathbb{T}^n_\theta)$ the Hilbert space completion of $C(\mathbb{T}^n_\theta)$ with respect to this inner product. Then the family $\{U^k;\ k \in \mathbb{Z}^n\}$ forms an orthonormal basis for $L_2(\mathbb{T}^n_\theta)$. This orthonormal basis plays the role of the standard orthonormal basis for $L_2(\mathbb{T}^n)$ utilized in Fourier analysis, i.e., every $x \in L_2(\mathbb{T}^n_\theta)$ can be uniquely written as follows.

$$x = \sum_{k\in\mathbb{Z}^n} x_k U^k, \qquad x_k := \left\langle x|U^k\right\rangle_{L_2(\mathbb{T}^n_\theta)}. \tag{10.2}$$

Furthermore, the GNS construction (see, e.g., [1]) associated with τ gives rise to the $*$-representations of $C(\mathbb{T}^n_\theta)$ and $L_\infty(\mathbb{T}^n_\theta)$ on the Hilbert space $L_2(\mathbb{T}^n_\theta)$.

The C^*-algebra $C(\mathbb{T}^n_\theta)$ admits a strongly continuous action of $\mathbb{R}^n$, denoted by $\alpha_s(x)$ for $s \in \mathbb{R}^n$ and $x \in C(\mathbb{T}^n_\theta)$. Hence, the triple $(C(\mathbb{T}^n_\theta), \mathbb{R}^n, \alpha)$ forms a C^*-dynamical system. For the elements U^k, $k \in \mathbb{Z}^n$, this action is given by

$$\alpha_s(U^k) = e^{is\cdot k} U^k, \qquad s \in \mathbb{R}^n.$$

The C^*-dynamical system structure on $C(\mathbb{T}^n_\theta)$ enables us to define the dense subalgebra $C^\infty(\mathbb{T}^n_\theta)$ of smooth elements of the action α, i.e., we define

$$C^\infty(\mathbb{T}^n_\theta) := \left\{x \in C(\mathbb{T}^n_\theta);\ \mathbb{R}^n \ni s \mapsto \alpha_s(x) \in C(\mathbb{T}^n_\theta) \text{ is a } C^\infty\text{-map}\right\}.$$

We also have the following characterization of the smooth elements in $C(\mathbb{T}^n_\theta)$ in terms of the Fourier series expansion given in (10.2).

$$C^\infty(\mathbb{T}^n_\theta) = \left\{x = \sum_{k\in\mathbb{Z}^n} x_k U^k;\ (x_k)_{k\in\mathbb{Z}^n} \in \mathcal{S}(\mathbb{Z}^n)\right\}.$$

Here $\mathcal{S}(\mathbb{Z}^n)$ denotes the space of rapidly decaying sequences indexed by $\mathbb{Z}^n$ with entries in $\mathbb{C}$. Furthermore, the action $\mathbb{R}^n \ni s \mapsto \alpha_s(x) \in C^\infty(\mathbb{T}^n_\theta)$ also enables us to define the derivations $\partial_1, \ldots, \partial_n$ on $C^\infty(\mathbb{T}^n_\theta)$. For $j = 1, \ldots, n$, we define $\partial_j : C^\infty(\mathbb{T}^n_\theta) \to C^\infty(\mathbb{T}^n_\theta)$ by letting

$$\partial_j(x) = \partial_{s_j}\alpha_s(x)\big|_{s=0}, \qquad x \in C^\infty(\mathbb{T}^n_\theta).$$

In particular, for the elements U^k, $k = (k_1, \ldots, k_n) \in \mathbb{Z}^n$, we obtain $\partial_j(U^k) = ik_j(U^k)$, $1 \leq j \leq n$.

10.2.2 Sobolev and L_p-Spaces

A detailed account on Sobolev spaces on noncommutative tori can be found in [18, 28, 30, 34].

Let us denote the L_2-version of Sobolev space of order $s \geq 0$ on $\mathbb{T}^n_\theta$ by $W^s_2(\mathbb{T}^n_\theta)$. This space consists of elements $x = \sum_{k\in\mathbb{Z}^n} x_k U^k$ in $L_2(\mathbb{T}^n_\theta)$ such that

$$\left((1+|k|^2)^{\frac{s}{2}} x_k\right)_{k\in\mathbb{Z}^n} \in \ell_2(\mathbb{Z}^n).$$

The space $W^s_2(\mathbb{T}^n_\theta)$ is a Hilbert space with the inner product,

$$\langle x|y\rangle_{W^s_2(\mathbb{T}^n_\theta)} := \sum_{k\in\mathbb{Z}^n} (1+|k|^2)^s x_k \overline{y_k},$$

$$x = \sum_{k\in\mathbb{Z}^n} x_k U^k,\, y = \sum_{k\in\mathbb{Z}^n} y_k U^k \in W^s_2(\mathbb{T}^n_\theta).$$

We shall denote the norm associated with this inner product by $\|\cdot\|_{W^s_2(\mathbb{T}^n_\theta)}$.

We apply the theory of noncommutative L_p-spaces (see, e.g., [22, 32]) to the von Neumann algebra $L_\infty(\mathbb{T}^n_\theta)$ to construct the L_p-spaces of $\mathbb{T}^n_\theta$. Recall that the elements of $L_\infty(\mathbb{T}^n_\theta)$ are represented as bounded linear operators on the Hilbert space $L_2(\mathbb{T}^n)$. We say that a closed and densely defined operator on $L_2(\mathbb{T}^n)$ is affiliated with $L_\infty(\mathbb{T}^n_\theta)$ if it commutes with the commutant of $L_\infty(\mathbb{T}^n_\theta)$ in $\mathscr{L}(L_2(\mathbb{T}^n))$. Given a positive operator x on $L_2(\mathbb{T}^n)$ affiliated with $L_\infty(\mathbb{T}^n_\theta)$ let $x = \int_0^\infty \lambda\, dE(\lambda)$ be its spectral representation and set

$$\tau(x) := \int_0^\infty \lambda\, d\tau(E(\lambda)).$$

For $1 \leq p < \infty$, the L_p-space of $\mathbb{T}^n_\theta$, denoted by $L_p(\mathbb{T}^n_\theta)$, consists of all elements x in the $*$-algebra of $L_\infty(\mathbb{T}^n_\theta)$-affiliated operators on $L_2(\mathbb{T}^n)$ such that

$$\|x\|_{L_p(\mathbb{T}^n_\theta)} := \tau\left(|x|^p\right)^{\frac{1}{p}} < \infty.$$

The space $L_p(\mathbb{T}^n_\theta)$ is a Banach space with the norm $\|\cdot\|_{L_p(\mathbb{T}^n_\theta)}$.

Furthermore, we have the following inclusion of Sobolev spaces into L_p-spaces. This is a particular case of the embedding theorem for Sobolev spaces on noncommutative tori stated in [34].

Proposition 10.1 (see [34, Theorem 6.6]) *Let $p > 2$ and set $s = n(\frac{1}{2} - \frac{1}{p})$. Then there is a continuous embedding,*

$$W^s_2(\mathbb{T}^n_\theta) \subset L_p(\mathbb{T}^n_\theta).$$

10.2.3 Jensen's Operator Inequality

Definition 10.1 Let $f(t)$ be a function on an interval $I \subset \mathbb{R}$. We say that $f(t)$ is *operator convex* if, for all $\lambda \in [0, 1]$ and for every selfadjoint operator A and B on a Hilbert space $\mathscr{H}$ such that $\operatorname{Sp} A \subset I$ and $\operatorname{Sp} B \subset I$, we have

$$f((1-\lambda)A + \lambda B) \leq (1-\lambda)f(A) + \lambda f(B).$$

In contrast, $f(t)$ is called *operator concave* if $-f(t)$ is operator convex.

Proposition 10.2 ([9, 14]; see also [23, Theorem 1.20]) *Let $\mathscr{H}$ and $\mathscr{H}'$ be Hilbert spaces. Suppose that $f(t)$ is an operator convex function on the interval I, $x \in \mathscr{L}(\mathscr{H})$ and $\operatorname{Sp} x \subset I$. Then we have*

$$\pi(f(x)) \geq f(\pi(x)), \tag{10.3}$$

for any positive normalized linear map $\pi : \mathscr{L}(\mathscr{H}) \to \mathscr{L}(\mathscr{H}')$.

Remark 10.1 The inequality (10.3) is of course reversed if we employ an operator concave function instead of an operator convex function.

10.3 The Proof of Theorem 10.1

In this section, we provide the proof of Theorem 10.1, the main result of this paper.

Proof of Theorem 10.1 Let $0 < x \in C^\infty(\mathbb{T}^n_\theta)$. Then, for any $\varepsilon > 0$, we have

$$\begin{aligned} \tau\left[x^2 \log\left(\frac{x^2}{\|x\|_{L_2}^2}\right)\right] &= \frac{1}{\varepsilon}\tau\left[x^2 \log\left(\frac{x^2}{\|x\|_{L_2}^2}\right)^{\varepsilon}\right] \\ &= \frac{\|x\|_{L_2}^2}{\varepsilon}\tau\left[\frac{x^2}{\|x\|_{L_2}^2}\log\left(\frac{x^{2\varepsilon}}{\|x\|_{L_2}^{2\varepsilon}}\right)\right]. \end{aligned} \tag{10.4}$$

Note that the map $\mathscr{L}(L_2(\mathbb{T}^n)) \ni u \mapsto \tau\left[\frac{|x|^2}{\|x\|_{L^2}^2}u\right] \in \mathbb{C}$ is a positive normalized linear map. Moreover, we know by [23, Example 1.7] that the logarithmic function $\log t$ is operator concave on $(0, \infty)$. Therefore, as we have $\operatorname{Sp}\left(x^{2\varepsilon}/\|x\|_{L_2}^{2\varepsilon}\right) \subset (0, \infty)$, it follows from Jensen's operator inequality (Proposition 10.2 and Remark 10.1) that we have

$$\begin{aligned}\tau\left[\frac{x^2}{\|x\|_{L_2}^2}\log\left(\frac{x^{2\varepsilon}}{\|x\|_{L_2}^{2\varepsilon}}\right)\right] &\le \log\tau\left(\frac{x^{2\varepsilon+2}}{\|x\|_{L_2}^{2\varepsilon+2}}\right)\\ &= (\varepsilon+1)\log\left\|\frac{x}{\|x\|_{L_2}}\right\|_{L_{2\varepsilon+2}}^2\\ &= (\varepsilon+1)\log\left(\frac{\|x\|_{L_{2\varepsilon+2}}^2}{\|x\|_{L_2}^2}\right).\end{aligned}\tag{10.5}$$

For all $0 < b, t \in \mathbb{R}$, we have

$$\log t \le bt - \log b - 1.$$

Combining this with the estimates (10.4) and (10.5) we obtain

$$\begin{aligned}\tau\left[x^2\log\left(\frac{x^2}{\|x\|_{L_2}^2}\right)\right] &\le \frac{\|x\|_{L_2}^2(\varepsilon+1)}{\varepsilon}\log\left(\frac{\|x\|_{L_{2\varepsilon+2}}^2}{\|x\|_{L_2}^2}\right)\\ &\le \frac{\|x\|_{L_2}^2(\varepsilon+1)}{\varepsilon}\left(b\frac{\|x\|_{L_{2\varepsilon+2}}^2}{\|x\|_{L_2}^2} - \log b - 1\right)\\ &= \frac{\varepsilon+1}{\varepsilon}\left(b\,\|x\|_{L_{2\varepsilon+2}}^2 - [\log b + 1]\,\|x\|_{L_2}^2\right).\end{aligned}\tag{10.6}$$

Let $\varepsilon > 0$ be such that $2\varepsilon + 2 = \frac{2n}{n-2s}$ for $0 < s < \frac{n}{2}$ and set $b = ea^2$ for $a > 0$. We know by Proposition 10.1 that there is $C(n, s) > 0$ depending only on n and s such that

$$\|x\|_{L_{2\varepsilon+2}}^2 \le C(n, s)\,\|x\|_{W_2^s}^2.$$

It then follows from this and (10.6) that

$$\begin{aligned}\tau\left[x^2\log\left(\frac{x^2}{\|x\|_{L_2}^2}\right)\right] &\le \frac{\frac{n}{n-2s}}{\frac{2s}{n-2s}}\left(ea^2\,\|x\|_{L_{2\varepsilon+2}}^2 - [\log(ea^2)+1]\,\|x\|_{L_2}^2\right)\\ &\le \frac{n}{2s}\left(ea^2C(n,s)\,\|x\|_{W_2^s}^2 - 2(\log a + 1)\,\|x\|_{L_2}^2\right)\\ &= \frac{nea^2}{2s}C(n,s)\,\|x\|_{W_2^s}^2 - \frac{n}{s}(\log a + 1)\,\|x\|_{L_2}^2.\end{aligned}$$

This completes the proof. □

Acknowledgments The author wishes to thank Michael Ruzhansky for helpful discussions related to the topic of this paper.

The research for this article is financially supported by the FWO Odysseus 1 grant G.0H94.18N: Analysis and Partial Differential Equations and by the Methusalem programme of the Ghent University Special Research Fund (BOF) (Grant number 01M01021).

References

1. Arveson, W.: An invitation to C^*-Algebras. Springer, Berlin (1981)
2. Bellissard, J., van Elst, A., Schulz-Baldes, H.: The noncommutative geometry of the quantum Hall effect. J. Math. Phys. **35**, 5373–5451 (1994)
3. Bodineau, T., Helffer, B.: The log-Sobolev inequalities for unbounded spin systems. J. Funct. Anal. **166**(1), 168–178 (1999)
4. Bourne, C., Carey, A.L., Rennie, A.: A non-commutative framework to topological insulators. Rev. Math. Phys. **28**(2), 1650004, 51pp. (2016)
5. Brannan, M., Gao, L., Junge, M.: Complete logarithmic Sobolev inequalities via Ricci curvature bounded below. Adv. Math. **394**, 108129, 60pp. (2022)
6. Chatzakou, M., Ruzhansky, M.: Revised logarithmic Sobolev inequalities of fractional order (2023), 8pp. arXiv:2302.05126
7. Chatzakou, M., Kassymov, A., Ruzhansky, M.: Logarithmic Sobolev inequalities on Lie groups (2021), 35pp. arXiv:2106.15652
8. Chen, Z., Xu, Q., Yin, Z.: Harmonic analysis on quantum tori. Commun. Math. Phys. **322**(3), 755–805 (2013)
9. Choi, M.D.: A Schwarz inequality for positive linear maps on C^*-algebras. Illinois J. Math. **18**, 565–574 (1974)
10. Connes, A.: C*-algèbres et géométrie differentielle. C. R. Acad. Sc. Paris, sér. A **290**, 599–604 (1980)
11. Connes, A.: Noncommutative geometry. Academic Press, San Diego (1994)
12. Connes, A., Douglas, M.R., Schwarz, A.: Noncommutative geometry and matrix theory: compactification on tori. J. High Energy Phys. **2**, 3, 35pp. (1998)
13. Cotsiolis, A., Tavoularis, N.: Best constants for Sobolev inequalities for higher order fractional derivatives. J. Math. Anal. Appl. **295**(1), 225–236 (2004)
14. Davis, C.: A Schwartz inequality for convex operator functions. Proc. Am. Math. Soc. **8**, 42–44 (1957)
15. Gross, L.: Logarithmic Sobolev inequalities. Am. J. Math. **97**(4), 1061–1083 (1975)
16. Gross, L.: Logarithmic Sobolev inequalities on Lie groups. Illinois J. Math. **36**(3), 447–490 (1992)
17. Ha, H., Lee, G., Ponge, R.: Pseudodifferential calculus on noncommutative tori, I. Oscillating integrals. Int. J. Math. **30**, 1950033, 74pp. (2019)
18. Ha, H., Lee, G., Ponge, R.: Pseudodifferential calculus on noncommutative tori, II. Main properties. Int. J. Math. **30**, 1950034, 73pp. (2019)
19. Khalkhali, M., Sadeghi, S.: On logarithmic Sobolev inequality for the noncommutative two torus. J. Pseudo-Differ. Oper. Appl. **8**, 453–484 (2017)
20. McDonald, E., Ponge, R.: Cwikel estimates and negative eigenvalues of Schrödinger operators on noncommutative tori. J. Math. Phys. **61**, 0435031, 37pp. (2021)
21. McDonald, E., Ponge, R.: Dixmier trace formulas and negative eigenvalues of Schrödinger operators on curved noncommutative tori. Adv. Math. **412**, 108815, 57pp. (2023)
22. Nelson, E.: Notes on non-commutative integration. J. Funct. Anal. **15**, 103–116 (1974)
23. Pečarić, J., Furuta, T., Mićić Hot J., Seo, Y.: Mond-Pečarić method in operator inequalities: inequalities for bounded selfadjoint operators on a Hilbert Space. Monographs in Inequalities. Element, Zagreb (2005)

24. Polishchuk, A., Schwarz, A.: Categories of holomorphic vector bundles on noncommutative two-tori. Commun. Math. Phys. **236**(1), 135–159 (2003)
25. Prodan, E., Schulz-Baldes, H.: Bulk and boundary invariants for complex topological insulators: from K-theory to physics. Mathematical Physics Studies. Springer, Berlin (2016)
26. Rieffel, M.: Non-commutative tori-A case study of non-commutative differentiable manifolds. Contemporary Mathematics, vol. 105, pp. 191–211. American Mathematical Society, Providence (1990)
27. Rosen, J.: Sobolev inequalities for weighted spaces and supercontractivity. Trans. Am. Math. Soc. **222**, 367–376 (1976)
28. Rosenberg, J.: Noncommutative variations on Laplace's equation. Anal. PDE **1**, 95–114 (2008)
29. Seiberg, N., Witten, E.: String theory and noncommutative geometry. J. High Energy Phys. **9**, 32, 93pp. (1999)
30. Spera, M.: Sobolev theory for noncommutative tori. Rend. Sem. Mat. Univ. Padova **86**, 143–156 (1992)
31. Stroock, D.W., Zegarlinski, B.: The logarithmic Sobolev inequality for continuous spin systems on a lattice. J. Funct. Anal. **104**(2), 299–326 (1992)
32. Terp, M.: L^p-spaces associated with von Neumann algebras. Copenhagen University, Copenhagen (1981)
33. Weissler, F.: Logarithmic Sobolev inequalities and hypercontractivity estimates on the circle. J. Funct. Anal. **37**, 218–234 (1980)
34. Xiong, X., Xu, Q., Yin, Z.: Sobolev, Besov and Triebel-Lizorkin spaces on quantum tori. Mem. Am. Math. Soc. **252**(1203), 86pp. (2018)

Chapter 11
The Prabhakar Fractional q-Integral and q-Differential Operators

Azizbek Mamanazarov

Abstract In this short note, we discuss the joint work (Shaimardan et al., The Prabhakar fractional q-integral and q-differential operators, and their properties, 2022. arXiv:2212.08843) with E. Karimov and M. Ruzhansky. More specifically, we give the definitions and the main properties of the Prabhakar fractional q-integral and q-differential operators. More precisely, we present the semigroup property of the Prabhakar fractional q-integral operator, which allowed us to introduce the corresponding q-differential operator. Formulas for compositions of q-integral and q-differential operators are also presented. Moreover, we give the result on the boundness of the Prabhakar fractional q-integral operator in the class of q-integrable functions.

11.1 Introduction

Fractional calculus is the area of mathematical analysis that deals with the study and application of integrals and derivatives of arbitrary order. In recent decades, fractional calculus has become of increasing significance due to its applications in many fields of science and engineering. For instance, it has many applications in viscoelasticity, signal processing, electromagnetics, fluid mechanics, and optics. For more information on this research, we refer the reader to [1–6] and the references therein.

First, we would like to give some brief information about the classical Prabhakar fractional calculus. The theory of Prabhakar fractional calculus [7] has been studied more intensively in recent years, and, as a result, certain differential equations involving Prabhakar operators have become an intensive target, which is interesting both for their pure mathematical properties [8, 9] and for their real-world applications in topics such as viscoelasticity, anomalous dielectrics, and

A. Mamanazarov (✉)
Fergana State University, Fergana, Uzbekistan

M. Ruzhansky, K. Van Bockstal (eds.), *Extended Abstracts 2021/2022*,
Research Perspectives Ghent Analysis and PDE Center 2,
https://doi.org/10.1007/978-3-031-42539-4_11

options pricing [10–12]. The origin of the q-difference calculus can be traced back to the works [13] and [14]. Recently, the fractional q-difference calculus has been proposed by Alsalam [15] and Agarwal [16]. Nowadays, new developments in the theory of fractional q-difference calculus have been addressed extensively by several researchers (see [17, 18] and the references therein).

In the present work, our aim is to introduce and study some properties of Prabhakar fractional q-integral and differential operators.

11.2 Preliminaries

First, we recall some elements of the q-calculus for the sequel. For more information, we note the works [19, 20] and the references therein. From now on, we assume that $0 < q < 1$ and $0 \leq a < b < \infty$. Let $\alpha \in \mathbb{R}$. A q-real number $[\alpha]_q$ is defined by

$$[\alpha]_q = \frac{1-q^\alpha}{1-q}.$$

The q-shifted factorial is defined by

$$(a;q)_n = \begin{cases} 1, & n = 0; \\ (1-a)(1-aq)\ldots\left(1-aq^{n-1}\right), & n \in \mathbb{N}. \end{cases}$$

The q-analogue of the factorial is

$$[n]_q! = [1]_q[2]_q[3]_q\ldots[n]_q = \frac{(q;q)_n}{(1-q)^n}, n \in \mathbb{N}, \quad [0]_q! = 1.$$

For the q-binomial coefficients, we have the following formula

$$\begin{bmatrix} n \\ k \end{bmatrix}_q = \frac{(1-q^n)(1-q^{n-1})\ldots(1-q)^{n-k+1}}{(q;q)_k} = \frac{[n]_q!}{[n-k]_q![k]_q!}.$$

Moreover, the q-analogue of the power $(a-b)_q^k$ is defined by

$$(a-b)_q^0 = 1, \quad (a-b)_q^k = \prod_{i=0}^{k-1}\left(a - bq^i\right), \quad k \in \mathbb{N}.$$

There is the following relationship between them:

$$(a-b)_q^0 = 1; \quad (a-b)_q^k = a^k (b/a; q)_k, \quad a \neq 0, \quad k \in \mathbb{N},$$

as well as

$$(a-b)_q^\alpha = a^\alpha \frac{(b/a; q)_\infty}{(q^\alpha b/a; q)_\infty}, \quad (a; q)_\alpha = \frac{(a; q)_\infty}{(aq^\alpha; q)_\infty},$$

$$(a; q)_\infty = \prod_{i=0}^{\infty} \left(1 - aq^i\right).$$

For $x > 0$, the q-analogue of the gamma function is defined by

$$\Gamma_q(x) = \frac{(q; q)_\infty}{(q^x; q)_\infty} (1-q)^{1-x}.$$

It has the following property

$$\Gamma_q(x+1) = [x]_q \Gamma_q(x).$$

The (Jackson) q-derivative of a function $f(x)$ is defined by

$$\left(D_q f\right)(x) = \frac{f(x) - f(qx)}{x(1-q)}, \quad (x \neq 0)$$

and q-derivatives $D_q^n f$ of higher order are defined inductively as follows:

$$D_q^0 f = f, \quad D_q^n f = D_q\left(D_q^{n-1} f\right) \quad (n = 1, 2, 3, ...).$$

Moreover,

$$D_q[(x-b)_q^\alpha] = [\alpha]_q (x-b)_q^{\alpha-1},$$

$$D_q[(a-x)_q^\alpha] = -[\alpha]_q (a-qx)_q^{\alpha-1}.$$

The q-integral (Jackson integral) is defined by

$$\left(I_{q,0+} f\right)(x) = \int_0^x f(t)\, d_q t = x(1-q) \sum_{k=0}^{\infty} f\left(xq^k\right) q^k,$$

and

$$\left(I_{q,a+}f\right)(x) = \int\limits_{a}^{x} f\left(t\right) d_q t = \int\limits_{0}^{x} f\left(t\right) d_q t - \int\limits_{0}^{a} f\left(t\right) d_q t.$$

For the n-th order integral operator $I_{q,a}^{n}$ we have

$$(I_{q,a+}^{0} f)(x) = f(x),\ (I_{q,a+}^{n} f)(x) = I_{q,a+}\left(I_{q,a+}^{n-1} f\right)(x) \qquad (n = 0, 1, 2, \cdots).$$

Between q-integral and q-derivative operators, we have the following relations:

$$\left(D_q I_{q,a+} f\right)(x) = f\left(x\right),\ \left(I_{q,a+} D_q f\right)(x) = f\left(x\right) - f\left(a\right).$$

For $\alpha, \beta > 0$ and $z \in \mathbb{R}$, a q-analogue of the Mittag–Leffler function is defined as follows ([20]):

$$e_{\alpha,\beta}\left(z; q\right) = \sum_{n=0}^{\infty} \frac{z^n}{\Gamma_q\left(\alpha n + \beta\right)}, \qquad \left(\left|z(1-q)^{\alpha}\right| < 1\right). \tag{11.1}$$

Definition 11.1 ([21]) Let $\alpha, \beta, \gamma, z \in \mathbb{R}$ such that $\alpha, \beta > 0$. Then the q-Prabhakar function $e_{\alpha,\beta}^{\gamma}\left(z; q\right)$ is defined by

$$e_{\alpha,\beta}^{\gamma}\left(z; q\right) = \sum_{n=0}^{\infty} \frac{(\gamma)_{n,q} z^n}{\Gamma_q\left(\alpha n + \beta\right)}, \qquad |z(1-q)^{\alpha}| < 1,$$

where

$$(\gamma)_{n,q} := \frac{(q^{\gamma}; q)_n}{(q; q)_n}.$$

Lemma 11.1 ([20]) *Let α and β be two complex numbers. Then*

$$(q^{\alpha+\beta}; q)_n = \sum_{k=0}^{n} \begin{bmatrix} n \\ k \end{bmatrix}_q q^{k\beta} (q^{\alpha}; q)_k (q^{\beta}; q)_{n-k}, \quad n = 0, 1, 2, \cdots.$$

Proposition 11.1 *Let $\gamma, \sigma \in \mathbb{C}$. Then the following equality is valid*

$$\sum_{k=0}^{n} (\gamma)_{n-k,q}\, q^{\gamma k} \left(\sigma\right)_{k,q} = \left(\gamma + \sigma\right)_{n,q}, \qquad (n = 0, 1, 2, \cdots). \tag{11.2}$$

Now, we introduce a generalised q-Prabhakar function.

Definition 11.2 Let $\alpha, \beta, \gamma, \omega, \delta, z, s \in \mathbb{R}$ be such that $\alpha, \beta > 0$ and $s < z$. Then the generalized q-Prabhakar function $e^{\gamma}_{\alpha,\beta}$ is defined by

$$e^{\gamma}_{\alpha,\beta}\left[\omega(z-s)^{\delta}_{q};q\right] := \sum_{n=0}^{\infty} \frac{(\gamma)_{n,q}\omega^{n}(z-s)^{\delta n}_{q}}{\Gamma_q(\alpha n+\beta)}, \tag{11.3}$$

where $|\omega(z-s)^{\delta}_{q}| < (1-q)^{-\alpha}$.

We note that (11.3) can be considered a generalisation of some known functions. For instance, if $\delta = \omega = 1, s = 0$ then from (11.3) we get Definition 11.1 of q-Prabhakar function. And also when $\gamma = 0$ and $\delta = \omega = 1$ then from (11.3) we get formula (11.1) for the q-Mittag-Leffler function.

Now, we give basic concepts of the q-fractional calculus.

Definition 11.3 ([17]) The Riemann-Liouville q-fractional integral $I^{\alpha}_{q,a+}$ of order $\alpha > 0$ is defined by

$$\left(I^{\alpha}_{q,a+}f\right)(x) = \frac{1}{\Gamma_q(\alpha)} \int_a^x (x-qt)^{\alpha-1}_{q} f(t)\, d_q t.$$

Definition 11.4 ([17]) The Riemann-Liouville q-fractional differential operator $D^{\alpha}_{q,a+}f$ of order $\alpha > 0$ is defined by

$$\left(D^{\alpha}_{q,a+}f\right)(x) = \left(D^{\lceil\alpha\rceil}_{q,a+} I^{\lceil\alpha\rceil-\alpha}_{q,a+} f\right)(x),$$

where $\lceil\alpha\rceil$ denotes the smallest integer greater or equal to α.

For $1 \le p < \infty$ the space $L^p_q[a,b]$ is defined by Shaimardan et al. [18]

$$L^p_q[a,b] = \left\{ f : [a,b] \to \mathbb{C} : \left(\int_a^b |f(x)|^p\, d_q x \right)^{1/p} < \infty \right\}.$$

Definition 11.5 ([20]) A function $f : [a,b] \to \mathbb{R}$ is called q-absolutely continuous if $\exists \varphi \in L^1_q[a,b]$ such that

$$f(x) = f(a) + \int_a^x \varphi(t) d_q t$$

for all $x \in [a,b]$.

The set of all q-absolutely continuous functions defined on $[a,b]$ is denoted by $AC_q[a,b]$. Moreover, $AC^n_q[a,b]$ $(n \in \mathbb{N})$ is the space of real-valued functions $f(x)$

which have q-derivatives up to order $n-1$ on $[a,b]$ such that $D_q^{n-1}f \in AC_q[a,b]$, i.e.

$$AC_q^n[a,b] = \left\{ f : [a,b] \to \mathbb{R};\ (D_q^{n-1}f)(x) \in AC_q[a,b] \right\}.$$

11.3 The Definitions and the Main Properties of the Prabhakar Fractional q-Integral and q-Differential Operators

In this section, using Definition 11.2 of the generalised q-Prabhakar function, we introduce the Prabhakar fractional q-integral and q-differential operators.

Definition 11.6 Let $f \in L_q^1[a,b]$ and $\alpha, \beta, \gamma, \omega \in \mathbb{R}$ be such that $\alpha, \beta > 0$. Then the Prabhakar fractional q-integral operator is defined by

$$\left({}^P I_{q,a+}^{\alpha,\beta,\gamma,\omega} f \right)(x) := \int_a^x (x-qt)_q^{\beta-1}\, e_{\alpha,\beta}^{\gamma} \left[\omega \left(x - q^{\beta} t \right)_q^{\alpha} ; q \right] f(t) d_q t. \quad (11.4)$$

From here and for the rest of the paper we denote $\omega' = q^{\gamma}\omega$.

Proposition 11.2 *Let $\alpha, \beta, \gamma, \mu, \sigma, \omega \in \mathbb{R}$ be such that $\alpha, \beta, \mu > 0$ and $x, s \in \mathbb{R}^+$, $x > s$. Then*

$${}^P I_{q,qs+}^{\alpha,\beta,\gamma,\omega} \left\{ g_{\sigma,\omega'}^{\alpha,\mu}(x,s) \right\} = g_{\sigma+\gamma,\omega}^{\alpha,\mu+\beta}(x,s),$$

where

$$g_{\sigma,\omega'}^{\alpha,\mu}(x,s) := (x-qs)_q^{\mu-1}\, e_{\alpha,\mu}^{\sigma} \left[\omega' \left(x - q^{\mu} s \right)_q^{\alpha} ; q \right]. \quad (11.5)$$

Proof By using (11.4), we have

$${}^P I_{q,qs+}^{\alpha,\beta,\gamma,\omega} \left\{ g_{\sigma,\omega'}^{\alpha,\mu}(x,s) \right\} = \int_{qs}^x g_{\gamma,\omega}^{\alpha,\beta}(x,t) g_{\sigma,\omega'}^{\alpha,\mu}(x,s) d_q t.$$

Hence considering (11.3) and using Definition 11.3, after some evaluations, we obtain

$${}^P I_{q,qs+}^{\alpha,\beta,\gamma,\omega} \left\{ g_{\sigma,\omega'}^{\alpha,\mu}(x,s) \right\} = \sum_{n=0}^{\infty} (\gamma)_{n,q} \omega^n \sum_{k=0}^{\infty} (\sigma)_{k,q} q^{\gamma k} \omega^k \frac{(x-qs)_q^{\alpha k + \alpha n + \beta + \mu - 1}}{\Gamma_q(\alpha n + \alpha k + \beta + \mu)}.$$

Using the Cauchy product formula ([22]) and then taking (11.2) into account and also the expansion (11.3) of the generalized q-Prabhakar function, we derive

$$
\begin{aligned}
{}^{P}I^{\alpha,\beta,\gamma,\omega}_{q,qs+}\left\{g^{\alpha,\mu}_{\sigma,\omega'}(x,s)\right\} &= \sum_{n=0}^{\infty}\frac{(x-qs)_q^{\alpha n+\beta+\mu-1}\omega^n}{\Gamma_q(\alpha n+\beta+\mu)}\sum_{k=0}^{n}(\gamma)_{n-k,q}q^{\gamma k}(\sigma)_{k,q} \\
&= \sum_{n=0}^{\infty}\frac{(\gamma+\sigma)_{n,q}\omega^n}{\Gamma_q(\alpha n+\beta+\mu)}(x-qs)_q^{\alpha n+\beta+\mu-1} \\
&= g^{\alpha,\mu+\beta}_{\sigma+\gamma,\omega}(x,s),
\end{aligned}
$$

which completes the proof. □

Lemma 11.2 *Let $f \in L_q^p[a,b]$ and $\alpha, \beta, \gamma, \omega, \mu, \sigma \in \mathbb{R}$ be such that $\alpha, \beta, \mu > 0$. Then the following relation*

$$
\left({}^{P}I^{\alpha,\beta,\gamma,\omega}_{q,a+}\,{}^{P}I^{\alpha,\mu,\sigma,\omega'}_{q,a+}f\right)(x) = \left({}^{P}I^{\alpha,\beta+\mu,\gamma+\sigma,\omega}_{q,a+}f\right)(x) \tag{11.6}
$$

holds for all $x \in [a,b]$. In particular

$$
\left({}^{P}I^{\alpha,\beta,\gamma,\omega}_{q,a+}\,{}^{P}I^{\alpha,\mu,-\gamma,\omega'}_{q,a+}f\right)(x) = \left(I^{\beta+\mu}_{q,a+}f\right)(x). \tag{11.7}
$$

Proof By Definition 11.4 of the Prabhakar fractional q-integral operator and taking into account notation (11.5), we have

$$
\left({}^{P}I^{\alpha,\beta,\gamma,\omega}_{q,a+}\,{}^{P}I^{\alpha,\mu,\sigma,\omega'}_{q,a+}f\right)(x) = \int_a^x g^{\alpha,\beta}_{\gamma,\omega}(x,t)\int_a^t g^{\alpha,\mu}_{\sigma,\omega'}(x,s)f(s)d_qs\,d_qt.
$$

Hence, by changing the order of integration and using Definition 11.6, we get

$$
\left({}^{P}I^{\alpha,\beta,\gamma,\omega}_{q,a+}\,{}^{P}I^{\alpha,\mu,\sigma,\omega'}_{q,a+}f\right)(x) = \int_a^x {}^{P}I^{\alpha,\beta,\gamma,\omega}_{q,qs+}\,g^{\alpha,\mu}_{\sigma,\omega'}(x,s)]f(s)d_qs.
$$

Applying Proposition 11.2 and taking into account Definition 11.6, we obtain

$$
\left({}^{P}I^{\alpha,\beta,\gamma,\omega}_{q,a+}\,{}^{P}I^{\alpha,\mu,\sigma,\omega'}_{q,a+}f\right)(x) = \int_a^x g^{\alpha,\beta+\mu}_{\gamma+\sigma,\omega}(x,s)f(s)d_qs = \left({}^{P}I^{\alpha,\beta+\mu,\gamma+\sigma,\omega}_{q,a+}f\right)(x).
$$

By putting $\sigma = -\gamma$ and taking $e^0_{\alpha,\beta}(z) = 1/\Gamma_q(\beta)$ into account from the last one can easily obtain (11.7). The proof of Lemma 11.2 is complete. □

For the boundness of the Prabhakar fractional q-integral operator, we give the following result without proof.

Lemma 11.3 *Let* $\alpha, \beta, \gamma, \omega \in \mathbb{R}$ *be such that* $\alpha, \beta > 0$ $|\gamma| < 1$, $|\omega(b-q^{\beta+1}a)_q^{\alpha}| < (1-q)^{\alpha}$ *and* $1 \leq p < \infty$. *Then the Prabhakar fractional* q*-integral operator* ${}^{P}I_{q,a+}^{\alpha,\beta,\gamma,\omega}$ *is bounded in* $L_q^p[a,b]$*:*

$$\left\| {}^{P}I_{q,a+}^{\alpha,\beta,\gamma,\omega} f \right\|_{L_q^p[a,b]} \leq M \, \|f\|_{L_q^p[a,b]},$$

where

$$M = (b-qa)_q^{\beta} e_{\alpha,\beta+1}[(b-q^{\beta+1}a)_q^{\alpha}; q]. \tag{11.8}$$

Now, we give the definition of the Prabhakar fractional q-differential operator.

Definition 11.7 Let $f \in L_q^1[a,b]$, ${}^{P}I_{q,a+}^{\alpha,n-\beta,-\gamma,\omega} f \in AC_q^n[a,b]$ and $\alpha, \beta, \gamma, \delta \in \mathbb{R}$ with $\alpha > 0$ and $\beta > 0$. Then the Prabhakar fractional q-differential operator ${}^{P}D_{q,+a}^{\alpha,\beta,\gamma,\omega}$ is defined by

$$\left({}^{P}D_{q,a+}^{\alpha,\beta,\gamma,\omega} f \right)(x) := \left(D_{q,a+}^{n} \, {}^{P}I_{q,a+}^{\alpha,n-\beta,-\gamma,\omega} f \right)(x),$$

where $n = \lceil \beta \rceil$, $\lceil \alpha \rceil$ denotes the smallest integer greater or equal to β.

Theorem 11.1 *Let* $\alpha, \beta, \gamma, \omega \in \mathbb{R}$ *with* $\alpha > 0$ *and* $\beta > 0$. *Then for any function* $f \in L_q^1[a,b]$ *the following equality is valid*

$$\left({}^{P}D_{q,a+}^{\alpha,\beta,\gamma,\omega} \, {}^{P}I_{q,a+}^{\alpha,\beta,\gamma,\omega'} f \right)(x) = f(x).$$

Proof Using Definition 11.7 and formula (11.6) and also Lemma 11.2, we have

$$\begin{aligned} {}^{P}D_{q,a+}^{\alpha,\beta,\gamma,\omega} \left({}^{P}I_{q,a+}^{\alpha,\beta,\gamma,\omega'} f \right)(x) &= D_{q,a+}^{n} \left({}^{P}I_{q,a+}^{\alpha,n-\beta,-\gamma,\omega} \, {}^{P}I_{q,a+}^{\alpha,\beta,\gamma,\omega'} f \right)(x) \\ &= D_{q,a+}^{n} \left(I_{q,a+}^{n} f \right)(x) = f(x). \end{aligned}$$

The proof is complete. □

In the classical case, Prabhakar fractional integral operators' semi-group property is commutative, but in the q-calculus case this property is non-commutative. To deal with this problem we need to introduce the following operator, which affects only one parameter of the Prabhakar fractional q-operators. We introduce the operator $\Lambda_q^{n\gamma,\omega}$ defined by setting

$$\Lambda_q^{n\gamma,\omega} \omega := q^{n\gamma} \omega, \quad n \in \mathbb{N}.$$

For example, $\Lambda_q^{n\gamma,\omega} f(\delta,\omega) = f(\delta, q^{n\gamma}\omega)$. Using this operator, we present some other properties of q-Prabhakar operators.

Theorem 11.2 *Let* $f \in L_q^1[a,b]$, ${}^P I_{q,a+}^{\alpha,1-\beta,-\gamma,\omega} f \in AC_q[a,b]$ *and* $\alpha, \beta, \gamma, \omega \in \mathbb{R}$ *with* $\alpha > 0$, $0 < \beta \leq 1$. *Then*

$$\left({}^P I_{q,a+}^{\alpha,\beta,\gamma,\omega'}\, {}^P D_{q,a+}^{\alpha,\beta,\gamma,\omega} f\right)(x) = f(x) - g_{\gamma,\omega'}^{\alpha,\beta}(x, a/q)\left({}^P I_{q,a+}^{\alpha,1-\beta,-\gamma,\omega} f\right)(a+).$$

11.4 A Cauchy Type Problem Associated with q-Prabhakar Differential Operator

Let us consider the following Cauchy-type problem with Prabhakar fractional q-differential operator:

$$\left({}^P D_{q,a+}^{\alpha,\beta,\gamma,\omega} y\right)(x) = f(x,y), \tag{11.9}$$

$$\left({}^P I_{q,a+}^{\alpha,1-\beta,-\gamma,\omega} y\right)(a+) = \xi_0, \tag{11.10}$$

where $\alpha, \beta, \gamma, \omega, \xi_0 \in \mathbb{R}$ are such that $\alpha > 0$, $0 < \beta \leq 1$, $\xi_0 \neq 0$.

To prove the existence and uniqueness of a solution to problem (11.9)–(11.10), first the equivalence of the considered problem to the q-Volterra integral equation is shown. Then the existence of the solution of the q-Volterra integral equation is proved by using the Banach fixed point theorem. We proved the following theorem.

Theorem 11.3 *Let* $f(\cdot,\cdot) : [a,b] \times \mathbb{R} \to \mathbb{R}$ *be a function such that* $f(\cdot, y(\cdot)) \in L_q^1[a,b]$ *for all* $y \in L_q^1[a,b]$. *Then* y *satisfies the relations (11.9) and (11.10) if and only if* y *satisfies the following* q*-Volterra integral equation:*

$$y(x) = {}^P I_{q,a+}^{\alpha,\beta,\gamma,\omega'} f(x,y) + \xi_0 g_{\gamma,\omega'}^{\alpha,\beta}(x, a/q).$$

The following result is obtained under the conditions of Theorem 11.3 and Lemma 11.3.

Theorem 11.4 *Let* G *be an open set in* $\mathbb{R}$. *Let* $f(\cdot,\cdot) : [a,b] \times G \to \mathbb{R}$ *be a function such that* $f(\cdot, y(\cdot)) \in L_q^1[a,b]$ *for all* $y \in G$, *and for all* $x \in (a,b]$ *and for all* $y_1, y_2 \in G$, *it satisfies*

$$|f(x,y_1) - f(x,y_2)| \leq A\,|y_1 - y_2|,$$

where $A > 0$ *does not depend on* $x \in [a, b]$ *and* $y_1, y_2 \in L_q^1[a, b]$. *Then there exists a unique solution* $y \in L_q^1[a, b]$ *to the problem (11.9)–(11.10).*

Acknowledgments The research was initiated during the visit of the authors to the intercontinental research center "Analysis and PDE" (Ghent University, Belgium), supported by the FWO Odysseus 1 grant G.0H94.18N: Analysis and partial differential equations and by the Methusalem programme of the Ghent University Special Research Fund (BOF) (Grant number 01M01021).

References

1. Meral, F.C., Royston, T.J., Magin, R.: Fractional calculus in viscoelasticity: an experimental study. Commun. Nonlinear Sci. Numer. Simul. **15**(4), 939–945
2. Mainardi, F.: Fractional Calculus and Waves in Linear Visco-Elasticity an Introduction to Mathematical Models. Imperial College Press, London (2010)
3. Machado, J.A.T.: A probabilistic interpretation of the fractional-order differentiation. Fract. Calc. Appl. Anal. **6**(1), 7380 (2003)
4. Kilbas, A.A., Srivastava, H.M., Trujillo, J.J.: Theory and Applications of Fractional Differential Equations. Elsevier, Boston (2006)
5. Hilfer, R.: Applications of Fractional Calculus in Physics. World Scientific, Singapore/River Edge (2000)
6. Lazopoulos, K.A.: Non-local continuum mechanics and fractional calculus. Mech. Res. Commun. **33**(6), 753–757 (2006)
7. Giusti, A., Colombaro, I., Garra, R., et al.: A practical guide to Prabhakar fractional calculus. Fract. Calc. Appl. Anal. **23**(1), 9–54 (2020)
8. Restrepo, J.E., Suragan, D.: Oscillatory solutions of fractional integro-differential equations II. Math. Methods Appl Sci. **44**(8), 7262–7274 (2021)
9. Eshaghi, S., Ghaziani, R.K., Ansari, A.: Stability and dynamics of neutral and integro-differential regularized Prabhakar fractional differential systems. Comput Appl. Math. **39**(4) (2020)
10. Colombaro, I., Giusti, A., Vitali, S.: Storage and dissipation of energy in Prabhakar viscoelasticity. Mathematics **6**(2), 15 (2018)
11. Garrappa, R., Maione, G.: Fractional Prabhakar derivative and applications in anomalous dielectrics: a numerical approach. In: Babiarz, A, Czornik, A, Klamka, J, Niezabitowski, M. (eds.) Theory and Applications of Non-integer Order Systems. Springer, Cham (2017)
12. Tomovski, Ž., Dubbeldam, J.L.A., Korbel, J.: Applications of Hilfer–Prabhakar operator to option pricing financial model. Fract. Calc. Appl. Anal. **23**(4), 996–1012 (2020)
13. Jackson, F.H.: On q-functions and a certain difference operator. Trans. R. Soc. Edin. **46**(2), 253–281 (1908)
14. Carmichael, R.D.: The general theory of linear q-difference equations. Am. J. Math. **34**(2), 147–168 (1912)
15. Al-Salam, W.A.: Some fractional q-integrals and q-derivatives. Proc. Edin. Math. Soc. **15**(2), 135–140 (1966)
16. Agarwal, R.P.: Certain fractional q-integrals and q-derivatives, Proc. Camb. Phil. Soc. **66**(2), 365–370 (1969)
17. Rajkovic, P.M., Marinkovic, S.D., Stankovic, M.S.: On q-analogues of Caputo derivative and Mittag–Leffler function. Fract. Calc. Appl. Anal. **10**(4), 359–373 (2007)
18. Shaimardan, S., Persson, L.E., Tokmagambetov, N.S.: Existence and uniqueness of some Cauchy-type problems in fractional q-difference calculus. Filomat **34**(13), 4429–4444 (2020). https://doi.org/10.2298/FIL2013429S.

19. Cheung, P., Kac, V.: Quantum Calculus. Edwards Brothers, Ann Arbor (2000)
20. Annaby, M.H., Mansour, Z.S.: q-Fractional Calculus and Equations. Springer, Heidelberg (2012)
21. Nadeem, R., Usman, T., Nisar, K.S., et al.: A new generalization of Mittag-Leffler function via q-calculus. Adv. Differ. Equ. 695 (2020). https://doi.org/10.1186/s13662-020-03157-z
22. Canuto, C., Tabacco, A.: Mathematical Analysis II, 2nd edn. Springer, Berlin (2015)

Chapter 12
A Note on Boundedness Properties of Pseudo-Differential Operators on Rank One Symmetric Spaces of Noncompact Type

Tapendu Rana

Abstract In this short note, we discuss the boundedness properties of the multiplier and pseudo-differential operators. More specifically, we first discuss multiplier and pseudo-differential operators theories in the classical setting. Then, we discuss multiplier results in symmetric space and their similarity and distinction with the corresponding results in Euclidean spaces. Finally, using the Helgason inversion formula, we define the pseudo-differential operators in noncompact type symmetric spaces and discuss its boundedness results.

12.1 Introduction

Partial differential equations, quantum physics, and signal analysis have all served as significant sources of inspiration for the systematic study of pseudo-differential operators. The early research on this topic in the 1960s, as examined, for instance, by Hörmander [11, 12] and Kohn Nirenberg [14], was strongly influenced by elliptic and hypoelliptic equations. Let us recall a pseudo-differential operator associated with a symbol $a(x, \xi)$ is an operator given by

$$a(x, D)f(x) = \int_{\mathbb{R}^n} e^{2\pi i x \cdot \xi} a(x, \xi)\mathcal{F}f(\xi)\, d\xi,$$

where $\mathcal{F}f$ denotes the Euclidean Fourier transform defined by,

$$\mathcal{F}f(\xi) = \int_{\mathbb{R}^n} f(x)e^{-2\pi i x \cdot \xi}\, dx.$$

T. Rana (✉)
Department of Mathematics: Analysis, Logic and Discrete Mathematics, Ghent University, Ghent, Belgium
e-mail: tapendu.rana@ugent.be

M. Ruzhansky, K. Van Bockstal (eds.), *Extended Abstracts 2021/2022*, Research Perspectives Ghent Analysis and PDE Center 2,
https://doi.org/10.1007/978-3-031-42539-4_12

The symbol $a(x,\xi)$ is assumed to be smooth in both space and frequency variables and satisfies certain growth conditions. An example of such a pseudo-differential operator is any classical differential operator $p(x,D)=\sum_{|\alpha|\le m} a_\alpha(x)\frac{\partial^\alpha}{\partial x^\alpha}$, which can be represented as

$$p(x,D)f(x)=\int_{\mathbb{R}^n} e^{2\pi i x\cdot\xi}\, p(x,\xi)\mathcal{F}f(\xi)\,d\xi,$$

where the symbol of this operator is $p(x,\xi)=\sum_{|\alpha|\le m} a_\alpha(x)\xi^\alpha$. The most widely used class of symbols are $\mathcal{S}^m_{\rho,\delta}$ introduced by Hörmander [12], which consists all $a\in C^\infty(\mathbb{R}^n\times\mathbb{R}^n)$ with

$$\left|\partial_x^\beta\partial_\xi^\alpha a(x,\xi)\right|\le C_{\alpha,\beta}(1+|\xi|)^{m-\rho|\alpha|+\delta|\beta|},$$

where $m\in\mathbb{R}, 0\le\rho,\delta\le 1$. One of the most important problems for pseudo-differential operators is whether they are bounded on the Lebesgue spaces, which has been extensively studied. In 1966, Hörmander proved that operators with symbols belongs to $\mathcal{S}^{n(\rho-1)/2}_{\rho,\delta}$ are bounded on $L^p(\mathbb{R}^n)$ for $1<p<\infty$, where $0\le\delta<\rho<1$ (see [12]). However, the most important result in this direction, which is relevant for us, is the following

Theorem 12.1 ([20, Chap VI, Theorem 1]) *Let $a\in\mathcal{S}^0_{1,0}$ and $a(x,D)$ be the associated pseudo-differential operator. Then the followings are true:*

1. *For $1<p<\infty$, the operator $a(x,D)$ extends to a bounded operator on $L^p(\mathbb{R}^n)$ to itself.*
2. *For $p=1$, there is the weak-type estimate*

$$meas\{x\in\mathbb{R}^n : |a(x,D)f(x)|>\lambda\}\le C_n\|f\|_{L^1(\mathbb{R}^n)}/\lambda,$$

for all $\lambda>0$.

Remark 12.1 We would like to mention that for the pseudo-differential operator, one cannot weaken the regularity assumption on the symbol $a(x,\xi)$, having singularity near $\xi=0$. In particular, if the symbol a satisfies the following simpler-looking condition

$$\left|\partial_x^\beta\partial_\xi^\alpha a(x,\xi)\right|\le C_{\alpha,\beta}\,|\xi|^{-|\alpha|},$$

then the corresponding operator may not be bounded. We refer the interested reader to the discussion [20, page 267].

If the symbol $a(x,\xi)$ is independent in x variable, say $a(x,\xi)=m(\xi)$, then the associated operator

$$m(D)f(x)=\int_{\mathbb{R}^n} e^{2\pi i x\cdot\xi} m(\xi)\mathcal{F}f(\xi)\,d\xi,$$

is called a multiplier operator. The boundedness of the multiplier operator is well studied on $\mathbb{R}^n$. In this regard, let us recall the celebrated Hörmander-Mihlin multiplier theorem on $\mathbb{R}^n$.

Theorem 12.2 *Let $m : \mathbb{R}^n \to \mathbb{C}$ satisfies the following Hörmander-Mihlin differential inequalities:*

$$\left| \frac{d^j}{d\xi^j} m(\xi) \right| \leq A_j |\xi|^{-j},$$

for all $\xi \in \mathbb{R}^n \setminus \{0\}$ and for all multi indices j with $0 \leq |j| \leq [\frac{n}{2}] + 1$. Then the followings are true:

1. *The multiplier operator $m(D)$ is a bounded operator on $L^p(\mathbb{R}^n)$, for all $1 < p < \infty$.*
2. *For $p = 1$, the multiplier operator $m(D)$ is a bounded operator from $L^1(\mathbb{R}^n)$ to $L^{1,\infty}(\mathbb{R}^n)$.*

In this note, we will discuss the boundedness properties of multiplier and pseudo-differential operators on rank one symmetric spaces of noncompact type. But before that let us discuss some basic preliminaries which are required to describe harmonic analysis on symmetric space of noncompact type.

12.2 Preliminaries

In this section, we describe the necessary preliminaries regarding semisimple Lie groups and harmonic analysis on Riemannian symmetric spaces, which are standard, and can be found, for example, in [7, 9, 10]. Let G be a noncompact connected semisimple real rank one Lie group with finite center, with its Lie algebra $\mathfrak{g}$. Let θ be a Cartan involution of $\mathfrak{g}$ and $\mathfrak{g} = \mathfrak{k} + \mathfrak{p}$ be the associated Cartan decomposition. Let $K = \exp \mathfrak{k}$ be a maximal compact subgroup of G and let $\mathbb{X} = G/K$ be an associated symmetric space with origin $\mathbf{0} = \{K\}$. Let $\mathfrak{a}$ be a maximal abelian subspace of $\mathfrak{p}$. Since the group G is of real rank one, dim $\mathfrak{a} = 1$. Let Σ be the set of nonzero roots of the pair $(\mathfrak{g}, \mathfrak{a})$, and let W be the associated Weyl group. For rank one case, it is well known that either $\Sigma = \{-\alpha, \alpha\}$ or $\{-2\alpha, -\alpha, \alpha, 2\alpha\}$, where α is a positive root and the Weyl group W associated to Σ is {-Id, Id}, where Id is the identity operator. Let $\mathfrak{a}^+ = \{H \in \mathfrak{a} : \alpha(H) > 0\}$ be a positive Weyl chamber, and let Σ^+ be the corresponding set of positive roots. In our case, $\Sigma^+ = \{\alpha\}$ or $\{\alpha, 2\alpha\}$. For any root $\beta \in \Sigma$, let $\mathfrak{g}_\beta$ be the root space associated to β. Let

$$\mathfrak{n} = \sum_{\beta \in \Sigma^+} \mathfrak{g}_\beta,$$

and let

$$\overline{\mathfrak{n}} = \theta(\mathfrak{n}).$$

Also, let

$$N = \exp \mathfrak{n}, \overline{N} = \exp \overline{\mathfrak{n}}.$$

The group G has an Iwasawa decomposition

$$G = K(\exp \mathfrak{a})N,$$

and a Cartan decomposition

$$G = K(\exp \mathfrak{a}^+)K.$$

This decomposition is unique. For each $g \in G$, we denote $H(g) \in \mathfrak{a}$ and $g^+ \in \overline{\mathfrak{a}^+}$ are the unique elements such that

$$g = k \exp H(g)n, k \in K, n \in N,$$

and

$$g = k_1 \exp(g^+)k_2, k_1, k_2 \in K.$$

We also have another Iwasawa decomposition

$$G = \overline{N}(\exp \mathfrak{a})K.$$

Let H_0 be the unique element in $\mathfrak{a}$ such that $\alpha(H_0) = 1$ and through this we identify $\mathfrak{a}$ with $\mathbb{R}$ as $t \leftrightarrow tH_0$ and $\mathfrak{a}_+ = \{H \in \mathfrak{a} \mid \alpha(H) > 0\}$ is identified with the set of positive real numbers. We also identify $\mathfrak{a}^*$ and its complexification $\mathfrak{a}^*_{\mathbb{C}}$ with $\mathbb{R}$ and $\mathbb{C}$ respectively by $t \leftrightarrow t\alpha$ and $z \leftrightarrow z\alpha$, $t \in \mathbb{R}$, $z \in \mathbb{C}$. Let $A = \exp \mathfrak{a} = \{a_t := \exp(tH_0) \mid t \in \mathbb{R}\}$ and $A^+ = \{a_t \mid t > 0\}$. Let $m_1 = \dim \mathfrak{g}_\alpha$ and $m_2 = \dim \mathfrak{g}_{2\alpha}$ where $\mathfrak{g}_\alpha$ and $\mathfrak{g}_{2\alpha}$ are the root spaces corresponding to α and 2α. As usual, then $\rho = \frac{1}{2}(m_1 + 2m_2)\alpha$ denotes the half sum of the positive roots. By abuse of notation, we will denote $\rho(H_0) = \frac{1}{2}(m_1 + 2m_2)$ by ρ.

Let dg, dk, dn and $d\overline{n}$ be the Haar measures on the groups G, K, N and $\overline{N}$ respectively. We normalise dk such that $\int_K dk = 1$. We have the following integral formulae corresponding to the Iwasawa and Cartan decomposition respectively, which holds for any integrable function f:

$$\int_G f(g)dg = \int_{\overline{N}} \int_{\mathbb{R}} \int_K f(\overline{n}a_t k)e^{2\rho t} dk dt d\overline{n},$$

and

$$\int_G f(g)dg = \int_K \int_{\mathbb{R}^+} \int_K f(k_1 a_t k_2)\Delta(t)\, dk_1\, dt\, dk_2.$$

where $\Delta(t) = (2\sinh t)^{m_1+m_2}(2\cosh t)^{m_2}$.

12.2.1 Fourier Transform

For a sufficiently nice function f on $\mathbb{X}$, its (Helgason) Fourier transform $\widetilde{f}$ is a function defined on $\mathbb{R} \times K$ given by

$$\widetilde{f}(\lambda, k) = \int_G f(g)e^{(i\lambda-\rho)H(g^{-1}k)}dg, \qquad \lambda \in \mathbb{R},\ k \in K,$$

whenever the integral exists [10, p. 199].

It is known that if $f \in L^1(\mathbb{X})$ then $\widetilde{f}(\lambda, k)$ is a continuous function of $\lambda \in \mathbb{R}$, for almost every $k \in K$. If in addition $\widetilde{f} \in L^1(\mathbb{R} \times K, |c(\lambda)|^{-2}\, d\lambda\, dk)$ then the following Fourier inversion holds,

$$f(gK) = |W|^{-1} \int_{\mathbb{R}\times K} \widetilde{f}(\lambda, k)\, e^{-(i\lambda+\rho)H(g^{-1}k)}\, |c(\lambda)|^{-2} d\lambda\, dk,$$

for almost every $gK \in \mathbb{X}$ [10, Chapter III, Theorem 1.8, Theorem 1.9], where $c(\lambda)$ is the Harish Chandra's c-function given by

$$c(\lambda) = \frac{2^{\rho-i\lambda}\Gamma(\frac{m_1+m_2+1}{2})\Gamma(i\lambda)}{\Gamma(\frac{\rho+i\lambda}{2})\Gamma(\frac{m_1+2}{4}+\frac{i\lambda}{2})}.$$

It is normalised such that $c(-i\rho) = 1$.

Moreover, $f \mapsto \widetilde{f}$ extends to an isometry of $L^2(\mathbb{X})$ onto $L^2(\mathbb{R} \times K, |c(\lambda)|^{-2}\, d\lambda\, dk)$, see [10, Chapter III, Theorem 1.5].

A function f is called K-biinvariant if

$$f(k_1 x k_2) = f(x) \text{ for all } x \in G, k_1, k_2 \in K.$$

We denote the set of all K-biinvariant functions by $\mathcal{F}(G//K)$. Let $\mathbb{D}(G/K)$ be the algebra of G-invariant differential operators on G/K. The elementary spherical functions ϕ are C^∞ functions and are joint eigenfunctions of all $D \in \mathbb{D}(G/K)$ for some complex eigenvalue $\lambda(D)$. That is

$$D\phi = \lambda(D)\phi,\ D \in \mathbb{D}(G/K).$$

They are parametrised by $\lambda \in \mathbb{C}$. The algebra $\mathbb{D}(G/K)$ is generated by the Laplace-Beltrami operator L. Then we have, for all $\lambda \in \mathbb{C}$, ϕ_λ is a C^∞ solution of

$$L\phi = -(\lambda^2 + \rho^2)\phi.$$

For any $\lambda \in \mathbb{C}$ the elementary spherical function ϕ_λ has the following integral representation

$$\phi_\lambda(x) = \int_K e^{-(i\lambda+\rho)H(xk)}\,dk \text{ for all } x \in G.$$

The spherical transform $\widehat{f}$ of a suitable K-biinvariant function f is defined by the formula:

$$\widehat{f}(\lambda) = \int_G f(x)\phi_\lambda(x^{-1})\,dx.$$

It is easy to check that for suitable K-biinvariant function f on G, its (Helgason) Fourier transform $\widetilde{f}$ reduces to the spherical transform $\widehat{f}$.

Since we are now equipped with the basic definitions and Helgason fourier transform on the noncompact type symmetric space, let us define the pseudo-differential operators and discuss their boundedness properties in the next section.

12.3 Results on Boundedness of Pseudo-Differential Operators on $\mathbb{X}$

We begin by discussing the case of multipliers in rank one symmetric spaces, that is when the symbol is independent of the space variable. Let $\mathbb{X} = G/K$ be the associated symmetric space, where G be a real rank one noncompact connected semisimple Lie group with finite center and K be a maximal compact subgroup of G. Corresponding to a multiplier function m on $\mathbb{R}$, the associated multiplier operator T_m on $\mathbb{X}$ is defined by

$$T_m(f)(x) = \int_{\mathbb{R}}\int_K m(\lambda)\widetilde{f}(\lambda, k)e^{-(i\lambda+\rho)H(x^{-1}k)}|c(\lambda)|^{-2}\,d\lambda,$$

where $\widetilde{f}$ is the (Helgason) Fourier transform of f and $c(\lambda)$ is the Harish-Chandra c-function. Clerc and Stein [5], Stanton and Tomas [19], Anker and Lohoué [2], Taylor [21], Anker [1] and Giulini et al. [8] have all investigated the boundedness of the multiplier operator on the symmetric spaces. Let us elaborate. The early work was completed in [5]. There, it became apparent that m has to be holomorphic and bounded inside the strip S_p if T_m is to be a continuous operator on $L^p(G/K)$, where

$$S_p = \left\{ \lambda \in \mathbb{C} : |\Im\lambda| \le \left| \frac{2}{p} - 1 \right| \rho \right\} \quad \text{for } p \in [1, \infty].$$

Conversely, the authors in [5] also gave a sufficient condition when G is complex. Similar results were obtained later when G is real rank one [19] and when G is a normal real form [2]. In 1990, Anker [1] improved and generalized previous results of [2, 5, 19], and [21] by proving the following multiplier theorem on $\mathbb{X} = G/K$:

Theorem 12.3 (Anker) *Let* $1 \le p < \infty$, $v = |\frac{1}{p} - \frac{1}{2}|$ *and* $N = [v \dim \mathbb{X}] + 1$. *Let* $m : \mathbb{R} \to \mathbb{C}$ *extends to an even holomorphic function on* S_p°, $\frac{\partial^i}{\partial \lambda^i} m$ $(i = 0, 1, \cdots, N)$ *extends continuously to* S_p *and satisfies*

$$\sup_{\lambda \in S_p} (1 + |\lambda|)^{-i} \left| \frac{\partial^i}{\partial \lambda^i} m(\lambda) \right| < \infty,$$

for $i = 0, 1, \cdots, N$. *Then the followings are true:*

1. *The associated multiplier operator* T_m *is a bounded operator on* $L^p(\mathbb{X})$ *for all* $p \in (1, \infty)$.
2. *For* $p = 1$, *the operator* T_m *extends to a bounded operator from* $L^1(\mathbb{X})$ *to* $L^{1,\infty}(\mathbb{X})$.

For technical reasons, it was necessary for the author to assume regularity assumptions on the boundary. Nonetheless, Anker mentioned that it should be possible to allow m to have a singularity at the boundary points $\pm i\rho_p := i \left| \frac{2}{p} - 1 \right| \rho$; and m will be still a L^p multiplier on $\mathbb{X}$. Later Ionescu [13] improved the theorem above by replacing the continuity of the multiplier m on the boundary with that of singularity condition at $\pm i\rho_p$.

Theorem 12.4 (Ionescu) *Let* $1 < p < \infty$. *Let* $m : \mathbb{R} \to \mathbb{C}$ *extends to an even holomorphic function on* S_p° *and satisfies*

$$\left| \frac{\partial^j}{\partial \lambda^j} m(\lambda) \right| \le A_j \left[|\lambda + i\rho_p|^{-j} + |\lambda - i\rho_p|^{-j} \right], \lambda \in S_p,$$

for $j = 0, 1, \cdots, \left[\frac{\dim \mathbb{X} + 1}{2} \right] + 2$. *Then the associated multiplier operator* T_m *is a bounded operator on* $L^p(\mathbb{X})$.

Additionally, the authors in [18] have examined the multiplier operator's boundedness for fixed K-types on $\mathrm{SL}(2, \mathbb{R})$. Now, it only makes sense to ask: What happens if we swap out the multiplier $m(\lambda)$ for the symbol $\sigma(x, \lambda)$? What requirements must the function $x \mapsto \sigma(x, \lambda)$ meet if the symbol's corresponding operator is required to be bounded? In our joint work with Pusti [17], we investigated the L^p-

boundedness problem for $p > 1$ of pseudo-differential operators in the context of rank one symmetric spaces of noncompact type. Here we are going to state the L^1 boundedness result for the pseudo-differential operator. Before we state our result, let us formally define the pseudo-differential operator on symmetric space.

Let $\sigma : \mathbb{X} \times \mathbb{R} \to \mathbb{C}$ be a suitable function. We can also think $g \mapsto \sigma(g, \lambda)$ is a right K-biinvariant function on G. We define the pseudo-differential operator associated with the symbol σ by

$$\Psi_\sigma f(x) = \int_{\mathbb{R}} \int_K \sigma(x, \lambda) \tilde{f}(\lambda, k) e^{-(i\lambda+\rho)H(x^{-1}k)} |c(\lambda)|^{-2} d\lambda\, dk,$$

for any smooth compactly supported function f on $\mathbb{X}$. Then we have the following L^1 boundedness result of the operator Ψ_σ.

Theorem 12.5 *Suppose $\sigma : G/K \times S_1 \to \mathbb{C}$ be a smooth function such that*

1. *For each $x \in \mathbb{X}$, $\lambda \mapsto \sigma(x, \lambda)$ is an even holomorphic function on S_1°.*
2. *The function $(g, \lambda) \mapsto \sigma(g, \lambda)$ satisfies the differential inequalities:*

$$\left| \frac{\partial^\beta}{\partial s^\beta} \frac{\partial^\alpha}{\partial \lambda^\alpha} \sigma(g a_s, \lambda) \right| \leq C_{\alpha,\beta} (1 + |\lambda|)^{-\alpha-\delta},$$

for all $\alpha = 0, 1, \cdots, \left[\frac{\dim \mathbb{X}+1}{2}\right] + 1$; $\beta = 0, 1, 2$; for all $g \in G, s \in \mathbb{R}, \lambda \in S_1$, and for some $\delta>0$.

Then the operator Ψ_σ extends to a bounded operator from $L^1(\mathbb{X})$ to itself.

For the boundedness of the pseudo-differential operators on nilpotent Lie groups and stratified Lie groups, we refer to [3, 5, 6] and references therein. Recently, the calculus of pseudo-differential operators on discrete spaces has also gained popularity, see for e.g. [4, 15] due to their natural connection with various problems of quantum ergodicity and in the discretization of continuous problems. To be more explicit, Molahajloo studied the pseudo-differential operators on $\mathbb{Z}$ (see [16]). For a higher dimensional analogue of the above result and a comprehensive pseudo-differential calculus on $\mathbb{Z}^n$, we refer to [4] and the references therein.

Acknowledgments The author is supported by the following project: Title: BOFMET20210006 01, Ugent-WE16-Wisk.: Analyse, Logica and Discrete Wisk. Adres: Kriijgslaan 281 S8, 9000 Gent.

References

1. Anker, J.-P.: L^p Fourier multipliers on Riemannian symmetric spaces of the noncompact type. Ann. Math. **132**(3), 597–628 (1990)
2. Anker, J.-P., Lohoué, N.: Multiplicateurs sur certains espaces symétriques. Am. J. Math. **108**(6), 1303–1353 (1986)
3. Bagchi, S., Thangavelu, S.: On Hermite pseudo-multipliers. J. Funct. Anal. **268**(1), 140–170 (2015)
4. Botchway, L.N.A., Kibiti, P.G., Ruzhansky, M.: Difference equations and pseudo-differential operators on $\mathbb{Z}^n$. J. Funct. Anal. **278**(11), 108473 (2020)
5. Clerc, J.-L., Stein, E.M.: L^p multipliers for noncompact symmetric spaces. Proc. Natl. Acad. Sci. USA **71**, 3911–3912 (1974)
6. Epperson, J.: Hermite multipliers and pseudo-multipliers. Proc. Am. Math. Soc. **124**(7), 2061–2068 (1996)
7. Gangolli, R., Varadarajan, V.S.: Harmonic Analysis of Spherical Functions on Real Reductive Groups. Springer-Verlag, Berlin (1988)
8. Giulini, S., Mauceri, G., Meda, S.: L^p multipliers on noncompact symmetric spaces. J. Reine Angew. Math. **482**, 151–175 (1997)
9. Helgason, S.: Groups and Geometric Analysis, Integral Geometry, Invariant Differential Operators, and Spherical Functions. Mathematical Surveys and Monographs. American Mathematical Society, Providence (2000)
10. Helgason, S.: Geometric Analysis on Symmetric Spaces. Mathematical Surveys and Monographs. American Mathematical Society, Providence (2008)
11. Hörmander, L.: Pseudo-differential operators. Commun. Pure Appl. Math. **18**, 501–517 (1965)
12. Hörmander, L.: Pseudo-differential operators and non-elliptic boundary problems. Ann. Math. **83**(2), 129–209 (1966)
13. Ionescu, A.D.: Fourier integral operators on noncompact symmetric spaces of real rank one. J. Funct. Anal. **174**(2), 274–300 (2000)
14. Kohn, J.J., Nirenberg L.: An algebra of pseudo-differential operators. Commun. Pure Appl. Math. **18**, 269–305 (1965)
15. Masson, E.L.: Pseudo-differential calculus on homogeneous trees. Ann. Henri Poincaré. **15**(9), 1697–1732 (2014)
16. Molahajloo, S.: Pseudo-differential operators on $\mathbb{Z}$. In: Pseudo-Differential Operators: Complex Analysis and Partial Differential Equations. Operator Theory: Advances and Applications, vol. 205, pp. 213–221. Birkhäuser Verlag, Basel (2010)
17. Pusti, S., Rana, T.: L^p boundedness of pseudo-differential operators on rank one Riemannian symmetric spaces of noncompact type. Preprint (2022). https://doi.org/10.48550/arXiv.2204.12327
18. Ricci, F., Wróbel, B.: Spectral multipliers for functions of fixed K-type on $L^p(\mathrm{SL}(2,\mathbb{R}))$. Math. Nachr. **293**(3), 554–584 (2020)
19. Stanton, R.J., Tomas, P.A.: Expansions for spherical functions on noncompact symmetric spaces. Acta. Math. **140**, 251–276 (1978)
20. Stein, E.M.: Harmonic Analysis: Real Variable Methods, Orthogonality, and Oscillatory Integrals. Princeton University Press, Princeton (1993)
21. Taylor, M.E.: L^p-estimates on functions of the Laplace operator. Duke Math. J. **58**(3), 773–793 (1989)

Chapter 13
L^p-L^q Norms of Spectral Multipliers

David Rottensteiner

Abstract This presentation is based on the joint work with M. Ruzhansky, which has recently been published in Rottensteiner and Ruzhansky (Arch. Math. (Basel) 120:507–520, 2023). We present an update on the multiplier theorems by Akylzhanov and Ruzhansky (J. Funct. Anal. 278(3):108324, 2020) by extending relevant L^p-L^q norm estimates to spectral multipliers of left-invariant weighted subcoercive operators on unimodular Lie groups. In particular, this includes spectral multipliers of Laplacians, sub-Laplacians and Rockland operators. As an application, we obtain, e.g., time asymptotics for the L^p-L^q norms of the heat kernels and Sobolev-type embeddings.

13.1 Introduction

Given a connected unimodular Lie group G, let us denote by λ_G and ρ_G the left and right regular representations on $L^2(G)$, respectively, and let us denote the associated left and right group von Neumann algebras by

$$\mathrm{VN_L}(G) := \lambda_G(G)'' \quad \text{and} \quad \mathrm{VN_R}(G) := \rho_G(G)'',$$

respectively. Then any left-invariant operator $T \in \mathcal{L}(L^2(G))$, i.e., any one that satisfies

$$T\lambda_G(x) = \lambda_G(x)T$$

for all $x \in G$, is a member of $\mathrm{VN_R}(G)$ by the double commutant theorem, and by the Schwartz kernel theorem, there exists a unique distribution $K_T \in \mathcal{D}'(G)$ such

D. Rottensteiner (✉)
Department of Mathematics: Analysis, Logic and Discrete Mathematics, Ghent University, Gent, Belgium
e-mail: david.rottensteiner@ugent.be

M. Ruzhansky, K. Van Bockstal (eds.), *Extended Abstracts 2021/2022*,
Research Perspectives Ghent Analysis and PDE Center 2,
https://doi.org/10.1007/978-3-031-42539-4_13

that

$$Tf(x) = f * K_T(x) = \int_G K_T(y^{-1}x)f(y)\,dy,$$

for all $x \in G$, $f \in \mathcal{D}(G)$. Any (possibly unbounded) positive self-adjoint operator $\mathcal{L}$ on $L^2(G)$ admits a spectral resolution $\mathcal{L} = \int_0^\infty \lambda dE_\lambda$. It is well known that, given a Borel measurable function $\varphi : \mathbb{R} \to \mathbb{C}$, the operator

$$\varphi(\mathcal{L}) = \int_0^\infty \varphi(\lambda)\,dE(\lambda)$$

is bounded on $L^2(G)$ if and only if the function φ is E-essentially bounded, and that the spectral multiplier $\varphi(\mathcal{L})$ is left-invariant if $\mathcal{L}$ is left-invariant.

In [1], it was shown that the L^p-L^q norms of $\varphi(\mathcal{L})$ depend essentially only on the growth of the spectral projections $\tau(E_{(0,s)}(|\mathcal{L}|))$ of $\mathcal{L}$ for any trace τ on the positive part $\mathrm{VN}_{\mathrm{R}}^+(G)$ of $\mathrm{VN}_{\mathrm{R}}(G)$. More precisely, the result [1, Thm. 1.2] asserts:

Theorem 13.1 ([1]) *Let G be a locally compact separable unimodular group and let $\mathcal{L}$ be a left-invariant operator on G. Assume that $\varphi : [0, \infty) \to \mathbb{R}$ is a monotonically decreasing continuous function such that*

$$\varphi(0) = 1,$$
$$\lim_{s\to\infty} \varphi(s) = 0.$$

Then

$$\|\varphi(|\mathcal{L}|)\|_{L^p(G)\to L^q(G)} \lesssim \sup_{s>0} \varphi(s)[\tau(E_{(0,s)}(|\mathcal{L}|))]^{\frac{1}{p}-\frac{1}{q}}, \tag{13.1}$$

for every $1 < p \leq 2 \leq q < \infty$.

As a consequence of Theorem 13.1, one obtains the following wide-ranging multiplier theorem (cf. [1, Cor. 1.3]):

Corollary 13.1 *Let G be a locally compact separable unimodular group and let $\mathcal{L}$ be a positive left-invariant operator on G. Suppose that there exists an $\alpha \in \mathbb{R}$ such that*

$$\tau(E_{(0,s)}(\mathcal{L})) \lesssim s^\alpha, \quad s \to \infty. \tag{13.2}$$

Then, for any $1 < p \leq 2 \leq q < \infty$, there is a constant $C = C(p, q) > 0$ such that

$$\|e^{-s\mathcal{L}}\|_{L^p(G)\to L^q(G)} \leq Cs^{-\alpha(\frac{1}{p}-\frac{1}{q})}, \quad s \to \infty. \tag{13.3}$$

Moreover, if

$$\gamma \geq \alpha\left(\frac{1}{p} - \frac{1}{q}\right),$$

then the embeddings

$$\|f\|_{L^q(G)} \leq C\|(1+\mathcal{L})^\gamma f\|_{L^p(G)}$$

hold for every $1 < p \leq 2 \leq q < \infty$.

The exponent $\alpha \in \mathbb{R}$ in (13.2) was explicitly computed in [1, §7] for several classes of groups and operators, e.g., for the sub-Laplacians on compact groups, the sub-Laplacian on the Heisenberg group $\mathbb{H}_n$ and certain higher order Rockland operators on $\mathbb{H}_n$. Furthermore, values of α for general positive Rockland operators on graded groups have been given in [6, §8]. While for all of these examples, the asymptotics were shown to be sharp, the proofs crucially rely on results that are particular for the specific settings and operators under consideration, e.g. [3, 9].

In the present extended abstract, we give a brief summary of the main notions and results of [7].

13.2 Weighted Subcoercive Operators and Spectral Estimates

In [10], ter Elst and Robinson proved Gaussian-type heat kernel estimates for subcoercive operators of an arbitrarily high order, previously only known for the heat kernels of sub-Laplacians and for Rockland operators on graded groups. We will briefly recall some fundamental facts and notions due to ter Elst and Robinson [10] and Martini [5], in order to justify the scope of [7]. For more details, confer [2, 4, 5, 8, 10].

13.2.1 Weighted Lie Algebras and Their Contractions

Let G be a d-dimensional connected Lie group with Lie algebra $\mathfrak{g}$. An algebraic basis $X_1, \ldots, X_{d'}$ of $\mathfrak{g}$ is a set of linearly independent elements that together with their multi-commutators span $\mathfrak{g}$. An algebraic basis $X_1, \ldots, X_{d'}$ is said to be a *weighted algebraic basis* if we associate to each vector X_j, $j = 1, \ldots, d'$, a so-called *weight* $w_j \in [1, \infty)$. We assume that $\bigcap_{j=1}^{d'} w_j\mathbb{N} \neq \emptyset$ and set

$$w = \min_{s\in[1,\infty)} \left\{s \in w_j\mathbb{N},\ \forall j \in \{1, ..., d'\}\right\}.$$

We denote by $J(d')$ the set of finite sequences of elements of $\{1, ..., d'\}$. For arbitrary $\alpha = (\alpha_1, ..., \alpha_n) \in J(d')$, let $|\alpha| = n$ denote the Euclidean length of α and denote by $\|\alpha\| := \sum_{j=1}^{d'} w_{\alpha_j}$ the weighted length of α. Moreover, we define

$$X^\alpha = X_{\alpha_1} \cdots X_{\alpha_n}$$

as an element of the universal enveloping algebra $U(\mathfrak{g})$ of $\mathfrak{g}$.

A given weighted algebraic basis defines a filtration $\{F_\lambda\}_{\lambda \in \mathbb{R}}$ on $\mathfrak{g}$,

$$F_\lambda := \operatorname{span}\big\{[\ldots [X_{\alpha_1}, X_{\alpha_2}], \ldots, X_{\alpha_n}] \mid \text{ non-empty } \alpha \in J(d'), \|\alpha\| \leq \lambda\big\},$$

which satisfies

$$[F_\lambda, F_\mu] \subseteq F_{\lambda+\mu}, \quad F_\lambda = \bigcap_{\mu>\lambda} F_\mu, \quad \bigcup_{\lambda \in \mathbb{R}} F_\lambda = \mathfrak{g}.$$

A weighted algebraic basis of $\mathfrak{g}$ is said to be *reduced* if for all $\lambda \in \mathbb{R}$

$$\operatorname{span}\big\{X_{\alpha_j} \mid w_j = \lambda\big\} \bigcap F_\lambda^- = \{0\}, \text{ where } F_\lambda^- := \bigcup_{\mu<\lambda} F_\mu = \{0\}.$$

Given a weighted basis and a filtration, there always exists a reduced weighted basis $X_1, \ldots, X_{d''}$ of $\mathfrak{g}$ which defines the same filtration. A Lie algebra equipped with a reduced weighted basis is said to be *weighted.* Any such Lie algebra admits an associated homogeneous Lie algebra $\mathfrak{g}_*$ together with a decomposition

$$\mathfrak{g}_* = W_{\lambda_1} \oplus \ldots \oplus W_{\lambda_k}$$

into a direct sum of eigenspaces of the homogeneous dilations, defined by $W_{\lambda_j} := F_{\lambda_j} \neq F_{\lambda_j}^-$, for weights $1 \leq \lambda_1 < \ldots < \lambda_k$ determined by $F_{\lambda_j}/F_{\lambda_j}^-$, $j = 1, \ldots, k$. The weights $w_1, \ldots, w_{d''}$ are among the weights $\lambda_1 < \ldots < \lambda_k$ since $X_1, \ldots, X_{d''}$ is reduced. If $\bar{X}_j$ denotes the projection onto W_{w_j} of $X_j \in F_{w_j}$, then $\bar{X}_1, \ldots, \bar{X}_{d''}$ is an adapted basis of $\mathfrak{g}_*$ with the same weights $w_1, \ldots, w_{d''}$. The homogeneous Lie algebra $\mathfrak{g}_*$ equipped with the adapted basis $\bar{X}_1, \ldots, \bar{X}_{d''}$ is called the contraction of the weighted Lie algebra $\mathfrak{g}_*$.

A *weighted Lie group* is a connected Lie group G whose Lie algebra $\mathfrak{g}$ is weighted. The *contraction* G_* of a weighted Lie group is the homogeneous Lie group whose Lie algebra is $\mathfrak{g}_*$.

Any weighted Lie group can be equipped with a control distance $d_* : G \times G \to [0, \infty)$ with the following properties: the left Haar measure of the metric balls $B_r(e)$ of radius r around the unit element $e \in G$ is given by $|B_r(e)| \sim r^{Q_*}$ for $r \leq 1$, where Q_* is the homogeneous dimension of the contraction G_*, and in the case that G is homogeneous the control modulus defined by $|x|_* := d_*(e, x)$ coincides with a homogeneous quasi-norm for small x; for $r \geq 1$, the metric coincides with the

Carnot-Carathéodory metric associated with the reduced (unweighted) basis vector fields $X_1, \ldots, X_{d''}$ on G, which necessarily satisfy Hörmander's condition. This control modulus is a connected distance. Note that for large radii r the volume growth is controlled by $|B_r(x)| \lesssim e^{\beta r}$ for some $\beta > 0$. For details, we refer to [4, 5].

13.2.2 Differential Operators

Let G be a weighted Lie group and fix a reduced weighted algebraic basis $X_1, ..., X_{d''}$ of $\mathfrak{g}$. A function $C : J(d'') \to \mathbb{C}$ such that $C(\alpha) = 0$ if $\|\alpha\| > m$ but $C(\alpha) \neq 0$ for at least one $\alpha \in J(d'')$ with $\|\alpha\| = m$ is said to be an m-th order *form*. The *principal part* of C is the form $P : J(d'') \to \mathbb{C}$ given by the sum of terms of C of order m:

$$P(\alpha) = \begin{cases} C(\alpha) & \text{if } \|\alpha\| = m, \\ 0 & \text{otherwise.} \end{cases}$$

A form C is said to be *homogeneous* if $C = P$. The *adjoint* C^+ of a form C is defined by $C^+(\alpha) := (-1)^{|\alpha|}\overline{C(\alpha_*)}$, where $\alpha_* := (\alpha_k, \ldots, \alpha_1)$ if $\alpha = (\alpha_1, \ldots, \alpha_k) \in J(d'')$. The form C is said to be *symmetric* if $C^+ = C$.

For a given form C, we can consider the m-th order operators

$$d\rho_G(C) = \sum_{\alpha \in J(d'')} C(\alpha) d\rho_G(X^\alpha)$$

with domain $D(d\rho_G(C)) = \bigcap_{\|\alpha\| \leq m} D(d\rho_G(X^\alpha))$. The associated semi-norms are defined by

$$N_s(f) = \max_{\substack{\alpha \in J(d''), \\ \|\alpha\| \leq s}} \left\| d\rho_G(X^\alpha) f \right\|_{L^2(G)}$$

for $s \in \mathbb{R}$ with $s \geq 0$. If we consider the right regular representation ρ_G on $L^p(G)$, $1 \leq p < \infty$, instead of $L^2(G)$, we denote the analogue of $D(d\rho_G(C))$ by $L^{p;m}(G)$ and set $\bigcap_{m=1}^{\infty} L^{p;m}(G) =: L^{p;\infty}(G)$.

An m-th order form C is said to be a *weighted subcoercive form*, with respect to $X_1, ..., X_{d''}$, if $m \in 2w\mathbb{N}$ and if $d\rho_G(C)$ satisfies a local *Gårding inequality*: there exist $\mu > 0$, $\nu \in \mathbb{R}$ and an open neighborhood V of the identity $e_G \in G$ such that

$$\mathrm{Re}(\langle f, d\rho_G(C) f \rangle) \geq \mu \big(N_{m/2}(f)\big)^2 - \nu \|f\|_{L^2(G)}^2$$

for all $f \in \mathcal{D}(G)$ with support $\mathrm{supp} f \subseteq V$.

We say that an operator $\mathcal{L} = d\rho_G(C)$ on G is *weighted subcoercive* if C is a weighted subcoercive form.

We recall that an operator $\mathcal{L} = d\rho_G(C) = d\rho_G(P)$ on a homogeneous group G is called Rockland if the principal form $C = P$ is homogeneous with respect to the given homogeneous dilations on G and $d\pi(C)$ is injective on the space of smooth vectors $\mathcal{H}^\infty$ for every nontrivial unitary irreducible representation $\pi \in \widehat{G}$. The existence of a Rockland operator on a homogeneous group G implies that the dilation weights $w_1, \ldots, w_d$ can be jointly rescaled to be positive integers and that $\mathfrak{g}$ possesses a gradation.

The following theorem collects several key properties of weighted subcoercive operators.

Theorem 13.2 ([5, 10]) *Let G be a weighted Lie group and C an m-th order form with principal part P. Then the following are equivalent:*

(i) $\mathcal{L} = d\rho_G(C)$ is a weighted subcoercive operator on G;
(ii) $d\rho_{G_*}(P + P^+)$ is a positive Rockland operator on the contraction G_*.

If these conditions are satisfied, then the following properties hold:

(a) The continuous semigroup $S = \{S_t\}_{t\geq 0}$ generated by $\mathcal{L}$ has a smooth kernel $k_t \in L^{1;\infty}(G) \cap C_0^\infty(G)$ such that

$$d\rho_G(X^\alpha)S_t f = \int_G (X^\alpha k_t)(x)\rho_G(x^{-1}) f \, dx$$

for all $\alpha \in J(d')$ and $f \in L^2(G)$.

(b) For all $\alpha \in J(d')$, there exist $b, c > 0$ and $\omega \geq 0$ such that

$$|k_t(x)| \leq c t^{-\frac{Q_*}{m}} \exp(\omega t) \exp\Big(-b\big(\frac{|x|_*^m}{t}\big)^{\frac{1}{m-1}}\Big), \quad x \in G, t > 0, \tag{13.4}$$

where Q_* is the homogeneous dimension of the contraction G_* and $|\cdot|_*$ is the control modulus associated with the control distance d_*.

(c) The function $k : \mathbb{R} \times G \to \mathbb{C}$ defined by $(t, x) \mapsto k_t(x)$ for $t > 0$ and $k_t = 0$ for $t \leq 0$ satisfies the heat equation

$$((\partial_t + d\rho_G(C))k_t)(x) = \delta(t)\delta(x)$$

as distributions, for all $(t, x) \in \mathbb{R} \times G$.

In [5], Martini developed the spectral theory of systems of pairwise commuting weighted subcoercive operators and proved estimates of the spectral projectors in terms of heat kernel estimates, which in the case of a single weighted subcoercive single operator yield the following (cf. [5, Prop. 3.11]):

Proposition 13.1 ([5]) *Let G be a connected unimodular Lie group with Lie algebra $\mathfrak{g}$. Suppose that $\mathcal{L}$ is an m-th order weighted subcoercive operator on G.*

Then

$$\tau(E_{(0,s)}(|\mathcal{L}|)) \lesssim s^{\frac{Q_*}{m}}, \quad s \to \infty, \tag{13.5}$$

where Q_* is the local dimension of G relative to the chosen weighted structure on $\mathfrak{g}$.

Combining Theorem 13.1 and Proposition 13.1, we obtain the main result of [7]:

Theorem 13.3 *Let G be a connected unimodular Lie group with Lie algebra $\mathfrak{g}$ and let $\mathcal{L}$ be an m-th order positive weighted subcoercive operator on G. Assume that $\varphi : [0, \infty) \to \mathbb{R}$ is a monotonically decreasing continuous function such that*

$$\varphi(0) = 1,$$
$$\lim_{s\to\infty} \varphi(s) = 0.$$

Then, for any $1 < p \leq 2 \leq q < \infty$, there is a constant $C = C(p, q) > 0$ such that

$$\|\varphi(|\mathcal{L}|)\|_{L^p(G)\to L^q(G)} \leq C \sup_{s>0} \varphi(s)\, s^{\frac{Q_*}{m}(\frac{1}{p}-\frac{1}{q})}, \quad s \to \infty, \tag{13.6}$$

where Q_* is the local dimension of G relative to the chosen weighted structure on $\mathfrak{g}$. In particular,

$$\|e^{-s\mathcal{L}}\|_{L^p(G)\to L^q(G)} \leq C s^{-\frac{Q_*}{m}(\frac{1}{p}-\frac{1}{q})}, \quad s \to \infty. \tag{13.7}$$

Moreover, if

$$\gamma \geq \frac{Q_*}{m}\left(\frac{1}{p} - \frac{1}{q}\right),$$

then the Sobolev norm estimates

$$\|f\|_{L^q(G)} \leq C\|(1 + \mathcal{L})^\gamma f\|_{L^p(G)} \tag{13.8}$$

hold for every $1 < p \leq 2 \leq q < \infty$.

In particular, this holds true for every left-invariant hypoelliptic (positive) sub-Laplacian $\mathcal{L}$ on G, in which case Q_* equals the Hausdorff dimension of G and $m = 2$, as well as for every positive Rockland operator $\mathcal{L}$ on a graded group G, in which case Q_* equals the homogeneous dimension of G and m equals the homogeneous order of $\mathcal{L}$.

Let us emphasize that this shows that the hypothesis (13.2) is automatically satisfied for all weighted subcoercive operators on a unimodular Lie group. Moreover, the estimate in (13.5) is sharp.

Acknowledgments The author would like to thank Jordy van Velthoven for useful discussions and Alessio Martini for pointing out the link between spectral traces and heat kernels estimates.

The author is supported by the FWO Odysseus 1 grant G.0H94.18N: Analysis and Partial Differential Equations and the FWO Senior Research Grant G022821N: Niet-commutatieve wavelet analyse.

References

1. Akylzhanov, R., Ruzhansky, M.: L^p-L^q multipliers on locally compact groups. J. Funct. Anal. **278**(3), 108324 (2020)
2. Dungey, N., ter Elst, A.F.M., Robinson, D.W.: Analysis on Lie groups with polynomial growth. Progress in Mathematics, vol. 214, p. viii+312. Birkhäuser Boston, Boston (2003)
3. Hassannezhad, A., Kokarev, G.: Sub-Laplacian eigenvalue bounds on sub-Riemannian manifolds. Ann. Sc. Norm. Super. Pisa Cl. Sci. **4**(5), 1049–1092 (2016)
4. Martini, A.: Algebras of differential operators on Lie groups and spectral multipliers. Ph.D. Thesis, Scuola Normale Superiore Pisa, 2010
5. Martini, A.: Spectral theory for commutative algebras of differential operators on Lie groups. J. Funct. Anal. **260**(9), 2767–2814 (2011)
6. Rottensteiner, D., Ruzhansky, M.: Harmonic and anharmonic oscillators on the Heisenberg group. J. Math. Phys. **63**(11), 111509 (2022)
7. Rottensteiner, D., Ruzhansky, M.: An update on the L^p-L^q norms of spectral multipliers on unimodular Lie groups. Arch. Math. (Basel) **120**, 507–520 (2023)
8. ter Elst, A.F.M., Robinson, D.W.: Weighted strongly elliptic operators on Lie groups. J. Funct. Anal. **125**(2), 548–603 (1994)
9. ter Elst, A.F.M., Robinson, D.W.: Spectral estimates for positive Rockland operators. Australian Mathematical Society Lecture Series, vol. 9, pp. 195–213. Cambridge University Press, Cambridge (1997)
10. ter Elst, A.F.M., Robinson, D.W.: Weighted subcoercive operators on Lie groups. J. Funct. Anal. **157**(1), 88–163 (1998)

Chapter 14
Estimates for Oscillatory Integrals with Discontinuous Amplitude

Isroil A. Ikromov, Akbar R. Safarov, and Dilshodbek G. Khudoyberdiev

Abstract In this paper, we consider oscillatory integrals with analytic phase function and amplitude having a set of discontinuity points. We obtain estimates for such integrals that are sharp up to a positive number $\varepsilon > 0$.

14.1 Introduction

In this paper, we consider oscillatory integrals with analytic phase function and amplitude having a set of discontinuity points. We obtain estimates for such integrals that are sharp up to a positive number $\varepsilon > 0$.

It is well known that oscillatory integrals arise in many problems of analysis, mathematical physics and analytic number theory. It suffices to recall the trigonometric integrals studied by I.M. Vinogradov in [8] and the Fourier transform of compact domains considered in the classical works of I.M. Stein [7] and W. Rendoll [5], Svensson.

I. A. Ikromov
Uzbekistan Academy of Sciences, V.I.Romanovskiy Institute of Mathematics, Tashkent, Uzbekistan

Department of Mathematics, Samarkand State University, Samarkand, Uzbekistan
e-mail: ikromov1@rambler.ru

A. R. Safarov (✉)
Uzbek-Finnish Pedagogical Institute, Samarkand, Uzbekistan

Department of Mathematics, Samarkand State University, Samarkand, Uzbekistan
e-mail: safarov-akbar@mail.ru

D. G. Khudoyberdiev
Department of Mathematics and Statistics, Xi'an Jioatong University, Xi'an Shaanxi, P. R. China

Department of Mathematics, Uzbek-Finnish Pedagogical Institute, Spitamen Shokh Samarkand, Uzbekistan

M. Ruzhansky, K. Van Bockstal (eds.), *Extended Abstracts 2021/2022*,
Research Perspectives Ghent Analysis and PDE Center 2,
https://doi.org/10.1007/978-3-031-42539-4_14

The monograph by M. V. Fedoryuk [3] considers a phase having singularity and amplitude discontinuities. However, that monograph considers the case where the boundary of the domain is smooth and the phase has only nondegenerate critical points.

In this article, using the ideas of I.M. Stein [7], we obtained estimates of the oscillatory integrals with amplitudes whose discontinuities lie near the corner points. The estimates obtained are sharp up to an arbitrary positive number ε. In our approach, arbitrary analytic functions can be considered as the phase. Moreover, the existence of nonisolated critical points is allowed. We conclude with an example showing that the $\varepsilon > 0$ parameter cannot be eliminated, although it is possible to replace it with a logarithmic multiplier, but this issue has not yet been finally investigated.

14.2 A Model Case

In this section, we study oscillatory integrals with model amplitude. Consider the oscillatory integral

$$J(\lambda, s) = \int_{T(s)} a(x)\, e^{i\lambda\Phi(x,s)} dx, \tag{14.1}$$

where $a \in C_0^\infty(\mathbb{R}^n)$, $T(s)$-the family of some polygons depending on the parameter $s \in \mathbb{R}^m$, λ is a large real parameter and $\Phi(x, s)$—is a real-valued function.

If we write (14.1) as

$$J(\lambda, s) = \int_{\mathbb{R}^n} a(x)\, \chi_{T(s)}(x) e^{i\lambda\Phi(x,s)} dx,$$

here and further $\chi_{T(s)}$− characteristic function of the set T, then the amplitude function $a_1(x, s) := a(x)\, \chi_{T(s)}(x)$ has a discontinuity at the boundary points $y \in T(s)$, with $a(y) \neq 0$. Such integrals occur in many problems of analysis and mathematical physics.

The monograph [3] investigated such integrals in cases where the boundaries of the region (the set of discontinuities) are smooth and $\Phi(x, s)$ has nondegenerate critical points near the boundary. In the present section, we consider a more general case.

Moreover, if T is a domain with a smooth boundary and $\lambda\Phi(x, s) = \lambda(x, s)$, (where $s \in S^{n-1}$ is a unit vector) and (x, s) is the scalar product of vectors x and s, then (14.1) is the same as the Fourier transform of a characteristic function of the set T, case $a \equiv 1$.

We will restrict ourselves to the two-dimensional case, i.e. $n = 2$. In addition to the integral (14.1), let us consider the corresponding oscillatory integral with a smooth amplitude having a compact support:

$$J_0(\lambda, s) = \int_{\mathbb{R}^2} a(x, s)\, e^{i\lambda\Phi(x,s)} dx.$$

The basic condition:

Let $s \in K$ and for (η_1, η_2) consider a new phase function

$$\Phi_1(x, s, \eta) = \Phi(x, s) + x_1\eta_1 + x_2\eta_2.$$

Main assumption: suppose that $\Phi(x, s)$ is an analytic function such that for any critical point of the function $\Phi_1(x, s^0, \eta^0)$ where $\left(s^0, \eta^0\right) \in K \times \mathbb{R}^2$, and its oscillation exponent $\beta \le \beta_0$. We have the following statement.

Theorem 14.1 *If the phase function satisfies the main condition, then for any positive number* $\varepsilon > 0$ *the following estimate is valid*

$$|J(\lambda, s)| \le \frac{c_\varepsilon \, \|a(\bullet, s)\|_{C^3}}{\lambda^{\beta_0 - \varepsilon}}.$$

Proof Since $T(s)$ is a family of compact triangles, we can assume that there exists a rectangle P such that for any $s \in K$ there is an inclusion, $T(s) \subset P$. Let's choose $\psi \in C_0^\infty\left(\mathbb{R}^2\right)$ such, that $\psi(x) = 1$ as $x \in P$. Then, since $\chi_{T(s)}(x)\psi(x) = \chi_{T(s)}(x)$. The integral $J(\lambda, s)$ can be written in the form

$$J(\lambda, s) = \int_{\mathbb{R}^2} a(x)\, \chi_{T(s)}(x)\psi(x) e^{i\lambda\Phi(x,s)} dx.$$

Now let's apply the Plancherel formula [6]:

$$\int_{\mathbb{R}^2} f(x)\, \overline{g(x)} dx = \frac{1}{(2\pi)^2} \int_{\mathbb{R}^2} \widehat{f}(\xi)\, \overline{\widehat{g}(\xi)} d\xi,$$

where $\widehat{f}$-Fourier transform of a function f. As f, g, we will take respectively

$$f(x) = a(x)\, \chi_{T(s)}(x),\; g(x) = \psi(x) e^{i\lambda\Phi(x,s)}.$$

Using Hölder's inequality we obtain

$$|J(\lambda, s)| \le \frac{1}{(2\pi)^2} \left(\int_{\mathbb{R}^2} \left|\widehat{f}(\xi)\right|^p d\xi\right)^{\frac{1}{p}} \left(\int_{\mathbb{R}^2} \left|\widehat{g}(\xi)\right|^q d\xi\right)^{\frac{1}{q}},$$

where $\frac{1}{p}+\frac{1}{q}=1$. Since a is a finite smooth function with compact support, $\widehat{a}\in \mathbb{L}^1\left(\mathbb{R}^2\right)$ and $\widehat{\chi}_{T(s)}\in \mathbb{L}^p\left(\mathbb{R}^2\right)$ for any $p>1$ and, moreover, the inequality is valid:

$$\left\|\widehat{\chi}_{T(s)}\right\|_{\mathbb{L}^p(\mathbb{R}^2)}\leq c\left(\mu\left(T(s)\right)\right)^{\frac{1}{q}}$$

as $p>1$, where $\mu\left(T(s)\right)$ polygon area $T(s)$. Consequently, using Young's inequality [2] to $\widehat{f}\ (\xi)=\left(\widehat{a}*\widehat{\chi}_{T(s)}\right)(\xi)$ obtain:

$$\left\|\widehat{f}\right\|_{\mathbb{L}^p}\leq\left\|\widehat{a}\right\|_{\mathbb{L}^1}\left\|\widehat{\chi}_{T(s)}\right\|_{\mathbb{L}^p}\leq c\left\|\widehat{a}\right\|_{\mathbb{L}^1}\left(\mu\left(T(s)\right)\right)^{\frac{1}{q}},$$

where it is essentially used that $T(s)$ is a polygon and therefore its boundary has no curvature, i.e. the curvature is zero at ordinary points. Next, we will show how to get rid of this condition in the local version of the main result.

Thus, in order to obtain the desired estimate, it is sufficient to estimate $\mathbb{L}^q-$norm of the function $\widehat{g}$

$$\widehat{g}\ (\xi)=\int_{\mathbb{R}^2}\psi(x)e^{-i(\lambda\Phi(x,s)+x_1\xi_1+x_2\xi_2)}dx.$$

Let us estimate L^q-norm of function g by writing it as a sum of two functions

$$\widehat{g}\ (\xi)=\widehat{g}\ (\xi)\,\chi_{|\xi|\leq c|\lambda|}\,(\xi)+\widehat{g}\ (\xi)\left(1-\chi_{|\xi|\leq c|\lambda|}\,(\xi)\right)$$

where c is a large enough fixed positive number. Let's introduce the notation

$$g_1\,(\xi)=\widehat{g}\ (\xi)\,\chi_{|\xi|\leq c|\lambda|}\,(\xi)\,,\qquad g_2\,(\xi)=\widehat{g}\ (\xi)\left(1-\chi_{|\xi|\leq c|\lambda|}\,(\xi)\right).$$

Replacing ξ with $\lambda\eta$ we write g_1 as the following oscillatory integral:

$$g_1\,(\lambda\eta)=\int_{\mathbb{R}^2}\psi(x)e^{i\lambda(\Phi(x,s)-x_1\eta_1-x_2\eta_2)}dx.$$

Since the oscillation exponent, $\beta\leq\beta_0$, then, according to Karpushkin's theorem [4], for any fixed positive number δ the estimate is valid [3]:

$$|g_1\,(\lambda\eta)|\leq\frac{A\,\|a\|_{C^3}}{\lambda^{\beta_0-\delta}}.$$

Since Karpushkin's estimate is locally uniform, it is uniform with respect to any compact. Now, let us estimate the L^q- norm of function g_1:

$$\int_{\mathbb{R}^2} |g_1(\xi)|^q \, d\xi = \lambda^2 \int_{|\eta|\leq c} |g_1(\lambda\eta)|^q \, d\eta \leq \frac{\lambda^2 A^q}{\lambda^{(\beta_0-\delta)q}}.$$

Hence,

$$\|g_1\|_{\mathbb{L}^q(R^2)} \leq \frac{A}{\lambda^{\beta_0-\frac{2}{q}-\delta}}.$$

Now, consider the estimation of the integral g_2. Since the support of ψ and the set of changes s are compact, then $||\nabla_x\Phi(x,s)| \leq c$ at all (x,s) $\in$supp$(\psi)\times K$. Therefore, for $|\eta|\geq 2A$ the phase function has no critical points and the formula for integrating by parts formula shows that

$$|g_2(\lambda\eta)| \leq \frac{A\,\|a\|_{C^{N+1}}}{|\lambda\eta|^{N+1}} \leq \frac{A\,\|a\|_{C^{N+1}}}{\lambda\,|\lambda\eta|^N}.$$

Now, estimate L^q-norm

$$\int_{\mathbb{R}^2} |g_2(\xi)|^q \, d\xi = \frac{1}{\lambda^{q(N+1)}} \int_{|\eta|>c} \frac{d\eta}{|\eta|^{Nq}} \leq \frac{c_N}{\lambda^{q(N+1)}}.$$

So

$$\left(\int_{\mathbb{R}^2} |g_2(\xi)|^q \, d\xi\right)^{\frac{1}{q}} \leq \frac{c_N}{\lambda^{N+1}}.$$

Moreover, we can choose $N = 2$ as N and obtain $C_N \leq c\,\|a(\bullet,s)\|_{C^3}$. By Minkowski's inequality, we have

$$|J(\lambda,s)| \leq c\left\|\widehat{a}\right\|_{\mathbb{L}^1} (\mu(T(s)))^{\frac{1}{q}} \left(\frac{1}{\lambda^{\beta_0-2/q-\delta}} + \frac{1}{\lambda^{N+1}}\right) \leq \frac{c\left\|\widehat{a}\right\|_{\mathbb{L}^1} (\mu(T(s)))^{\frac{1}{q}}}{\lambda^{\beta_0-\frac{2}{q}-\delta}}.$$

Now, for any $\varepsilon > 0$, choosing q and $\delta > 0$ such that $\frac{2}{q}+\delta < \varepsilon$ come to the proof of Theorem 2.1. □

14.3 A More General Case

In this section, we consider the case when the boundary contains two transversally intersecting curves. For the sake of being defined, assume that the analytic curves γ_1 and γ_2 transversally intersect at the origin and the phase function $\Phi(x,0)$ at $x=0$ has a singularity, that is, $\nabla\Phi(0,0)=0$.

Consider the oscillatory integral

$$J(\lambda, s) = \int_{\mathbb{R}^2} \chi_{D(s)}(x) a(x, s) e^{i\lambda\Phi(x,s)} dx, \tag{14.2}$$

where $D(s)$–family of regions in $\mathbb{R}^2$ and $\{F_1(x, s) > 0, F_2(x, s) > 0\} = D(s) \cap U \times V$, (where $U \times V$ is a sufficiently small neighborhood of the origin in $\mathbb{R}^2 \times \mathbb{R}^m$), furthermore $\gamma_1, \gamma_2 \subset \partial D(0)$, where $\gamma_j = \{x : F_j(x, 0) = 0\}$, $j = 1, 2$.
The following theorem is true.

Theorem 14.2 *If the oscillation index of the critical point $x = 0$ of the function $\Phi(x, 0)$ is β, then there exists a neighbourhood of zero $U \times V \subset \mathbb{R}^2 \times \mathbb{R}^m$ such that, for any fixed positive number $\varepsilon > 0$, for any amplitude function $a \in C_0^\infty(U \times V)$ the following estimate is valid:*

$$|J(\lambda, s)| \leq \frac{A_\varepsilon \|a\|_{C^3}}{\lambda^{\beta-\varepsilon}}.$$

Proof In the integral (14.2) we change the variables $y_1 = F_1(x, s)$, $y_2 = F_2(x, s)$. Since γ_1 and γ_2 transversally intersect at the origin, then $F := (F_1, F_2)$ is a family of diffeomorphisms in the neighbourhood of zero. Therefore, we have

$$J(\lambda, s) = \int_{\mathbb{R}^2} \chi_{\{y_1>0, y_2>0\}}(y) a_1(y, s) e^{i\lambda\Phi(F^{-1}(y,s),s)} dy. \tag{14.3}$$

To the last integral (14.3) applying Theorem 14.1, we obtain:

$$|J(\lambda, s)| \leq \frac{A_\varepsilon}{\lambda^{\beta-\varepsilon}}.$$

This completes a proof of Theorem 14.2. □

14.4 Sharpness of Result

Let us show that generally speaking, the estimates obtained do not hold for $\varepsilon = 0$. Indeed, consider the integral

$$J(\lambda) = \int_0^1 \int_0^1 e^{i\lambda xy} dx dy. \tag{14.4}$$

Let us show that for the integral (14.4) the following estimate from below is valid:

$$|J(\lambda, s)| \geq \frac{\ln \lambda}{2\lambda},$$

for large positive values λ. We represent (14.4) as

$$J(\lambda) = \int_0^1 dx \int_0^1 e^{i\lambda xy} dy + \int_{\frac{2}{\lambda}}^1 dx \int_0^1 e^{i\lambda xy} dy = J_1 + J_2.$$

Obviously, for the integral J_1 following inequality holds true:

$$|J_1| \leq \frac{2}{\lambda}.$$

Consider the estimation of the integral J_2. So, we calculate the inner integral:

$$J_2 = \int_{\frac{2}{\lambda}}^1 dx \int_0^1 e^{i\lambda xy} dy = \int_{\frac{2}{\lambda}}^1 \left(\frac{e^{i\lambda xy}}{i\lambda xy}\Big|_0^1 \right) dx = \frac{1}{i\lambda} \int_{\frac{2}{\lambda}}^1 \frac{e^{i\lambda xy} - 1}{x} dx$$

$$= \frac{1}{i\lambda} \left(\int_{\frac{2}{\lambda}}^1 \frac{\cos \lambda x - 1}{x} dx + i \int_{\frac{2}{\lambda}}^1 \frac{\sin \lambda x}{x} dx \right)$$

$$= \frac{1}{i\lambda} \left(\ln \frac{2}{\lambda} + \int_{\frac{2}{\lambda}}^1 \frac{\cos \lambda x}{x} dx + i \int_{\frac{2}{\lambda}}^1 \frac{\sin \lambda x}{x} dx \right).$$

Obviously, $\int_{\frac{2}{\lambda}}^1 \frac{\cos \lambda x}{x} dx = \int_2^\lambda \frac{\cos x}{x} dx = O(1)$ implies $\int_{\frac{2}{\lambda}}^1 \frac{\sin \lambda x}{x} dx = \int_2^\lambda \frac{\sin x}{x} dx = O(1)$ since $\lambda \to +\infty$. Thus, for large λ, we have the following estimate from below:

$$|J(\lambda)| > \frac{\ln \lambda}{2\lambda}.$$

The last estimate shows that it is impossible to get rid of the positive number ε. The question remains whether ε can be replaced by a suitable logarithmic multiplier.

References

1. Arkhipov, G.I., Karatsuba, A.A., Chubarikov, V.N.: Trigonometric Integrals. Izv. Akad. Nauk SSSR Ser. Mat. **5**(43), 971–1003 (1979)
2. Agarwal, R.P.: A propos d'une note de M.Pierre Humbert. C. R. Acad. Sci. Paris **236**, 2031–2032 (1953)
3. Fedoryuk, M.V.: Saddle-Point Method. Nauka, Moscow (1977) [Russian]
4. Karpushkin, V.N.: Uniform estimates of oscillating integrals in $\mathbb{R}^2$. Dokl. Acad. Sci. USSR **1**(254), 28–31 (1980) [Russian]
5. Randol, B.: On the Fourier transform of the indicator function of a planar set. Trans. AMS. **139**, 271–278 (1969)
6. Stein, E.M.: Singular Integrals and Differentiability Properties of Functions. Princeton Mathematical Series, vol. 30. Princeton University Press (1973)

7. Stein, E.M.: Harmonic Analysis: Real-Valued Methods, Orthogonality, and Oscillatory Integrals. Princeton University Press (1993)
8. Vinogradov, I.M.: Method Trigonometric Sums in Number Theory. Moscow, Nauka (1980) [Russian]

Chapter 15
The Unitary Dual of the Heisenberg Group Over $\mathfrak{R}_p$

Juan Pablo Velasquez-Rodriguez

Abstract In this short note, we review the essentials of the representation theory of the Heisenberg group over a certain compact ring obtained as the direct product of infinite copies of the finite field $\mathbb{F}_p$, where p is a prime number, and we use it to calculate the spectrum of the Vladimirov-Taibleson operator on the group.

15.1 The Ring $\mathfrak{R}_p$

Let $\mathbb{F}_p \cong \mathbb{Z}/p\mathbb{Z}$ be the finite field with p elements, where p is a prime number that we fix for the rest of this exposition. In this note, we are interested in the representation theory of groups of nilpotent matrices over the ring $\mathfrak{R}_p$, which we define here as the set

$$\mathfrak{R}_p := \{x = (x^1, x^2, x^3, \ldots) \; : \; x^j \in \mathbb{F}_p, \quad \text{for all } j \in \mathbb{N}\} \cong \prod_{k=1}^{\infty} \mathbb{F}_p,$$

together with the component-wise addition and multiplication inherited from $\mathbb{F}_p$. We can give $\mathfrak{R}_p$ the structure of a pseudo-compact ring by endowing it with the topology defined by the distance function

$$d(x, y) = |x - y|_p := p^{-\mathbf{Ord}(x-y)}, \quad \mathbf{Ord}(x - y) := \min\{j \; : \; x^j - y^j \neq 0\}.$$

The representation theory of $\mathfrak{R}_p$ is fairly simple: the characters of the additive group of $\mathfrak{R}_p$ can be obtained as finite products of the characters of $\not\Bbbk F_p$. That is, every character of $\mathfrak{R}_p$ has the form

J. P. Velasquez-Rodriguez (✉)
Ghent University, Ghent, Belgium
e-mail: juanpablo.velasquezrodriguez@ugent.be

M. Ruzhansky, K. Van Bockstal (eds.), *Extended Abstracts 2021/2022*, Research Perspectives Ghent Analysis and PDE Center 2,
https://doi.org/10.1007/978-3-031-42539-4_15

$$\chi(x) = \chi_{\xi^1}(x^1)\chi_{\xi^2}(x^2)\ldots\chi_{\xi^n}(x^n),\ \ \xi^1,\ldots,\xi^n \in \mathbb{F}_p,\ \text{ where } \chi_s(x^j) := e^{\frac{2\pi i}{p}x^j s}.$$

In this way, we can identify the Pontryagin dual of the additive group of $\mathfrak{R}_p$ with the group

$$\widehat{\mathfrak{R}_p} := \{\xi = (\xi^1, \xi^2, \xi^3, \ldots) \in \mathfrak{R}_p\ :\ \xi^j = 0,\ \text{for } j \text{ large enough}\}.$$

We will write

$$\chi_\xi(x) := \chi_{\xi^1}(x^1)\chi_{\xi^2}(x^2)\ldots\chi_{\xi^n}(x^n),\ \ \xi = (\xi^1, \ldots, \xi^n, 0, 0, \ldots) \in \widehat{\mathfrak{R}_p},$$

where we can express $\chi_\xi(x)$ in terms of the complex exponential with the following notation:

$$\chi_\xi(x) = e^{\frac{2\pi i}{p}x*\xi},\ \ x * \xi := \sum_j x^j\xi^j \in \mathbb{F}_p.$$

We also introduce the following notations:

- $|\xi|_p := p^{\mathbf{Ord}^*(\xi)}$ for $\xi \in \widehat{\mathfrak{R}_p}$, where $\mathbf{Ord}^*(\xi) := \max\{j\ :\ \xi^j \neq 0\}$,
- $\|x\|_p := \max_{1\leq k\leq d} |x_k|_p$, for $x = (x_1, \ldots, x_d) \in \mathfrak{R}_p^d$,
- $\|\xi\|_p := \max_{1\leq k\leq d} |\xi_k|_p$, for $\xi = (\xi_1, \ldots, \xi_d) \in \widehat{\mathfrak{R}_p^d} = (\widehat{\mathfrak{R}_p})^d$,
- $(x, \xi) := x_1 * \xi_1 + \ldots + x_d * \xi_d$, for $x \in \mathfrak{R}_p^d$ and $\xi \in \widehat{\mathfrak{R}_p^d}$,
- $\chi_\xi(x) = e^{\frac{2\pi i}{p}(x,\xi)}$, for $x \in \mathfrak{R}_p^d$ and $\xi \in \widehat{\mathfrak{R}_p^d}$.

15.2 The Heisenberg Group Over $\mathfrak{R}_p$

Let $\mathbb{H}_d(\mathfrak{R}_p)$ denote the group of $(d + 2) \times (d + 2)$ uni-triangular matrices of the form

$$\mathbb{H}_d(\mathfrak{R}_p) = \left\{ (x, y, z) := \begin{bmatrix} 1 & x^t & z \\ 0 & I_d & y \\ 0 & 0 & 1 \end{bmatrix} \in GL_{d+2}(\mathfrak{R}_p)\ :\ x, y \in \mathfrak{R}_p^d,\ z \in \mathfrak{R}_p \right\}.$$

We call this group the *Heisenberg group over* $\mathfrak{R}_p$ and, similarly to $\mathfrak{R}_p$, it is isomorphic to the direct product of infinite copies of $\mathbb{H}_d(\mathbb{F}_p)$. To see this, we simply need to define the projections

$$P_j : \mathfrak{R}_p \to \mathbb{F}_p,\ \ P_j(x) = x^j,$$

which we extend to $\mathfrak{R}_p^d$ as $P_j(x_1, \ldots, x_d) = (P_j(x_1), \ldots, P_j(x_d)) \in \mathbb{F}_p^d$. Then it is easy to check that the map

$$Q : \mathbb{H}_d(\mathfrak{R}_p) \to \prod_{k=1}^{\infty} \mathbb{H}_d(\mathbb{F}_p),$$

defined as

$$\begin{aligned} Q(x, y, z) &= ((P_1(x), P_1(y), P_1(z)), (P_2(x), P_2(y), P_2(z)), \ldots) \\ &= ((x^1, y^1, z^1), (x^2, y^2, z^2), (x^3, y^3, z^3), \ldots), \end{aligned}$$

is a group isomorphism. So, in order in order to describe explicitly the unitary dual of $\mathbb{H}_d(\mathfrak{R}_p)$, we need to briefly recall the representation theory of the Heisenberg group over $\mathbb{F}_p$.

As it is well known, there are two different kinds of unitary irreducible representations of $\mathbb{H}_d(\mathbb{F}_p)$:

1. **Type I representations:** Representations which are trivial on the center $\mathcal{Z}(\mathbb{H}_d(\mathbb{F}_p))$ and therefore descend to a representation of the quotient group $\mathbb{H}_d(\mathbb{F}_p)/\mathcal{Z}(\mathbb{H}_d(\mathbb{F}_p)) \cong \mathbb{F}_p^{2d}$. These are simply the characters of $\mathbb{F}_p^{2d}$, which have the form

$$\chi_{(\xi^j, \eta^j)}(x^j, y^j, z^j) = e^{\frac{2\pi i}{p}(x^j \cdot \xi^j + y^j \cdot \eta^j)}, \quad x^j, \xi^j, y^j, \eta^j \in \mathbb{F}_p^d, \quad z^j \in \mathbb{F}_p.$$

2. **Type II representations:** Noncommutative representations π_{λ^j} induced by a non-trivial central character $\chi_{\lambda^j}(x^j, y^j, z^j) = e^{\frac{2\pi i}{p} z^j \lambda^j}$, $\lambda^j \in \mathbb{F}_p$.

Now, for the representations of $\mathbb{H}_d(\mathfrak{R}_p)$, we just need to consider all the possible finite tensor products of representations of $\mathbb{H}_d(\mathbb{F}_p)$ composed with the projection maps. That is, unitary irreducible representations of $\mathbb{H}_d(\mathfrak{R}_p)$ have the form

$$\pi(x, y, z) = \bigotimes_{i=1}^{n} \tilde{\pi}((P_{k_i}(x), P_{k_i}(y), P_{k_i}(z))),$$

where each $\tilde{\pi}$ is a unitary irreducible representation of $\mathbb{H}_d(\mathbb{F}_p)$. Notice that, even though the tensor product is noncommutative in general, Type I representations are unidimensional, so that they commute with the non-commutative representations and we can group them together. With this consideration, we can write down the unitary irreducible representations of $\mathbb{H}_d(\mathfrak{R}_p)$ as

$$\pi_{(\xi, \eta, \lambda)}(x, y, z) = \chi_{(\xi, \eta)}(x, y, z) \otimes \pi_\lambda(x, y, z), \quad \xi, \eta \in \widehat{\mathfrak{R}_p^d}, \ \lambda \in \widehat{\mathfrak{R}_p},$$

where

$$\pi_\lambda(x, y, z) := \bigotimes_{j=1}^{\infty} \pi_{\lambda^j}((P_j(x), P_j(y), P_j(z))), \quad dim(\pi_{(\xi,\eta,\lambda)}) = dim(\pi_\lambda),$$

and the following condition must hold for all $j \in \mathbb{N}$:

$$P_j(\lambda)P_j(\xi_1) = \ldots = P_j(\lambda)P_j(\xi_d)$$
$$= P_j(\lambda)P_j(\eta_1) = \ldots = P_j(\lambda)P_j(\eta_d) = 0. \qquad (15.1)$$

Summing up, the unitary dual $\widehat{\mathbb{H}}_d$ of $\mathbb{H}_d(\mathfrak{R}_p)$, or $\mathbb{H}_d$ for short, can be described as

$$\widehat{\mathbb{H}}_d = \{\pi_{(\xi,\eta,\lambda)} \ : \ (\xi, \eta, \lambda) \in \Gamma \subset \widehat{\mathfrak{R}_p}^{2d+1}\},$$

where Γ is just the collection of elements of $\widehat{\mathfrak{R}_p}^{2d+1}$ satisfying Condition (15.1).

15.3 Applications

With this description of the unitary dual of $\widehat{\mathbb{H}}_d$, the Fourier series of a function $f \in L^2(\mathbb{H}_d)$ takes the form

$$f(x, y, z) = \sum_{(\xi,\eta,\lambda)\in\Gamma} d_\lambda Tr[\pi_{(\xi,\eta,\lambda)}(x, y, z)\widehat{f}(\xi, \eta, \lambda)],$$

where

$$\widehat{f}(\xi, \eta, \lambda) = \int_{\mathbb{H}_d} f(x, y, z)\pi^*_{(\xi,\eta,\lambda)}(x, y, z)dx \cdot dy \cdot dz.$$

By using this representation, left invariant operators on $\mathbb{H}_d$ take the form

$$T_\sigma f(x, y, z) = \sum_{(\xi,\eta,\lambda)\in\Gamma} d_\lambda Tr[\pi_{(\xi,\eta,\lambda)}(x, y, z)\sigma(\xi, \eta, \lambda)\widehat{f}(\xi, \eta, \lambda)],$$

where the mapping

$$\sigma : \Gamma \to \bigcup_{(\xi,\eta,\lambda)\in\Gamma} \mathcal{H}_{(\xi,\eta,\lambda)}, \quad \sigma(\xi, \eta, \lambda) \in \mathcal{H}_{(\xi,\eta,\lambda)},$$

is called the symbol of the operator, and it is given by the expression

$$\sigma(\xi, \eta, \lambda) = \pi^*_{(\xi,\eta,\lambda)}(x, y, z)T_\sigma \pi_{(\xi,\eta,\lambda)}(x, y, z).$$

In particular, we want to consider the following left invariant operator, often called the Vladimirov-Taibleson operator of order $\alpha > 0$:

$$D^\alpha f(x, y, z) := \frac{1 - p^\alpha}{1 - p^{-(\alpha+(2d+1))}} \times \int_{\mathbb{H}_d} \frac{f((x, y, z)(a, b, c)^{-1}) - f((x, y, z))}{\|(a, b, c)\|_p^{\alpha+(2d+1)}} da \cdot db \cdot dc.$$

To calculate the symbol of this operator, let us define the sequence of compact open subgroups

$$G_n := \{(x, y, z) \in \mathbb{H}_d \ : \ \|(x, y, z)\|_p \leq p^{-n}\} = B(e, p^{-n}),$$

which makes $\mathbb{H}_d$ a compact constant-order Vilenkin group. If we fix a $(\xi, \eta, \lambda) \in \Gamma$ with $\|(\xi, \eta, \lambda)\|_p = p^m$, $m > n$, then $\pi_{(\xi,\eta,\lambda)}$ is a non-trivial unitary irreducible representation of G_n and therefore

$$\int_{G_n} \pi^*_{(\xi,\eta,\lambda)}(x, y, z)dx \cdot dy \cdot dz = 0_{d_\lambda}.$$

With this information, the calculation of the symbol of the operator is just a simple computation:

$$\begin{aligned}
\sigma_{D^\alpha}(\xi, \eta, \lambda) &= C_\alpha \int_{\mathbb{H}_d} \frac{\pi_{(\xi,\eta,\lambda)}(x, y, z) - I_{d_\lambda}}{\|(x, y, z)\|_p^{\alpha+(2d+1)}} dxdydz \\
&= C_\alpha \int_{\mathbb{H}_d \setminus G_m} \frac{\pi_{(\xi,\eta,\lambda)}(x, y, z) - I_{d_\lambda}}{\|(x, y, z)\|_p^{\alpha+(2d+1)}} dxdydz \\
&= C_\alpha p^{(m-1)(\alpha+(2d+1))} \int_{G_{m-1} \setminus G_m} \pi_{(\xi,\eta,\lambda)}(x, y, z)dxdydz \\
&\quad - C_\alpha \int_{\mathbb{H}_d \setminus G_m} \frac{I_{d_\lambda}}{\|(x, y, z)\|_p^{\alpha+(2d+1)}} dxdydz \\
&= C_\alpha\big(- p^{(m-1)\alpha}(p^{-(2d+1)}) - \frac{p^{\alpha m} - 1}{p^\alpha - 1}(1 - p^{-(2d+1)})\big) I_{d_\lambda} \\
&= -C_\alpha p^{m\alpha}\big(p^{-(\alpha+(2d+1)} + \frac{1 - p^{-(2d+1)}}{p^\alpha - 1} - p^{-\alpha m}\frac{1 - p^{-(2d+1)}}{p^\alpha - 1}\big) I_{d_\lambda}
\end{aligned}$$

$$= \Big(- C_\alpha p^{m\alpha} \Big(\frac{1 - p^{-(\alpha+(2d+1))}}{p^\alpha - 1} - p^{-\alpha m} \frac{1 - p^{-(2d+1)}}{p^\alpha - 1} \Big) \Big) I_{d_\lambda}$$
$$= \Big(p^{m\alpha} - C_\alpha \frac{1 - p^{-(2d+1)}}{1 - p^\alpha} \Big) I_{d_\lambda}$$
$$= \Big(\|(\xi, \eta, \lambda)\|_p^\alpha - C_\alpha \frac{1 - p^{-(2d+1)}}{1 - p^\alpha} \Big) I_{d_\lambda}.$$

This shows how the group Fourier transform diagonalises the Vladimirov-taibleson operator on $\mathbb{H}_d$, and its eigenvalues are proportional to $\|(\xi, \eta, \lambda)\|_p^\alpha$. Sometimes it is also convenient to consider the operator

$$\mathbb{D}^\alpha := \frac{1 - p^{-(2d+1)}}{1 - p^{-(\alpha+(2d+1))}} I + D^\alpha,$$

and we will denote by $\langle(\xi, \eta, \lambda)\rangle_{\mathbb{H}_d}^\alpha$ its eigenvalue associated to the representation $\pi_{(\xi,\eta,\lambda)} \in \widehat{\mathbb{H}}_d$. For these eigenvalues we have

$$\langle(\xi, \eta, \lambda)\rangle_{\widehat{\mathbb{H}}_d}^\alpha = \begin{cases} \frac{1-p^{-(2d+1)}}{1-p^{-(\alpha+(2d+1))}} & \text{if } \pi_{(\xi,\eta,\lambda)} \text{ is the identity representation;} \\ \|(\xi, \eta, \lambda)\|_p^\alpha & \text{if } \|(\xi, \eta, \lambda)\|_p > 1. \end{cases}$$

We will explore further some of the consequences of our calculations in a future note. For example, we refer the interested reader to [1], where Tichmarsh theorems for Fourier series on some classes of profinite groups are proven.

Acknowledgments The author is supported by the FWO Odysseus 1 grant G.0H94.18N: Analysis and Partial Differential Equations.

References

1. Velasquez-Rodriguez, J.P.: Titchmarsh Theorems for Hölder-Lipschitz functions on profinite groups (2023). e-prints, arXiv:2303.18073

Chapter 16
Critical Sobolev-Type Identities and Inequalities on Stratified Lie Groups

Yerkin Shaimerdenov and Nurgissa Yessirkegenov

Abstract In this note, we discuss the critical case of the extended Hardy inequality obtained by Badiale and Tarantello (Arch. Ration. Mech. Anal. 163:259–293, 2002). Moreover, weighted versions of such inequalities, the so-called weighted critical Sobolev-type inequalities, with sharp constants are established. Actually, we obtain weighted critical Sobolev-type identities. Furthermore, anisotropic versions of these identities with any homogeneous quasi-norm are presented. Finally, we discuss hypoelliptic versions of these results in the setting of stratified Lie groups.

16.1 Introduction

The classical Hardy inequality asserts that for any $f \in C_0^\infty(\mathbb{R}^n)$

$$\left\| \frac{f}{|x|} \right\|_{L^p(\mathbb{R}^n)} \leq \frac{p}{n-p} \|\nabla f\|_{L^p(\mathbb{R}^n)}, \quad 1 \leq p < n, \tag{16.1}$$

where ∇ is the standard gradient on $\mathbb{R}^n$, $|x|$ is Euclidean norm on $\mathbb{R}^n$ and the constant $\frac{p}{n-p}$ is known to be sharp.

In 2002, the inequality (16.1) was extended by Badiale and Tarantello in [1] as follows: Let $x = (x', x'') \in \mathbb{R}^k \times \mathbb{R}^{n-k}$, $2 \leq k \leq n$. Then there exists a positive constant $C_{n,k,p}$ such that

$$\left\| \frac{f}{|x'|} \right\|_{L^p(\mathbb{R}^n)} \leq C_{n,k,p} \|\nabla f\|_{L^p(\mathbb{R}^n)}, \quad 1 \leq p < k, \tag{16.2}$$

Y. Shaimerdenov · N. Yessirkegenov (✉)
SDU University, Kaskelen, Kazakhstan Institute of Mathematics and Mathematical Modeling, Almaty, Kazakhstan
e-mail: yerkin.shaimerdenov@sdu.edu.kz

M. Ruzhansky, K. Van Bockstal (eds.), *Extended Abstracts 2021/2022*, Research Perspectives Ghent Analysis and PDE Center 2,
https://doi.org/10.1007/978-3-031-42539-4_16

where ∇ is the standard Euclidean gradient on $\mathbb{R}^n$ and $|x'|$ is Euclidean norm on $\mathbb{R}^k$. Indeed, it is easy to note that the extended Hardy inequality (16.2) implies the Hardy inequality (16.1) when $k = n$, so in this case the best constant equals $C_{n,k,p} = \frac{p}{n-p}$ in (16.2). In general case, it was conjectured by Badiale and Tarantello that the best constant in (16.2) is given by $C_{k,p} = \frac{p}{k-p}$ for $1 < p < k$. This conjecture was proved in [10] and later in [6] providing an alternative method. The extension (16.2) was motivated by its application to a nonlinear elliptic equation in $\mathbb{R}^3$, which was proposed as a model for the dynamics of galaxies. There, the authors call such inequalities the cylindrical version of the Hardy inequality (16.1).

Then, in [8] an improved version of the extended Hardy inequality (16.2) was obtained in the form: Let $x = (x', x'') \in \mathbb{R}^k \times \mathbb{R}^{n-k}$. Then for any $f \in C_0^\infty(\mathbb{R}^n \backslash \{x' = 0\})$, we have

$$\frac{|k-2|}{2} \left\| \frac{f}{|x'|} \right\|_{L^2(\mathbb{R}^n)} \leq \left\| \frac{x' \cdot \nabla_k f}{|x'|} \right\|_{L^2(\mathbb{R}^n)}, \tag{16.3}$$

where $|x'|$ is Euclidean norm on $\mathbb{R}^k$, ∇_k is the Euclidean gradient on $\mathbb{R}^k$. Moreover, the constant $\frac{|k-2|}{2}$ is sharp whenever $k \neq 2$. Namely, by using Schwarz's inequality one can derive (16.2) from (16.3). For the horizontal versions on stratified Lie groups, we refer to [8].

In this note, we discuss the critical case $k = 2$ of the improved Hardy inequality (16.3). Actually, we present the following improved Hardy identity on $\mathbb{R}^2 \times \mathbb{R}^{n-2}$:

$$4 \left\| \frac{\log|x'|(x' \cdot \nabla_2) f}{|x'|} \right\|^2_{L^2(\mathbb{R}^n)} = \left\| \frac{f}{|x'|} \right\|^2_{L^2(\mathbb{R}^n)} + \left\| \frac{f}{|x'|} + \frac{2\log|x'|(x' \cdot \nabla_2) f}{|x'|} \right\|^2_{L^2(\mathbb{R}^n)} \tag{16.4}$$

for all $f \in C_0^\infty(\mathbb{R}^n \backslash \{x' = 0\})$. By dropping the non-negative remainder term on the right-hand side of (16.4), we obtain the critical case $k = 2$ of (16.3):

$$\left\| \frac{f}{|x'|} \right\|_{L^2(\mathbb{R}^n)} \leq 2 \left\| \frac{\log|x'|(x' \cdot \nabla_2) f}{|x'|} \right\|_{L^2(\mathbb{R}^n)}.$$

Moreover, the constant 2 is sharp. Furthermore, in this note we discuss such identities with more general weights (weighted Sobolev-type identities) in Sect. 16.2. Then, their hypoelliptic versions on stratified Lie groups will be presented in Sect. 16.3.

16.2 Critical Sobolev-Type Identity and Inequality on $\mathbb{R}^k \times \mathbb{R}^{n-k}$

In this section, we discuss weighted versions of the identity (16.4). Namely, the critical case $\alpha = k/2$ of the following Sobolev-type inequality from [8]: Let $x = (x', x'') \in \mathbb{R}^k \times \mathbb{R}^{n-k}$ and let $\alpha \in \mathbb{R}$. Then for any $f \in C_0^\infty(\mathbb{R}^n \backslash \{x' = 0\})$, we have

$$\frac{|k-2\alpha|}{2} \left\| \frac{f}{|x'|^\alpha} \right\|_{L^2(\mathbb{R}^n)} \leq \left\| \frac{x' \cdot \nabla_k f}{|x'|^\alpha} \right\|_{L^2(\mathbb{R}^n)}, \tag{16.5}$$

where $|x'|$ is Euclidean norm on $\mathbb{R}^k$, ∇_k is the Euclidean gradient on $\mathbb{R}^k$. Moreover, the constant $\frac{|k-2\alpha|}{2}$ is sharp whenever $\alpha \neq k/2$. For the case when $\alpha = 0$ and $k = n$ we refer to [5]. Thus, the critical case $\alpha = k/2$ of the inequality (16.5) takes the following form.

Theorem 16.1 (Critical Sobolev-Type Identity on $\mathbb{R}^k \times \mathbb{R}^{n-k}$) *Let $x = (x', x'') \in \mathbb{R}^k \times \mathbb{R}^{n-k}$, $1 \leq k \leq n$. Then for any complex-valued function $f \in C_0^\infty(\mathbb{R}^n \backslash \{x' = 0\})$ we have*

$$4 \left\| \frac{\log|x'|(x' \cdot \nabla_k) f}{|x'|^{\frac{k}{2}}} \right\|^2_{L^2(\mathbb{R}^n)} = \left\| \frac{f}{|x'|^{\frac{k}{2}}} \right\|^2_{L^2(\mathbb{R}^n)} + \left\| \frac{f}{|x'|^{\frac{k}{2}}} + \frac{2\log|x'|(x' \cdot \nabla_k) f}{|x'|^{\frac{k}{2}}} \right\|^2_{L^2(\mathbb{R}^n)}, \tag{16.6}$$

where $|x'|$ is the Euclidean norm on $\mathbb{R}^k$ and ∇_k is the standard gradient on $\mathbb{R}^k$.

When $k = n$, we have an anisotropic version of this identity with any homogeneous quasi-norm for the radial derivative operator.

Theorem 16.2 (Critical Sobolev-Type Identity with any Homogeneous Quasi-Norm) *Let $|\cdot|$ be any homogeneous quasi-norm on $\mathbb{R}^n$. Then for any complex-valued function $f \in C_0^\infty(\mathbb{R}^n \backslash \{0\})$ we have*

$$4 \left\| \frac{\log|x|}{|x|^{\frac{k}{2}-1}} \frac{df}{d|x|} \right\|^2_{L^2(\mathbb{R}^n)} = \left\| \frac{f}{|x|^{\frac{k}{2}}} \right\|^2_{L^2(\mathbb{R}^n)} + \left\| \frac{f}{|x|^{\frac{k}{2}}} + \frac{2\log|x|}{|x|^{\frac{k}{2}-1}} \frac{df}{d|x|} \right\|^2_{L^2(\mathbb{R}^n)}.$$

Remark 16.1 Actually, Theorem 16.2 can be conveniently formulated in the language of Folland and Stein's homogeneous Lie groups [4].

Remark 16.2 Note that when $k = 2$ the identity (16.6) implies (16.4).

Remark 16.3 By dropping the remainder term, we derive from (16.6) the critical case $\alpha = k/2$ of (16.5):

$$\left\| \frac{f}{|x'|^{\frac{k}{2}}} \right\|_{L^2(\mathbb{R}^n)} \leq 2 \left\| \frac{\log |x'|(x' \cdot \nabla_k) f}{|x'|^{\frac{k}{2}}} \right\|_{L^2(\mathbb{R}^n)} . \tag{16.7}$$

Here, the constant 2 is sharp.

Remark 16.4 By using Schwarz's inequality on the right-hand side of (16.7), we obtain

$$\left\| \frac{f}{|x'|^{\frac{k}{2}}} \right\|_{L^2(\mathbb{R}^n)} \leq 2 \left\| \frac{\log |x'| \nabla_k f}{|x'|^{\frac{k}{2}-1}} \right\|_{L^2(\mathbb{R}^n)} \leq 2 \left\| \frac{\log |x'| \nabla f}{|x'|^{\frac{k}{2}-1}} \right\|_{L^2(\mathbb{R}^n)} , \tag{16.8}$$

which implies the critical case of the inequality (16.2) from [1] when $k = 2$.

Remark 16.5 In (16.6), by taking $k = 2$ and using Schwarz's inequality on the left-hand side, we obtain the inequality (16.8) with the remainder term

$$\left\| \frac{f}{|x'|} + \frac{2 \log |x'|(x' \cdot \nabla_2) f}{|x'|} \right\|^2_{L^2(\mathbb{R}^n)} + \left\| \frac{f}{|x'|} \right\|^2_{L^2(\mathbb{R}^n)} \leq 4 \left\| \log |x'| \nabla f \right\|^2_{L^2(\mathbb{R}^n)} .$$

16.3 Critical Sobolev-Type Identity on Stratified Lie Groups

Here, we now discuss the above results in the setting of stratified Lie groups. Before stating the main results on stratified Lie groups, let us recall the necessary definitions and notations. We say that a Lie group $\mathbb{G} = (\mathbb{R}^n, \circ)$ is stratified (or a homogeneous Carnot group) if it satisfies the conditions:

- The space $\mathbb{R}^n$ admits the decomposition $\mathbb{R}^n = \mathbb{R}^N \times \cdots \times \mathbb{R}^{N_r}$ and the dilation $\delta_\lambda : \mathbb{R}^n \to \mathbb{R}^n$ given by

$$\delta_\lambda(x) = \delta_\lambda \left(x', x^{(2)}, \ldots, x^{(r)} \right) := \left(\lambda x', \lambda^2 x^{(2)}, \ldots, \lambda^r x^{(r)} \right)$$

  is an automorphism of the group $\mathbb{G}$ for every $\lambda > 0$, where $x' \equiv x^{(1)} \in \mathbb{R}^N$ and $x^{(k)} \in \mathbb{R}^{N_k}$ for $k = 2, \ldots, r$.
- Let N be as in above and let $X_1, \ldots, X_N$ be the left invariant vector fields on $\mathbb{G}$ such that $X_k(0) = \left. \frac{\partial}{\partial x_k} \right|_0$ for $k = 1, \ldots, N$. Then the iterated commutators of $X_1, \ldots, X_N$ span the Lie algebra of $\mathbb{G}$ for all $x \in \mathbb{R}^n$, that is,

$$\operatorname{rank} (\operatorname{Lie} \{X_1, \ldots, X_N\}) = n.$$

So, the triple $\mathbb{G} = (\mathbb{R}^n, \circ, \delta_\lambda)$ is a stratified Lie group.

Recall also that the left invariant vector fields $X_1, \ldots, X_N$ are called the (Jacobian) generators of $\mathbb{G}$. If r is a step of $\mathbb{G}$, then the homogeneous dimension of $\mathbb{G}$ is defined by

$$Q = \sum_{k=1}^{r} kN_k, \quad N_1 = N.$$

The Haar measure dx on $\mathbb{G}$ is the standard Lebesgue measure for $\mathbb{R}^n$ (see, e.g. [3, Proposition 1.6.6]). We refer to [2, 3] or [7] for more details on stratified Lie groups. The left invariant vector fields X_j have an explicit form and satisfy the divergence theorem, see e.g. [3, Section 3.1.5] and [7],

$$X_k = \frac{\partial}{\partial x'_k} + \sum_{\ell=2}^{r} \sum_{m=1}^{N_1} a_{k,m}^{(\ell)} \left(x', \ldots, x^{\ell-1}\right) \frac{\partial}{\partial x_m^{(\ell)}}. \tag{16.9}$$

We will also use the following notations:

$$\nabla_H := (X_1, \ldots, X_N)$$

for the horizontal gradient,

$$\mathrm{div}_H\, v := \nabla_H \cdot v$$

for the horizontal divergence, and

$$|x'| := \sqrt{x_1'^2 + \cdots + x_N'^2}$$

for the Euclidean norm on $\mathbb{R}^N$.

Now, we are ready to discuss our results on stratified Lie groups.

Theorem 16.3 (Critical Sobolev-Type Identity on Stratified Lie Groups) *Let $\mathbb{G}$ be a stratified Lie group with N being the dimension of the first stratum. Denote by x' the variables from the first stratum of $\mathbb{G}$. Then for any $f \in C_0^\infty(\mathbb{G}\backslash\{x' = 0\})$, we have*

$$4 \left\| \frac{\log|x'|(x' \cdot \nabla_H) f}{|x'|^{\frac{N}{2}}} \right\|_{L^2(\mathbb{G})}^2 = \left\| \frac{f}{|x'|^{\frac{N}{2}}} \right\|_{L^2(\mathbb{G})}^2 + \left\| \frac{f}{|x'|^{\frac{N}{2}}} + \frac{2\log|x'|(x' \cdot \nabla_H) f}{|x'|^{\frac{N}{2}}} \right\|_{L^2(\mathbb{G})}^2, \tag{16.10}$$

where $|x'|$ is the Euclidean norm on $\mathbb{R}^N$ and ∇_H is the horizontal gradient on $\mathbb{G}$.

Remark 16.6 As in Remark 16.3, dropping the remainder term implies the following L^2-critical Sobolev-type inequality:

$$\left\| \frac{f}{|x'|^{\frac{N}{2}}} \right\|_{L^2(\mathbb{G})} \leq 2 \left\| \frac{\log|x'|(x' \cdot \nabla_H) f}{|x'|^{\frac{N}{2}}} \right\|_{L^2(\mathbb{G})}.$$

This inequality can be viewed as the critical case of the Sobolev-type inequality on stratified Lie groups from [8]. We also refer to [9] and the references therein for the investigation of the best constant in hypoelliptic Sobolev inequalities.

Remark 16.7 In the Abelian case $\mathbb{G} = (\mathbb{R}^n, +)$, we have $N = n, \nabla_H = \nabla = (\partial_{x_1}, \ldots, \partial_{x_n})$, so (16.10) implies Theorem 16.2 with the Euclidean norm $|\cdot|$ on $\mathbb{R}^n$.

Remark 16.8 Applying Schwarz's inequality in (16.10), we obtain the following critical Hardy inequality with the remainder term on stratified Lie groups:

$$4 \left\| \log|x'| \frac{\nabla_H f}{|x'|^{\frac{N}{2}-1}} \right\|^2_{L^2(\mathbb{G})} \geq \left\| \frac{f}{|x'|^{\frac{N}{2}}} \right\|^2_{L^2(\mathbb{G})} + \left\| \frac{f}{|x'|^{\frac{N}{2}}} + \frac{2\log|x'|(x' \cdot \nabla_H) f}{|x'|^{\frac{N}{2}}} \right\|^2_{L^2(\mathbb{G})}.$$

It follows that

$$\left\| \frac{f}{|x'|^{\frac{N}{2}}} \right\|_{L^2(\mathbb{G})} \leq 2 \left\| \frac{\log|x'|}{|x'|^{\frac{N}{2}-1}} \nabla_H f \right\|_{L^2(\mathbb{G})}.$$

These inequalities give critical cases of the horizontal Hardy inequalities on stratified Lie groups from [6].

Acknowledgments This research is funded by the Committee of Science of the Ministry of Science and Higher Education of the Republic of Kazakhstan (Grant No. AP14871691).

References

1. Badiale, M., Tarantello, G.: A Sobolev-Hardy inequality with applications to a nonlinear elliptic equation arising in astrophysics. Arch. Ration. Mech. Anal. **163**, 259–293 (2002)
2. Bonfiglioli, A., Lanconelli, E., Uguzzoni, F.: Stratified Lie Groups and Potential Theory for Their Sub-Laplacians. Springer, Berlin/Heidelberg (2007)

3. Fischer, V., Ruzhansky, M.: Quantization on Nilpotent Lie Groups. Progress in Mathematics, vol. 314. Birkhäuser, Basel (2016)
4. Folland, G.B., Stein, E.M.: Hardy Spaces on Homogeneous Groups. Mathematical Notes, vol. 28. Princeton University Press/University of Tokyo Press, Princeton/Tokyo (1982)
5. Ozawa, T., Sasaki, H.: Inequalities associated with dilations. Commun. Contemp. Math. **11**, 265–277 (2009)
6. Ruzhansky, M., Suragan, D.: On horizontal Hardy, Rellich, Caffarelli-Kohn-Nirenberg and p-sub-Laplacian inequalities on stratified groups. J. Differ. Equ. **262**, 1799–1821 (2017)
7. Ruzhansky, M., Suragan, D.: Hardy Inequalities on Homogeneous Groups. Progress in Mathematics, vol. 537. Birkhäuser, Basel (2019)
8. Ruzhansky, M., Suragan, D., Yessirkegenov, N.: Caffarelli-Kohn-Nirenberg and Sobolev-type inequalities on stratified Lie groups. Nonlinear Differ. Equ. Appl. **24**, Article no. 56 (2017)
9. Ruzhansky, M., Tokmagambetov, N., Yessirkegenov, N.: Best constants in Sobolev and Gagliardo-Nirenberg inequalities on graded groups and ground states for higher order nonlinear subelliptic equations. Calc. Var. Partial. Differ. Equ. **59**, Article no. 175 (2020)
10. Secchi, S., Smets, D., Willen, M.: Remarks on a Hardy-Sobolev inequality. Acad. Sci. Paris, Ser. I **336**, 811–815 (2003)

Part II
Partial Differential Equations

Chapter 17
Anisotropic Picone Type Identities for General Vector Fields and Some Applications

Abimbola Abolarinwa

Abstract This extended abstract presents a generalisation of nonlinear Picone-type identities for anisotropic subelliptic p-Laplacian in the context of general vector fields, as proved in the forthcoming paper (Abolarinwa, Nonlinear Anisotropic Picone Type Identities for General Vector Fields and Applications, Submitted). Furthermore, several applications, which range from Hardy-type inequalities to Liouville-type and Sturmian-type comparison principles to first eigenvalue monotonicity, are highlighted. There is also an extension of these identities to subelliptic p-bi-Laplacian, and their applications to nonlinear eigenvalues in Abolarinwa (Nonlinear Anisotropic Picone Type Identities for General Vector Fields and Applications, Submitted).

17.1 Introduction and Preliminaries

17.1.1 Introduction

The classical Picone identity states that, for nonnegative differentiable functions u and v with $v \neq 0$, the following formula

$$|\nabla u|^2 + \frac{u^2}{v^2}|\nabla v|^2 - 2\frac{u}{v}\nabla u \nabla v = |\nabla u|^2 - \nabla\left(\frac{u^2}{v}\right)\nabla v \geq 0 \tag{17.1}$$

holds. Here, $|\cdot|$ and ∇ denote the length of a vector and the gradient of a function in the Euclidean space $\mathbb{R}^n$, respectively. The identity (17.1) has been extensively applied to the study of second-order elliptic equations and systems involving the Laplace operator. Several extensions and generalisation of Picone identity have been established in order to handle more general elliptic operators. For instance, in

A. Abolarinwa (✉)
Department of Mathematics, University of Lagos, Lagos, Nigeria

M. Ruzhansky, K. Van Bockstal (eds.), *Extended Abstracts 2021/2022*, Research Perspectives Ghent Analysis and PDE Center 2,
https://doi.org/10.1007/978-3-031-42539-4_17

order to study p-Laplace equations and eigenvalue problems involving p-Laplacian, Allegretto and Huang [2] extended (17.1) to the case of general $p > 1$ as follows: for $u \geq 0$, $v > 0$, then

$$|\nabla u|^p + (p-1)\frac{u^p}{v^p}|\nabla v|^p - p\frac{u^{p-1}}{v^{p-1}}|\nabla v|^{p-2}\nabla v \nabla u = \mathcal{R}_p(u, v), \tag{17.2}$$

where

$$\mathcal{R}_p(u, v) := |\nabla u|^p - \nabla\left(\frac{u^p}{v^{p-1}}\right)|\nabla v|^{p-2}\nabla v \geq 0.$$

Tyagi [22] and Bal [3] established nonlinear versions of (17.1) and (17.2), respectively, with several applications (see also [8, 21]).

The Picone identity for the p-biLaplacian is a generalisation of the identity to a class of fourth-order elliptic operators (biLaplacian) as established by Dunninger [6] and Dwivedi and Tyagi [5]. Jaroš [10] obtained a generalised nonlinear analogue of this and highlighted its applications to comparison results for a class of half-linear differential equations of fourth-order. Other interesting extensions of Picone type identities one can find [11, 12] (for Finsler p-Laplacian with application to Caccioppoli inequality), [15, 17] (for general vector fields and p-sub-Laplacian with applications to Grushin plane, Heisenberg group), [13] (for p-sub-Laplacian on Heisenberg group and applications to Hardy inequalities), [19] (for nonlinear Picone identities for anisotropic p-sub-Laplacian and p-biLaplacian with applications to horizontal Hardy inequalities and weighted eigenvalue problem on stratified Lie groups).

The main purpose of this work is to derive generalised nonlinear versions of Picone identities (as presented in [1]) for the anisotropic subelliptic p-Laplacian and p-biLaplacian for the general vector fields. The generalisation of Picone-type identities to subelliptic contexts is experiencing rapid attention nowadays owing to their numerous applications in the analysis of partial differential equations and other areas of mathematical analysis and applications.

17.1.2 Anisotropic Sub-Laplacian for General Vector Fields

The anisotropic Euclidean p-Laplacian is defined for C^2-functions as

$$\sum_{k=1}^{N} \frac{\partial}{\partial x_k}\left(\left|\frac{\partial f}{\partial x_k}\right|^{p_k-2}\frac{\partial f}{\partial x_k}\right)$$

with $p_k > 1$, $k = 1, \cdots, N$. Setting $p_k = 2$ and $p_k = p$ for all k, this operator reduces to the usual Laplacian and the pseudo-p-Laplacian, respectively. The anisotropic Laplacian plays crucial roles in several areas of mathematical theories and their applications in engineering and sciences. For example, it reflects the anisotropic characteristics of some reinforced materials [20], as well as explains the fluid dynamics in anisotropic media that have different conductivities in each direction [4]. Anisotropic Laplacian models also arise in image processing and computer vision [18].

Let M be an n-dimensional smooth manifold equipped with a volume form dx, and $\{X_k\}_{k=1}^N$, $n \geq N$, be a family of real vector fields defined on M. Consider the operator for sums of squares of vector fields given as

$$\mathcal{L} := \sum_{k=1}^{N} X_k^2,$$

which is a second-order differential operator, usually called a canonical sub-Laplacian. This operator is well known to be locally hypoelliptic if the commutators of the vector fields $\{X_k\}_{k=1}^N$ generate the tangent space of M as the Lie algebra, due to Hörmander's pioneering work. The operator is also well studied under the weaker assumption without hypoellipticity [14]. The horizontal differential operator X_k identifies each vector field X with its derivative in the direction k. We denote the horizontal gradients for general vector fields by

$$\nabla_X = (X_1, \cdots, X_N).$$

The anisotropic p-sub-Laplacian for $1 < p_k < \infty$ is defined by

$$\mathcal{L}_p f := \sum_{k=1}^{N} X_k(|X_k f|^{p_k-2} X_k f),$$

while the anisotropic p-sub-biLaplacian for $1 < p_k < \infty$ is defined by

$$\mathcal{L}_p^2 f := \sum_{k=1}^{N} X_k^2(|X_k^2 f|^{p_k-2} X_k^2 f).$$

There are numerous examples of submanifolds in which vector fields can be defined. For example, we list, among others, the Carnot groups, Heisenberg groups, Engel groups, and the Grushin plane (which does not even possess a group structure). Interested readers can see the recent open access book [14] for more examples and detailed discussions on the sub-Laplacian and its various extensions in each case. In the case $M = \mathbb{R}^n$, then dx is the Lebesgue measure, $\nabla_X = \nabla$ and $\mathcal{L} = \Delta$ are the usual Euclidean gradient and Laplacian, respectively.

17.2 Anisotropic Picone Identities for Subelliptic p-Laplacian

Here, we present a generalised nonlinear Picone identity for the subelliptic p-Laplacian, $\mathcal{L}_p := \sum_{k=1}^{N} X_k(|X_k u|^{p_k-1} X_k u)$, $p_k > 1$, which extends the results of Feng and Cui [9] (for the Euclidean anisotropic Laplacian) and Feng [8] (for the Euclidean p-Laplacian). We refer to Remark 17.1 for other special cases of our result. An analogue of generalised second-order Picone-type identities for general vector fields, that is, the extension of Theorem 17.1 to anisotropic p-subelliptic bi-Laplacian, is proved and its applications to nonlinear eigenvalue problems are discussed in [1].

Theorem 17.1 *Let M be an n-dimensional smooth manifold and Ω any domain in M such that u and v are nonconstant real-valued differentiable functions a.e. in Ω. Suppose further that functions $g_k, f_k : \mathbb{R} \to (0, \infty)$ are C^1-functions for $k = 1, \cdots, N$, satisfying the following properties:*

$$\begin{aligned} &g_k(u) > 0, \quad g_k'(u) > 0 \ \text{ for } u > 0, \quad x \in \Omega; \\ &g_k(u) = 0, \quad g_k'(u) = 0 \ \text{ for } u = 0, \quad x \in \partial\Omega; \\ &f_k(v) > 0, \quad f_k'(v) > 0 \ \text{ for } v > 0, \quad x \in \Omega; \end{aligned}$$

and for $1 < p_k < \infty$

$$\sum_{k=1}^{N} \frac{g_k(u) f_k'(v)}{[f_k(v)]^2} |X_k v|^{p_k} \geq \sum_{k=1}^{N} (p_k - 1) \left[\frac{g_k'(u)}{p_k f_k(v)} |X_k v|^{p_k-1} \right]^{\frac{p_k}{p_k-1}}. \tag{17.3}$$

Define the quantities $R(u, v)$ and $L(u, v)$ as follows:

$$R(u, v) := \sum_{k=1}^{N} |X_k u|^{p_k} - \sum_{k=1}^{N} X_k \left(\frac{g_k(u)}{f_k(v)} \right) |X_k v|^{p_k-2} X_k v, \tag{17.4}$$

$$L(u, v) := \sum_{k=1}^{N} |X_k u|^{p_k} - \sum_{k=1}^{N} \frac{g_k'(u)}{f_k(v)} |X_k v|^{p_k-2} X_k v X_k u + \sum_{k=1}^{N} \frac{g_k(u) f_k'(v)}{[f_k(v)]^2} |X_k v|^{p_k}, \tag{17.5}$$

where $p_k > 1$, $k = 1, \cdots, N$. Then

$$R(u, v) = L(u, v) \geq 0. \tag{17.6}$$

Moreover, $L(u, v) = 0$ a.e. in Ω if and only if

$$u = \alpha v \quad a.e \; in \; \Omega \quad for \quad \alpha \in \mathbb{R}, \tag{17.7}$$

$$|X_k u| = \left[\frac{g_k'(u)}{p_k f_k(v)} |X_k v|^{p_k - 1} \right]^{\frac{1}{p_k - 1}}, \quad k = 1, \cdots, N, \tag{17.8}$$

and

$$\frac{g_k(u) f_k'(v)}{[f_k(v)]^2} = (p_k - 1) \left[\frac{g_k'(u)}{p_k f_k(v)} \right]^{\frac{p_k}{p_k - 1}}, \quad k = 1, \cdots, N. \tag{17.9}$$

Remark 17.1 Theorem 17.1 generalises many known results. Examples:

(i) If $M = \mathbb{R}^n$, and we allow $p_k = 2$, $g_k = u^2$ and $f_k(y) = y$ for $k = 1, \cdots, N$ in (17.4) and (17.5). Then, we obtain the classical Picone identity (17.1) for the usual Euclidean Laplacian.
(ii) If $M = \mathbb{R}^n$, and we allow $p_k = 2$ and $g_k = u^2$, $k = 1, \cdots, N$ in (17.4) and (17.5). Then, we arrive at Tyagi's result [22], which is the nonlinear analogue of (17.1).
(iii) If we allow $p_k = p$ and $g_k = u^p$ for each k in (17.4) and (17.5), we obtain a special case proved by Bal [3] in the Euclidean setting and a special case proved by Suragan and Yessirkegenov [19] in the setting of stratified Lie groups.
(iv) The case $p_k = p$, $g_k = u^p$ and $f_k(y) = y^{p-1}$ for each k in (17.4) and (17.5) was proved by Allegretto and Huang [2] (for $M = \mathbb{R}^n$ and Euclidean p-Laplacian), by Niu, Zhang and Wang [13] (for the Heisenberg group and horizontal p-sub-Laplacian) and by Ruzhansky et al. [17] (for general vector fields).
(v) The case $g_k = u^{p_k}$ for $k = 1, \cdots, N$ in (17.4) and (17.5) was presented by Suragan and Yessirkegenov in [19, Theorem 1.3] (see also [14]).
(vi) The case $g_k = u^{p_k}$ and $f_k(y) = y^{p_k - 1}$ for $k = 1, \cdots, N$ in (17.4) and (17.5) was proved by Ruzhansky et al. [15, 16] (for general vector fields and stratified Lie groups).

Proof By direct computation, we have

$$X_k \left(\frac{g_k(u)}{f_k(v)} \right) = \frac{g_k'(u) X_k u}{f_k(v)} - \frac{g_k(u) f_k'(v) X_k v}{[f_k(v)]^2}. \tag{17.10}$$

Substituting (17.10) into the expression $R(u, v)$ gives the identity

$$R(u, v) = L(u, v).$$

Next is to show that $L(u, v) \geq 0$. First, we rewrite $L(u, v)$ as follows

$$L(u, v) := \mathscr{A}_1(u, v) \;+\; \mathscr{A}_2(u, v) \;+\; \mathscr{A}_3(u, v),$$

where

$$\mathscr{A}_1(u, v) = \sum_{k=1}^{N} p_k \left(\frac{1}{p_k} |X_k u|^{p_k} + \frac{p_k - 1}{p_k} \left[\frac{g_k'(u)|X_k v|^{p_k-1}}{p_k f_k(v)} \right]^{\frac{p_k}{p_k-1}} \right) - \sum_{k=1}^{N} \frac{g_k'(u)|X_k u||X_k v|^{p_k-1}}{f_k(v)},$$

$$\mathscr{A}_2(u, v) = \sum_{k=1}^{N} \frac{g_k'(u)|X_k v|^{p_k-2}}{f_k(v)} \Big(|X_k v||X_k u| - X_k v X_k u \Big),$$

$$\mathscr{A}_3(u, v) = \sum_{k=1}^{N} \frac{g_k(u) f_k'(v)|X_k v|^{p_k}}{[f_k(v)]^2} - \sum_{k=1}^{N} (p_k - 1) \left[\frac{g_k'(u)|X_k v|^{p_k-1}}{p_k f_k(v)} \right]^{\frac{p_k}{p_k-1}}.$$

By the application of Young inequality, we can prove that $\mathscr{A}_1(u, v) \geq 0$. Considering the inequality $|X_k v||X_k u| - X_k v X_k u \geq 0$ and the fact that $\frac{g_k'(u)|X_k v|^{p_k-2}}{f_k(v)} > 0$, (since $g_k'(u) > 0,\ f(v) > 0$ in Ω), we obtain $\mathscr{A}_2(u, v) \geq 0$. Clearly, the condition (17.3) implies that $\mathscr{A}_3(u, v) \geq 0$. Putting all these considerations together, we obtain that $L(u, v) \geq 0$ a.e. in Ω.

By the above analysis, we infer that $L(u, v) = 0$ holds if and only if we have $|X_k v||X_k u| - X_k v X_k u = 0$, equality in Young's inequality, and there is equality in (17.3). Therefore, it follows from (17.7) that there exists a positive constant α such that $u = \alpha v$, which implies $\mathscr{A}_2 = 0$. By (17.8) we have that $\mathscr{A}_1(u, v) = 0$. We can show that $\mathscr{A}_3(u, v) = 0$ by (17.9). Putting these considerations together, it follows that $L(u, v) = \mathscr{A}_1(u, v) + \mathscr{A}_2(u, v) + \mathscr{A}_3(u, v) = 0$, and we conclude that $L(u, v) = 0$ if and only if (17.7)–(17.9) hold.

Indeed, if $u = 0$, the conclusion follows. If $u \neq 0$ the conclusion holds from the above procedure. □

17.3 Some Applications

Here we give some results which can be obtained by the application of the anisotropic Picone identities.

17.3.1 Functional Setting

Let $\Omega \subset M$ be an open domain. Recall the classical spaces $\mathcal{D}^{1,p}(\Omega)$, $1 \leq p < \infty$, defined by $\mathcal{D}^{1,p}(\Omega) := \{u \in L^p(\Omega) : |\nabla_X u| \in L^p(\Omega)\}$ with respect to the norm $\|u\|_{\mathcal{D}^{1,p}(\Omega)} := \left(\int_\Omega |u|^p dx + \int_\Omega |\nabla_X u|^p dx\right)^{\frac{1}{p}}$. We define the anisotropic functional spaces, $\mathcal{D}^{1,p_k}(\Omega)$, $k = 1, \cdots, N$, by

$$\mathcal{D}^{1,p_k}(\Omega) := \{u \in \mathcal{D}^{1,1}(\Omega) : | X_k u| \in L^{p_k}(\Omega)\}$$

with respect to the norm $\|u\|_{\mathcal{D}^{1,p_k}(\Omega)} := \int_\Omega |u| dx + \sum_{k=1}^N \left(\int_\Omega |X_k u|^{p_k} dx\right)^{\frac{1}{p_k}}$. Consider the anisotropic functional

$$\mathbb{J}_{p_k}(u) := \sum_{k=1}^N \left(\int_\Omega |X_k u|^{p_k} dx\right)^{\frac{1}{p_k}},$$

then we also define the anisotropic functional class $\mathcal{D}_0^{1,p_k}(\Omega)$ to be the closure of $C_0^\infty(\Omega)$ with respect to the norm generated by the functional $\mathbb{J}_{p_k}(u)$. For more details on anisotropic functional spaces, see [7]. It is well known that $\mathcal{D}^{1,p_k}(\Omega)$ and $\mathcal{D}_0^{1,p_k}(\Omega)$ are separable and reflexive Banach spaces.

17.3.2 Nonlinear Anisotropic p-Sub-Laplacian Equation

Consider the nonlinear anisotropic p-sub-Laplacian equation

$$\begin{aligned} -\sum_{k=1}^N X_k(|X_k u|^{p_k-2} X_k u) &= \sum_{k=1}^N F(p_k, x, u), & x \in \Omega, \\ u &> 0, & x \in \Omega, \\ u &= 0, & x \in \partial\Omega. \end{aligned} \tag{17.11}$$

By the (weak)-solution of (17.11), we refer to a positive function $u \in \mathcal{D}_0^{1,p_k}(\Omega)$ satisfying

$$\sum_{k=1}^N \int_\Omega |X_k u|^{p_k-2} X_k u X_k \phi dx = \sum_{k=1}^N \int_\Omega F(p_k, x, u)\phi dx$$

for all $\phi \in C_0^\infty(\Omega)$.

17.3.3 Applications

In the rest of this section, g_k and f_k are assumed to be $C^1(\mathbb{R})$-functions, except if stated otherwise.

17.3.3.1 Anisotropic Sturmian Comparison Principle

Here, we present a Sturmian comparison principle to the weighted anisotropic p-sub-Laplacian equation with singular terms.

Proposition 17.1 *Let $h_1(x)$ and $h_2(x)$ be two continuous weight functions such that $h_1(x) < h_2(x)$ in the bounded open domain $\Omega \subset M$. If there exists a positive function $u \in \mathcal{D}_0^{1,p_k}(\Omega)$ solving*

$$
\begin{aligned}
-\sum_{k=1}^{N} X_k(|X_k u|^{p_k-2} X_k u) &= \sum_{k=1}^{N} h_1(x)\frac{g_k(u)}{u}, && x \in \Omega, \\
g_k(u) > 0, \quad u > 0, & && x \in \Omega, \\
g_k(u) = 0, \quad u = 0, & && x \in \partial\Omega.
\end{aligned}
\tag{17.12}
$$

Then any nontrivial solution v of the weighted anisotropic p-sub-Laplacian equation

$$
\begin{aligned}
-\sum_{k=1}^{N} \frac{g_k(u)}{f_k(v)} X_k(|X_k v|^{p_k-2} X_k v) &= \sum_{k=1}^{N} h_2(x) g_k(u), && x \in \Omega, \\
g_k(u) > 0, \quad u > 0, & && x \in \Omega,
\end{aligned}
$$

must change sign.

Corollary 17.1 *Suppose that the hypotheses of Proposition 17.1 hold. If $u \in \mathcal{D}_0^{1,p_k}(\Omega)$ is a positive solution of* (17.12), *then any nontrivial solution*

$$
-X_k(|X_k v|^{p_k-2} X_k v) = h_2(x) f_k(v), \quad x \in \Omega, \tag{17.13}
$$

changes sign.

17.3.3.2 Liouville Type Principle

The next application is the proof of a Liouville-type result for anisotropic p-sub-Laplacian using anisotropic Picone identities.

Proposition 17.2 *Suppose $p_k > 1$, $\eta_0 > 0$ is a constant and $f_k : \mathbb{R}^+ \to \mathbb{R}^+$ is a C^1-function satisfying $f_k(s) > 0$, $f_k'(s) > 0$ for $s > 0$ in Ω. Then, the anisotropic p-sub-Laplacian inequality*

$$-\sum_{k=1}^{N} X_k(|X_k v|^{p_k-2} X_k v) \geq \eta_0 \sum_{k=1}^{N} f_k(v), \qquad x \in \Omega,$$

$$v = 0, \qquad x \in \partial\Omega,$$

has no solution in $\mathcal{D}_0^{1,p_k}(\Omega)$.

17.3.3.3 Anisotropic Quasilinear System with Singular Nonlinearities

We can show that Theorem 17.1 yields a linear relation between u and v solving anisotropic quasilinear system with singular nonlinearities. Given the following system of anisotropic p-sub-Laplace equations

$$\begin{aligned}
-\sum_{k=1}^{N} X_k(|X_k u|^{p_k-2} X_k u) &= \sum_{k=1}^{N} f_k(v), && x \in \Omega,\\
-\sum_{k=1}^{N} X_k(|X_k v|^{p_k-2} X_k v) &= \sum_{k=1}^{N} \frac{[f_k(v)]^2 u}{g_k(u)}, && x \in \Omega,\\
g_k(u) > 0,\ f_k(v) > 0,\ u > 0,\ v > 0,& && x \in \Omega,\\
g_k(u) = 0,\ f_k(v) = 0,\ u = 0,\ v = 0,& && x \in \partial\Omega.
\end{aligned} \tag{17.14}$$

Proposition 17.3 *Let $(u, v) \in \mathcal{D}_0^{1,p_k}(\Omega) \times \mathcal{D}_0^{1,p_k}(\Omega)$ be a pair of (weak)-solutions to* (17.14). *Then $u = \alpha v$ a.e. in Ω, where $\alpha \in \mathbb{R}$ is a constant.*

17.3.3.4 Weighted Anisotropic Hardy Inequalities

In this part of the paper, we discuss the application of Theorem 17.1 to the derivation of a generalised anisotropic Hardy inequality for the general vector fields, from which several Hardy-type inequalities could be deduced. This is done in the spirit of [15].

Proposition 17.4 *Let $\Omega \subset M$ be an open bounded domain. Suppose that a function $v \in \mathcal{D}_0^{1,p_k}(\Omega)$, $p_k > 1$, $k = 1, \cdots, N$, satisfies*

$$\begin{aligned} -X_k(|X_k v|^{p_k-2} X_k v) &\geq H_k(x) f_k(v), && x \in \Omega, \\ f_k(v) &> 0,\ v > 0, && x \in \Omega, \\ f_k(v) &= 0,\ v = 0, && x \in \partial\Omega, \end{aligned}$$

where $f_k : \mathbb{R}^+ \to \mathbb{R}^+$ is a C^1-function and $H_k(x)$ is a nonnegative weight function, $k = 1, \cdots, N$. Then for any nonnegative $u \in C_0^1(\Omega)$ with $g_k(u) > 0$, where $g_k(u)$ is a C^1-function, there holds

$$\sum_{k=1}^{N} \int_\Omega |X_k u|^{p_k} dx \geq \sum_{k=1}^{N} \int_\Omega H_k(x) g_k(u) dx.$$

Acknowledgments The author gratefully acknowledges the grant support from the IMU-Simons African Fellowship Grant and EMS-Simons for African program. He also thanks Professor Michael Ruzhansky for the useful discussion on this project. Parts of the results were presented during the Summer School "Singularities in Science and Engineering" on 22-31 August 2022 at Ghent Analysis and PDE Center, Ghent University.

References

1. Abolarinwa, A.: Nonlinear Anisotropic Picone Type Identities for General Vector Fields and Applications (Submitted)
2. Allegretto, W., Huang, Y.X.: A Picone's identity for the p-Laplacian and applications. Nonl. Anal. Theory Meth. Appl. **32**(7), 819–830 (1998)
3. Bal, K.: Generalized Picone's identity and its applications. Electron. J. Differ. Equ. **2013**(243), 1–6 (2013)
4. Bear, J.: Dynamics of Fluids in Porous Media. Elsevier, New York (1972)
5. Dwivedi, G., Tyagi, J.: Some remarks on the qualitative questions for biharmonic equations. Taiwanese J. Math. **19**(6), 1743–1758 (2015)
6. Dunninger, D.R.: A Picone integral identity for a class of fourth order elliptic differential inequalities. Atti Accad. Naz. Lincei Rend. Cl. Sci. Fis. Mat. Natur. **50**(8), 630–641 (1971)
7. Evans, L.C.: Partial Differential Equations, 2nd edn. American Mathematical Society, Providence (2010)
8. Feng, T.: A new nonlinear Picone identity and applications. Math. Appl. **30**(2), 278–283 (2017)
9. Feng, T., Cui, X.: Anisotropic Picone identities and anisotropic Hardy inequalities. J. Inequal. Appl. **2017**, 16 (2017)
10. Jaroš, J.: Picones identity for the p-biharmonic operator with applications. Electron. J. Differ. Equ. **2011**(122), 1–6 (2011)
11. Jaroš, J.: A-harmonic Picone's identity with applications. Ann. Mat. **194**(3), 719–729 (2015)
12. Jaroš, J.: Caccioppolli estimates through an anisotropic Picone's identity. Proc. Am. Math. Soc. **143**(3), 1137–1144 (2015)

13. Niu, P., Zhang, H., Wang, Y.: Hardy type and Rellich type inequalities on the heisenberg group. Proc. Am. Math. Soc. **129**(12), 3623–3630 (2001)
14. Ruzhansky, M., Suragan, D.: Hardy inequalities on homogeneous groups. Progress in Mathematics, vol. 327, 588 pp. Birkhäuser, Basel (2019)
15. Ruzhansky, M., Sabitbek, B., Suragan, D.: Weighted anisotropic Hardy and Rellich type inequalities for general vector fields. Nonlinear Differ. Eqn. Appl. **26**, 13 (2019)
16. Ruzhansky, M., Sabitbek, B., Suragan, D.: Weighted L^p-Hardy and L^p-Rellich inequalities with boundary terms on stratified Lie groups. Rev. Mat. Complut. **32**(1), 19–35 (2019)
17. Ruzhansky, M., Sabitbek, B., Suragan, D.: Principal frequency of p-versions of sub-Laplacians for general vector fields. Z. Anal. Anwend. **40**(1), 97–109 (2021)
18. Sapiro, G.: Geometric Partial Differential Equations and Image Analysis. Cambridge University Press, Cambridge (2001)
19. Surugan, D., Yessirkegenov, N.: Generalised nonlinear Picone identities for p-sub-Laplacians and p-biharmonic operators and applications. Adv. Oper. Theory **6**(53), 1–17 (2021)
20. Tang, Q.: Regularity of minimizer of nonisotropic integrals of the calculus of variations. Ann. Mat. Pura Appl. **164**, 77–87 (1993)
21. Tirayaki, A.: Generalized nonlinear Picone's identity for the p-Laplacian and its applications. Electron. J. Differ. Equ. **2016**(269), 1–7 (2016)
22. Tyagi, J.: A nonlinear Picone's identity and its applications. Appl. Math. Lett. **26**(6), 624–626 (2013)

Chapter 18
An Equivalence Between the Neumann Problem and Its Boundary Domain Integral Equation Systems for Stokes Equations

Mulugeta A. Dagnaw and Habtamu Z. Alemu

Abstract For a compressible viscous fluid with variable viscosity coefficient, the Neumann boundary value problem for the steady-state Stokes system of partial differential equations is taken into account in a three-dimensional bounded domain. This problem can be reduced to two different systems of Boundary-Domain Integral Equations (BDIEs) by using parametrix. The Neumann boundary value problem related to the Stokes system and the reduced BDIE systems are equivalent.

18.1 Introduction

The boundary integral equations and hydrodynamic potential theory for the Stokes system with constant viscosity have been thoroughly explored by a number of authors, including [3–5, 10, 11]. Neumann BVP for Stokes equations in the two-dimensional case and the incompressible and compressible mixed BVP Stokes system with variable viscosity in three dimensions have both been studied in terms of the boundary-domain integral equation system [1, 8, 9], but not for the Neumann BVP Stokes system with variable viscosity in 3D. In the case of constant viscosity, fundamental solutions for both velocity and pressure are available in analytical form. However, such fundamental solutions are not available for PDEs with variable viscosity. Therefore, the parametrix (Levi function, see e.g. [8, 9]) is used in order to derive and investigate the BDIE systems for the corresponding variable-coefficient BVPs. In [1, 8, 9], authors transformed BVPs with variable coefficients for the Stokes problem defined on a bounded domain to BDIE systems. In this work, we shall derive and analyse BDIE systems for Neumann BVP for compressible three-

M. A. Dagnaw (✉) · H. Z. Alemu
Department of Mathematics, Injibara University, Injibara, Ethiopia
e-mail: mulugeta.alemayehu@inu.edu.et

M. Ruzhansky, K. Van Bockstal (eds.), *Extended Abstracts 2021/2022*,
Research Perspectives Ghent Analysis and PDE Center 2,
https://doi.org/10.1007/978-3-031-42539-4_18

dimensional Stokes equations with variable viscosity in appropriate Bessel potential spaces.

18.2 Formulation of the Boundary Value Problem

Let $\Omega = \Omega^+$ be a bounded and simply connected domain and let $\Omega^- := \mathbb{R}^3 \setminus \overline{\Omega}^+$. We will assume that the boundary $S := \partial\Omega$ is simply connected, closed and infinitely differentiable, $S \in C^\infty$.

Let $\boldsymbol{v}$ be the velocity vector field; p the pressure scalar field and $\mu \in C^\infty(\Omega)$ be the variable kinematic viscosity of the fluid such that $\mu(\boldsymbol{x}) > c > 0$, we can define the Stokes operator as

$$\begin{aligned}\mathcal{A}_j(p, \boldsymbol{v})(\boldsymbol{x}) :&= \frac{\partial}{\partial x_i}\sigma_{ji}(p, \boldsymbol{v})(\boldsymbol{x}) \\ &= \frac{\partial}{\partial x_i}\left(\mu(\boldsymbol{x})\left(\frac{\partial v_j}{\partial x_i} + \frac{\partial v_i}{\partial x_j} - \frac{2}{3}\delta_i^j \operatorname{div}\boldsymbol{v}\right)\right) - \frac{\partial p}{\partial x_j}, \quad j, i \in \{1, 2, 3\},\end{aligned}$$

where δ_i^j is Kronecker symbol. Here and henceforth, we assume the Einstein summation in repeated indices from 1 to 3. We also denote the Stokes operator as $\boldsymbol{\mathcal{A}} = \{\mathcal{A}_j\}_{j=1}^3$. Occasionally, we may use the following notation for derivative operators: $\partial_j = \partial_{x_j} := \dfrac{\partial}{\partial x_j}$ with $j = 1, 2, 3$; $\nabla := (\partial_1, \partial_2, \partial_3)$.

For a compressible fluid $\operatorname{div}\boldsymbol{v} = g$, which gives the following stress tensor operator and the Stokes operator, respectively, to

$$\sigma_{ji}(p, \boldsymbol{v})(\boldsymbol{x}) = -\delta_i^j p(\boldsymbol{x}) + \mu(\boldsymbol{x})\left(\frac{\partial v_i(\boldsymbol{x})}{\partial x_j} + \frac{\partial v_j(\boldsymbol{x})}{\partial x_i} - \frac{2}{3}\delta_i^j g\right),$$

$$\mathcal{A}_j(p, \boldsymbol{v})(\boldsymbol{x}) = \frac{\partial}{\partial x_i}\left(\mu(\boldsymbol{x})\left(\frac{\partial v_j}{\partial x_i} + \frac{\partial v_i}{\partial x_j} - \frac{2}{3}\delta_i^j g\right)\right) - \frac{\partial p}{\partial x_j}, \quad j, i \in \{1, 2, 3\}.$$

In what follows $H^s(\Omega) = H_2^s(\Omega)$, $H^s(\partial\Omega)$ are the Bessel potential spaces, where s is a real number, see e.g. [6]. We recall that H^s coincide with the Sobolev-Slobodetski spaces W_2^s for any non-negative s. We denote by $\widetilde{H}^s(\Omega)$ the subspace of $H^s(\mathbb{R}^3)$, $\widetilde{H}^s(\Omega) = \{g : g \in H^s(\mathbb{R}^3), \mathrm{supp}(g) \subset \overline{\Omega}\}$; similarly, $\widetilde{H}^s(S_1) = \{g : g \in H^s(\partial\Omega), \mathrm{supp}(g) \subset \overline{S_1}\}$ is the Sobolev space of functions having support in $S_1 \subset \partial\Omega$. We will also use the notations $\mathbf{H}^s(\Omega) = [H^s(\Omega)]^3$, $\mathbf{L}^2(\Omega) = \left[L^2(\Omega)\right]^3$, $\mathbf{H}^1_{\mathcal{R}}(\Omega) = \mathbf{H}^1(\Omega)/\mathcal{R} = \{\mathbf{v} \in \mathbf{H}^1(\Omega) : \int_\Omega \mathbf{v} \cdot \mathbf{w} d\mathbf{x} = 0, \quad \text{for all} \quad \mathbf{w} \in \mathcal{R}\}$, $\boldsymbol{\mathcal{D}}(\Omega) = [\mathcal{D}(\Omega)]^3$ for three dimensional vector space. Where $\mathcal{R} = \{a + b \times \mathbf{x};\ a \text{ and } b \quad \text{are constant vectors}\}$ the space of rigid body motions.

We will also make use of the following space, see e.g. [1, 2, 8].

$$\mathbf{H}^{s,0}(\Omega;\mathcal{A}) := \{(p,\mathbf{v}) \in H^{s-1}(\Omega) \times \mathbf{H}^{s}(\Omega) : \boldsymbol{\mathcal{A}}(p,\mathbf{v}) \in \mathbf{L}^{2}(\Omega)\}$$

endowed with the norm

$$\|(p,v)\|^{2}_{\mathbf{H}^{s,0}(\Omega;\mathcal{A})} := \|p\|^{2}_{H^{s-1}(\Omega)} + \|\mathbf{v}\|^{2}_{\mathbf{H}^{s}(\Omega)} + \|\boldsymbol{\mathcal{A}}(p,\mathbf{v})\|^{2}_{\mathbf{L}^{2}(\Omega)}.$$

Let us define the space

$$\mathbf{H}^{1,0}_{\mathcal{R}}(\Omega;\mathcal{A}) := \{(p,\mathbf{v}) \in L^{2}(\Omega) \times \mathbf{H}^{1}_{\mathcal{R}}(\Omega) : \boldsymbol{\mathcal{A}}(p,\mathbf{v}) \in \mathbf{L}^{2}(\Omega)\}$$

endowed with the norm

$$\|(p,\mathbf{v})\|^{2}_{\mathbf{H}^{1,0}_{\mathcal{R}}(\Omega;\mathcal{A})} := \|p\|^{2}_{L_{2}(\Omega)} + \|\mathbf{v}\|^{2}_{\mathbf{H}^{1}_{\mathcal{R}}(\Omega)} + \|\boldsymbol{\mathcal{A}}(p,\mathbf{v})\|^{2}_{\mathbf{L}^{2}(\Omega)}.$$

For sufficiently smooth functions $(p,\mathbf{v}) \in H^{s-1}(\Omega^{\pm}) \times \mathbf{H}^{s}(\Omega^{\pm})$ with $s > 3/2$, we can define the classical traction operators, $\mathbf{T}^{c\pm} = \{T^{c\pm}_{j}\}^{3}_{j=1}$ on the boundary $\partial\Omega$ as

$$T^{c\pm}_{j}(p,\mathbf{v})(\mathbf{x}) := \left[\gamma^{\pm}\sigma_{ij}(p,\mathbf{v})(\mathbf{x})\right] n_{i}(\mathbf{x}), \tag{18.1}$$

where $n_i(\mathbf{x})$ denote components of the unit outward normal vector $\mathbf{n(x)}$ to the boundary $\partial\Omega$ of the domain and $\gamma^{\pm}$ is the trace operator from inside and outside Ω.

The traction operator (18.1) can be continuously extended to the canonical traction operator $\mathbf{T}^{\pm} : \mathbf{H}^{1,0}(\Omega^{\pm};\boldsymbol{\mathcal{A}}) \to \mathbf{H}^{-\frac{1}{2}}(\partial\Omega)$ defined in the weak form similar to [1, 8, 9] as

$$\langle \mathbf{T}^{\pm}(p,\mathbf{v}),\mathbf{w}\rangle_{\partial\Omega} := \pm\int_{\Omega^{\pm}} \left[\boldsymbol{\mathcal{A}}(p,\mathbf{v})(\gamma^{-1}\mathbf{w}) + E((p,\mathbf{v}),\gamma^{-1}\mathbf{w})\right] d\mathbf{x},$$
$$(p,\mathbf{v}) \in \mathbf{H}^{1,0}(\Omega^{\pm};\boldsymbol{\mathcal{A}}),\ \forall \mathbf{w} \in \mathbf{H}^{\frac{1}{2}}(\partial\Omega).$$

Here the operator $\gamma^{-1} : \mathbf{H}^{\frac{1}{2}}(\partial\Omega) \to \mathbf{H}^{1}(\mathbb{R}^{3})$ denotes a continuous right inverse of the trace operator $\gamma : \mathbf{H}^{1}(\mathbb{R}^{3}) \to \mathbf{H}^{\frac{1}{2}}(\partial\Omega)$, and the bilinear form E is defined as

$$E((p,\mathbf{v}),\mathbf{u})(\mathbf{x}) := \frac{\mu(\mathbf{x})}{2}\left(\frac{\partial u_i(x)}{\partial x_j} + \frac{\partial u_j(x)}{\partial x_i}\right)\left(\frac{\partial v_i(x)}{\partial x_j} + \frac{\partial v_j(x)}{\partial x_i}\right)$$
$$-\frac{2}{3}\mu(x)\mathrm{div}\mathbf{v}(\mathbf{x})\mathrm{div}\mathbf{u}(\mathbf{x}) - p(\mathbf{x})\mathrm{div}\mathbf{u}(\mathbf{x}).$$

Furthermore, if $(p, \mathbf{v}) \in \mathbf{H}^{1,0}(\Omega; \mathcal{A})$ and $\mathbf{u} \in \mathbf{H}^1(\Omega)$, the following first Green identity holds (see e.g. [1, 2, 7, 8] and [9])

$$\langle \mathbf{T}^+(p, \mathbf{v}), \gamma^+\mathbf{u}\rangle_{\partial\Omega} := \int_\Omega [\mathcal{A}(p, \mathbf{v})\mathbf{u} + E((p, \mathbf{v}), \mathbf{u})(\mathbf{x})]\, d\mathbf{x}. \tag{18.2}$$

Applying the identity (18.2) to the pairs $(p, \mathbf{v}) \in \mathbf{H}^{1,0}(\Omega; \mathcal{A})$ and $(q, \mathbf{u}) \in \mathbf{H}^{1,0}(\Omega; \mathcal{A})$ with exchanged roles and subtracting the one from the other, we arrive at the second Green identity (see e.g. [1, 6–9])

$$\int_\Omega \left[\mathcal{A}_j(p, \mathbf{v})u_j - \mathcal{A}_j(q, \mathbf{u})v_j + q\mathrm{div}\mathbf{v} - p\mathrm{div}\mathbf{u}\right] dx$$
$$= \int_{\partial\Omega} \left[T_j(p, \mathbf{v})u_j - T_j(q, \mathbf{u})v_j\right] dS_{\mathbf{x}}.$$

We shall derive and investigate the BDIE systems for the following Neumann boundary value problem: Given the functions $\boldsymbol{\psi}_0 \in \mathbf{H}^{-\frac{1}{2}}(\partial\Omega)$, $g \in L_2(\Omega)$ and $\mathbf{f} \in \mathbf{L}_2(\Omega)$, find a couple of functions $(p, \boldsymbol{v}) \in L_2(\Omega) \times \mathbf{H}^1(\Omega)$ satisfying

$$\mathcal{A}(p, v)(x) = \mathbf{f}(\mathbf{x}), \quad \mathbf{x} \in \Omega, \tag{18.3a}$$

$$\mathrm{div}\boldsymbol{v}(\mathbf{x}) = g(\mathbf{x}), \quad \mathbf{x} \in \Omega, \tag{18.3b}$$

$$\mathbf{T}^+(p(\mathbf{x}), \mathbf{v}(\mathbf{x})) = \boldsymbol{\psi}_0(\mathbf{x}), \quad \mathbf{x} \in \partial\Omega. \tag{18.3c}$$

Theorem 18.1 *The Neumann BVP (18.3a)–(18.3c) has a unique solution in the space* $L_2(\Omega) \times \boldsymbol{H}^1_{\mathcal{R}}(\Omega)$.

Proof Let (p_1, v_1) and (p_2, v_2) be in $L_2(\Omega) \times \mathbf{H}^1(\Omega)$ that satisfy the BVP (18.3a)–(18.3c). Then $(p, \mathbf{v}) := (p_2, v_2) - (p_1, v_1)$ also belongs to $L_2(\Omega) \times \mathbf{H}^1(\Omega)$ and satisfy the following homogeneous Neumann BVP

$$\boldsymbol{\mathcal{A}}(p, \mathbf{v})(\mathbf{x}) = 0, \quad \mathbf{x} \in \Omega, \tag{18.4a}$$

$$\mathrm{div}\mathbf{v}(\mathbf{x}) = 0, \quad \mathbf{x} \in \Omega, \tag{18.4b}$$

$$T^+(p(\mathbf{x}), \mathbf{v}(\mathbf{x})) = 0, \quad \mathbf{x} \in \partial\Omega. \tag{18.4c}$$

Applying the first Green identity (18.2) to $(p, \mathbf{v})$ and $\mathbf{u} = \mathbf{v}$ and taking in to account (18.4), we obtain

$$\int_\Omega \frac{\mu(\mathbf{x})}{2}\left(\frac{\partial v_i(\mathbf{x})}{\partial x_j} + \frac{\partial v_j(\mathbf{x})}{\partial x_i}\right)^2 d\mathbf{x} = 0$$

as $\mu(\mathbf{x}) > 0$. Thus $\mathbf{v} = a + b \times \mathbf{x} \in \mathcal{R}$, where a and b are constant vectors. But $\mathbf{v} \in \mathbf{H}^1_{\mathcal{R}}(\Omega)$ and so $\mathbf{v} = 0$. Hence $v_1 = v_2$. Keeping in mind the Neumann condition (18.4c), we have $p = 0$. □

18.3 Parametrix and Parametrix-Based Hydrodynamic Potentials

18.3.1 Parametrix and Remainder

The operator $\boldsymbol{\mathcal{A}}$ becomes the constant-coefficient Stokes operator $\mathring{\boldsymbol{\mathcal{A}}}$ when $\mu = 1$. The fundamental solution defined by the pair of distributions $(\mathring{q}^k, \mathring{\mathbf{u}}^k)$, where $\mathring{u}^k_j$ represents the components of the incompressible velocity fundamental solution and $\mathring{q}^k$ represents the components of the pressure fundamental solution (see e.g. [5, 10, 11]):

$$\mathring{q}^k(\boldsymbol{x}, \boldsymbol{y}) = \frac{(x_k - y_k)}{4\pi|\boldsymbol{x} - \boldsymbol{y}|^3},$$

$$\mathring{u}^k_j(\boldsymbol{x}, \boldsymbol{y}) = -\frac{1}{8\pi}\left\{\frac{\delta^k_j}{|\boldsymbol{x} - \boldsymbol{y}|} + \frac{(x_j - y_j)(x_k - y_k)}{|\boldsymbol{x} - \boldsymbol{y}|^3}\right\}, \quad j, k \in \{1, 2, 3\}.$$

Therefore, $(\mathring{q}^k, \mathring{\boldsymbol{u}}^k)$ satisfy

$$\mathring{\mathcal{A}}_j(\mathring{q}^k, \mathring{\boldsymbol{u}}^k)(\boldsymbol{x}) = \sum_{i=1}^{3} \frac{\partial^2 \mathring{u}^k_j}{\partial x_i^2} - \frac{\partial \mathring{q}^k}{\partial x_j} = \delta^k_j \delta(\boldsymbol{x} - \boldsymbol{y}).$$

Let us denote $\mathring{\sigma}_{ij}(p, \boldsymbol{v}) := \sigma_{ij}(p, \boldsymbol{v})|_{\mu=1}$. Then, in the particular case $\mu = 1$, the stress tensor $\mathring{\sigma}_{ij}(\mathring{q}^k, \mathring{\boldsymbol{u}}^k)(\boldsymbol{x} - \boldsymbol{y})$ reads as

$$\mathring{\sigma}_{ij}(\mathring{q}^k, \mathring{\boldsymbol{u}}^k)(\boldsymbol{x} - \boldsymbol{y}) = \frac{3}{4\pi} \frac{(x_i - y_i)(x_j - y_j)(x_k - y_k)}{|\boldsymbol{x} - \boldsymbol{y}|^5},$$

and the boundary traction becomes

$$\begin{aligned}
\mathring{T}_i(\boldsymbol{x}; \mathring{q}^k, \mathring{\boldsymbol{u}}^k)(\boldsymbol{x}, \boldsymbol{y}) &:= \mathring{\sigma}_{ij}(\mathring{q}^k, \mathring{\boldsymbol{u}}^k)(\boldsymbol{x} - \boldsymbol{y})\, n_j(\boldsymbol{x}) \\
&= \frac{3}{4\pi} \frac{(x_i - y_i)(x_j - y_j)(x_k - y_k)}{|\boldsymbol{x} - \boldsymbol{y}|^5} n_j(\boldsymbol{x}).
\end{aligned}$$

Let us define a pair of functions $(q^k, \boldsymbol{u}^k)_{k=1,2,3}$ as

$$q^k(\boldsymbol{x}, \boldsymbol{y}) = \frac{\mu(\boldsymbol{x})}{\mu(\boldsymbol{y})}\mathring{q}^k(\boldsymbol{x}, \boldsymbol{y}) = \frac{\mu(\boldsymbol{x})}{\mu(\boldsymbol{y})}\frac{x_k - y_k}{4\pi|\boldsymbol{x} - \boldsymbol{y}|^3}, \quad j, k \in \{1, 2, 3\}, \tag{18.5}$$

$$u_j^k(\boldsymbol{x}, \boldsymbol{y}) = \frac{1}{\mu(\boldsymbol{y})}\mathring{u}_j^k(\boldsymbol{x}, \boldsymbol{y}) = -\frac{1}{8\pi\mu(\boldsymbol{y})}\left\{\frac{\delta_j^k}{|\boldsymbol{x} - \boldsymbol{y}|} + \frac{(x_j - y_j)(x_k - y_k)}{|\boldsymbol{x} - \boldsymbol{y}|^3}\right\}. \tag{18.6}$$

Then,

$$\sigma_{ij}(\boldsymbol{x}; q^k, \boldsymbol{u}^k)(\boldsymbol{x}, \boldsymbol{y}) = \frac{\mu(\boldsymbol{x})}{\mu(\boldsymbol{y})}\mathring{\sigma}_{ij}(\mathring{q}^k, \mathring{\boldsymbol{u}}^k)(\boldsymbol{x} - \boldsymbol{y}),$$

$$T_i(\boldsymbol{x}; q^k, \boldsymbol{u}^k)(\boldsymbol{x}, \boldsymbol{y}) := \sigma_{ij}(\boldsymbol{x}; q^k, \boldsymbol{u}^k)(\boldsymbol{x}, \boldsymbol{y})\, n_j(\boldsymbol{x}) = \frac{\mu(\boldsymbol{x})}{\mu(\boldsymbol{y})}\mathring{T}_i(\boldsymbol{x}; \mathring{q}^k, \mathring{\boldsymbol{u}}^k)(\boldsymbol{x}, \boldsymbol{y}).$$

Substituting (18.5) and (18.6) in the Stokes system with variable coefficient gives

$$\mathcal{A}_j(\boldsymbol{x}; q^k, \boldsymbol{u}^k)(\boldsymbol{x}, \boldsymbol{y}) = \delta_j^k \delta(\boldsymbol{x} - \boldsymbol{y}) + R_{kj}(\boldsymbol{x}, \boldsymbol{y}),$$

where

$$R_{kj}(\boldsymbol{x}, \boldsymbol{y}) = \frac{1}{\mu(\boldsymbol{y})}\frac{\partial\mu(\boldsymbol{x})}{\partial x_i}\mathring{\sigma}_{ij}(\mathring{q}^k, \mathring{\boldsymbol{u}}^k)(\boldsymbol{x} - \boldsymbol{y}) = O(|\boldsymbol{x} - \boldsymbol{y}|)^{-2})$$

is a weakly singular remainder. This implies that $(q^k, \boldsymbol{u}^k)$ is a parametrix of the operator $\boldsymbol{\mathcal{A}}$.

18.3.2 Volume and Surface Potentials

Definition 18.1 The parametrix-based Newton-type and the Remainder vector potential operators are defined as ($\mathbf{y} \in \mathbb{R}^3$)

$$[\boldsymbol{\mathcal{U}}\rho]_k(\mathbf{y}) = \mathcal{U}_{kj}\rho_j(\mathbf{y}) := \int_\Omega u_j^k(\mathbf{x}, \mathbf{y})\rho_j(\mathbf{x})d\mathbf{x},$$

$$[\boldsymbol{\mathcal{R}}\rho]_k(\mathbf{y}) = \mathcal{R}_{kj}\rho_j(\mathbf{y}) := \int_\Omega R_{kj}(\mathbf{x}, \mathbf{y})\rho_j(\mathbf{x})d\mathbf{x},$$

for the velocity $\mathbf{v}$, and ($\mathbf{y} \in \mathbb{R}^3$)

$$[\mathcal{Q}\rho]_j(\mathbf{y}) = \mathcal{Q}_j\rho(\mathbf{y}) := -\int_\Omega q^j(\mathbf{x},\mathbf{y})\rho(\mathbf{x})d\mathbf{x},$$

$$\mathcal{Q}\boldsymbol{\rho}(\mathbf{y}) = \boldsymbol{\mathcal{Q}}\cdot\boldsymbol{\rho}(\mathbf{y}) = \mathcal{Q}_j\rho_j(\mathbf{y}) := -\int_\Omega q^j(\mathbf{x},\mathbf{y})\rho_j(\mathbf{x})d\mathbf{x},$$

$$\begin{aligned}\mathcal{R}^\bullet\boldsymbol{\rho}(\mathbf{y}) &= -2\langle\partial_i\mathring{q}^j(.,\mathbf{y}),\rho_i\partial_j\mu\rangle_\Omega - 2\rho_i(\mathbf{y})\partial_i\mu(\mathbf{y})\\ &= -2v.p.\int_\Omega \frac{\partial\mathring{q}^j(\mathbf{x},\mathbf{y})}{\partial x_i}\frac{\partial\mu(\mathbf{x})}{\partial x_i}\rho_j(\mathbf{x})d\mathbf{x} - \frac{4}{3}\rho_j(\mathbf{y})\frac{\partial\mu(\mathbf{y})}{\partial y_j},\end{aligned} \tag{18.7}$$

the scalar Newton-type and remainder potentials for the pressure p, see e.g. [8, 9]. The integral in (18.7) is understood as a 3D strongly singular integral in the Cauchy sense.

Definition 18.2 For the velocity, the parametrix-based single-layer and double-layer potentials are defined for $\mathbf{y} \notin \partial\Omega$ as

$$[\mathbf{V}\boldsymbol{\rho}]_k(\mathbf{y}) = V_{kj}\rho_j(\mathbf{y}) := -\int_{\partial\Omega} u_j^k(\mathbf{x},\mathbf{y})\rho_j(\mathbf{x})dS_{\mathbf{x}},$$

$$[\mathbf{W}\boldsymbol{\rho}]_k(\mathbf{y}) = W_{kj}\rho_j(\mathbf{y}) := -\int_{\partial\Omega} T_j^+(\mathbf{x}; q^k,\mathbf{u}^k)(\mathbf{x},\mathbf{y})\rho_j(\mathbf{x})dS_{\mathbf{x}},$$

and for pressure in the variable coefficient Stokes system, the single-layer and double-layer potentials are defined for $\mathbf{y} \notin \partial\Omega$ as

$$\Pi^s\boldsymbol{\rho}(\mathbf{y}) = \Pi_j^s\rho_j(\mathbf{y}) := \int_{\partial\Omega} \mathring{q}^j(\mathbf{x},\mathbf{y})\rho_j(\mathbf{x})dS_{\mathbf{x}},$$

$$\Pi^d\boldsymbol{\rho}(\mathbf{y}) = \Pi_j^d\rho_j(\mathbf{y}) := -2\int_{\partial\Omega} \frac{\partial\mathring{q}^j(\mathbf{x},\mathbf{y})}{\partial\mathbf{n}(\mathbf{x})}\mu(\mathbf{x})\rho_j(\mathbf{x})dS_{\mathbf{x}}.$$

The corresponding boundary integral (pseudo-differential) operators of direct surface values of the single-layer potential and the double-layer potential, the traction of the single-layer potential and the double-layer potential are

$$[\boldsymbol{\mathcal{V}}\boldsymbol{\rho}]_k(\mathbf{y}) = -\int_{\partial\Omega} u_j^k(\mathbf{x},\mathbf{y})\rho_j(\mathbf{x})dS_{\mathbf{x}}, \quad \mathbf{y} \in \partial\Omega,$$

$$[\boldsymbol{\mathcal{W}}\boldsymbol{\rho}]_k(\mathbf{y}) = -\int_{\partial\Omega} T_j^+(\mathbf{x}; q^k,\mathbf{u}^k)(\mathbf{x},\mathbf{y})\rho_j(\mathbf{x})dS_{\mathbf{x}}, \quad \mathbf{y} \in \partial\Omega,$$

$$\left[\mathcal{W}'\rho\right]_k(\mathbf{y}) = -\int_{\partial\Omega} T_j^+(\mathbf{y}; q^k, \mathbf{u}^k)(\mathbf{x}, \mathbf{y})\rho_j(\mathbf{x})dS_{\mathbf{x}}, \quad \mathbf{y} \in \partial\Omega,$$

$$\mathcal{L}^{\pm}\rho(\mathbf{y}) := \mathbf{T}^{\pm}(\Pi^d\rho, \mathbf{W}\rho)(\mathbf{y}), \quad \mathbf{y} \in \partial\Omega,$$

where $\mathbf{T}^{\pm}$ are the traction operators, see e.g. [1, 8, 9].

The parametrix-based integral operators depending on the variable coefficient $\mu(\mathbf{x})$ can be expressed in terms of the corresponding integral operators for the constant coefficient case $\mu = 1$ as in [1, 8, 9]. This helps us to look at the mapping properties of the potentials.

18.4 The Third Green Identities

Theorem 18.2 *For any* $(p, v) \in \boldsymbol{H}^{1,0}(\Omega; \mathcal{A})$ *or* $\boldsymbol{H}^{1,0}_{\mathcal{R}}(\Omega; \mathcal{A})$ *the following third Green identities hold*

$$v + \mathcal{R}v - \boldsymbol{V}\boldsymbol{T}^+(p, v) + \boldsymbol{W}\gamma^+ v = \boldsymbol{\mathcal{U}}\boldsymbol{\mathcal{A}}(p, v) - \mathcal{Q}divv \quad in\ \Omega, \tag{18.8}$$

$$p + \mathcal{R}^{\bullet}v - \Pi^s\boldsymbol{T}^+(p, v) + \Pi^d\gamma^+ v = \mathring{\mathcal{Q}}\boldsymbol{\mathcal{A}}(p, v) + \frac{4}{3}\mu divv \quad in\ \Omega. \tag{18.9}$$

If the couple $(p, \mathbf{v}) \in \mathbf{H}^{1,0}(\Omega; \boldsymbol{\mathcal{A}})$ or $\mathbf{H}^{1,0}_{\mathcal{R}}(\Omega; \boldsymbol{\mathcal{A}})$ is a solution of the Stokes PDE (18.3a) with variable coefficient, then (18.8) and (18.9) gives

$$\mathbf{v} + \mathcal{R}\mathbf{v} - \mathbf{V}\mathbf{T}^+(p, \mathbf{v}) + \mathbf{W}\gamma^+\mathbf{v} = \mathcal{U}\mathbf{f} - \mathcal{Q}g, \quad \text{in } \Omega, \tag{18.10}$$

$$p + \mathcal{R}^{\bullet}\mathbf{v} - \Pi^s\mathbf{T}^+(p, \mathbf{v}) + \Pi^d\gamma^+\mathbf{v} = \mathring{\mathcal{Q}}\mathbf{f} + \frac{4}{3}\mu g, \quad \text{in } \Omega. \tag{18.11}$$

We will also need the trace and traction of the third Green identities (18.10) and (18.11) on $\partial\Omega$:

$$\frac{1}{2}\gamma^+\mathbf{v} + \mathcal{R}^+\mathbf{v} - \mathcal{V}\mathbf{T}^+(p, \mathbf{v}) + \mathcal{W}\gamma^+\mathbf{v} = \gamma^+\mathcal{U}\mathbf{f} - \gamma^+\mathcal{Q}g, \tag{18.12}$$

$$\frac{1}{2}\mathbf{T}^+(p, \mathbf{v}) + \mathbf{T}^+(\mathcal{R}^{\bullet}, \mathcal{R})\mathbf{v} - \mathcal{W}'\mathbf{T}^+(p, \mathbf{v}) + \mathcal{L}^+\gamma^+\mathbf{v}$$
$$= \mathbf{T}^+(\mathring{\mathcal{Q}}\mathbf{f} + \frac{4}{3}\mu g, \mathcal{U}f - \mathcal{Q}g). \tag{18.13}$$

One can prove the following two assertions that are instrumental for proof of equivalence of the BDIEs and the Neumann PDE (18.3).

Lemma 18.1 *Let* $\boldsymbol{v} \in \boldsymbol{H}^1(\Omega)$(*or* $\boldsymbol{H}^1_{\mathcal{R}}(\Omega)$), $p \in L^2(\Omega)$, $g \in L^2(\Omega)$, $\boldsymbol{f} \in \boldsymbol{L}^2(\Omega)$, $\boldsymbol{\Psi} \in \boldsymbol{H}^{-\frac{1}{2}}(\partial\Omega)$, $\boldsymbol{\Phi} \in \boldsymbol{H}^{\frac{1}{2}}(\partial\Omega)$ *satisfy equations*

$$\boldsymbol{v} + \mathcal{R}\boldsymbol{v} - \boldsymbol{V}\boldsymbol{\Psi} + \boldsymbol{W}\boldsymbol{\Phi} = \mathcal{U}\boldsymbol{f} - \mathcal{Q}g, \quad in\ \Omega,$$

$$p + \mathcal{R}^{\bullet}\boldsymbol{v} - \Pi^s\boldsymbol{\Psi} + \Pi^d\boldsymbol{\Phi} = \mathring{\mathcal{Q}}\boldsymbol{f} + \frac{4}{3}\mu g, \quad in\ \Omega.$$

Then $(p, \boldsymbol{v}) \in \boldsymbol{H}^{1,0}(\Omega; \boldsymbol{\mathcal{A}})$ *or* $\boldsymbol{H}^{1,0}_{\mathcal{R}}(\Omega; \boldsymbol{\mathcal{A}})$ *and solve the equations*

$$\boldsymbol{\mathcal{A}}(\boldsymbol{y}; p, \boldsymbol{v}) = \boldsymbol{f}, \ div\boldsymbol{v} = g.$$

Moreover, the following relations hold true:

$$\boldsymbol{V}(\boldsymbol{\Psi} - \boldsymbol{T}^+(p, \boldsymbol{v}))(\boldsymbol{y}) - \boldsymbol{W}(\boldsymbol{\Phi} - \gamma^+\boldsymbol{v})(\boldsymbol{y}) = 0, \quad \boldsymbol{y} \in \Omega,$$

$$\Pi^s(\boldsymbol{\Psi} - \boldsymbol{T}^+(p, \boldsymbol{v}))(\boldsymbol{y}) - \Pi^d(\boldsymbol{\Phi} - \gamma^+\boldsymbol{v})(\boldsymbol{y}) = 0, \quad \boldsymbol{y} \in \Omega.$$

Lemma 18.2

(i) Let either $\boldsymbol{\Psi}^* \in \boldsymbol{H}^{-\frac{1}{2}}(\partial\Omega)$. *If* $\boldsymbol{V}\boldsymbol{\Psi}^*(\boldsymbol{y}) = 0, \boldsymbol{y} \in \Omega$, *then* $\boldsymbol{\Psi}^* = 0$;
(ii) Let $\boldsymbol{\Phi}^* \in \boldsymbol{H}^{\frac{1}{2}}(\partial\Omega)$. *If* $\boldsymbol{W}\boldsymbol{\Phi}^*(\boldsymbol{y}) = 0, \boldsymbol{y} \in \Omega$, *then* $\boldsymbol{\Phi}^* = 0$.

18.5 BDIE Systems for Neumann BVP

We aim to obtain a segregated boundary-domain integral equation systems for Neumann BVP (18.3a)–(18.3c). We will use similar procedures as in [1]. We can reduce the BVP (18.3a)–(18.3c) to two different systems of Boundary-Domain Integral Equations for the unknowns $(p, \mathbf{v}, \varphi) \in \mathbf{H}^{1,0}_{\mathcal{R}}(\Omega; \boldsymbol{\mathcal{A}}) \times \mathbf{H}^{\frac{1}{2}}(\partial\Omega)$.

18.5.1 BDIE System (N1)

From the equations (18.10), (18.11) and its traction (18.13), we obtain

$$p + \mathcal{R}^{\bullet}\mathbf{v} + \Pi^d\varphi = G_0 \quad \text{in } \Omega, \tag{18.14a}$$

$$\mathbf{v} + \mathcal{R}\mathbf{v} + \mathbf{W}\varphi = \mathbf{G} \quad \text{in } \Omega, \tag{18.14b}$$

$$\mathbf{T}^+(\mathcal{R}^{\bullet}, \mathcal{R})\mathbf{v} + \mathcal{L}^+\varphi = \mathbf{T}^+(G_0, \mathbf{G}) - \boldsymbol{\psi}_0 \quad \text{on } \partial\Omega, \tag{18.14c}$$

where

$$G_0 := \mathring{\mathcal{Q}}\mathbf{f} + \frac{4}{3}\mu g + \Pi^s \boldsymbol{\psi}_0, \qquad \mathbf{G} := \boldsymbol{\mathcal{U}}\mathbf{f} - \boldsymbol{\mathcal{Q}}g + \mathbf{V}\boldsymbol{\psi}_0$$

and $(G_0, \mathbf{G}) \in L^2(\Omega) \times \mathbf{H}^1_{\mathcal{R}}(\Omega)$. We denote the right hand side of BDIE system (18.14a)–(18.14c) as

$$\mathcal{G}^1 := [G_0, \mathbf{G}, \mathbf{T}^+(G_0, \mathbf{G}) - \boldsymbol{\psi}_0]^T,$$

which implies $\mathcal{G}^1 \in L^2(\Omega) \times \mathbf{H}^1_{\mathcal{R}}(\Omega) \times \mathbf{H}^{-\frac{1}{2}}(\partial\Omega)$. In matrix form it can be written as $\mathcal{N}^1 \mathcal{X} = \mathcal{G}^1$, where

$$\mathcal{N}^1 = \begin{bmatrix} I & \mathcal{R}^\bullet & \Pi^d \\ 0 & I + \boldsymbol{\mathcal{R}} & \mathbf{W} \\ 0 & \mathbf{T}^+(\mathcal{R}^\bullet, \boldsymbol{\mathcal{R}}) & \mathcal{L}^+ \end{bmatrix}, \qquad \mathcal{G}^1 = \begin{bmatrix} G_0 \\ \mathbf{G} \\ \mathbf{T}^+(G_0, \mathbf{G}) - \boldsymbol{\psi}_0 \end{bmatrix}$$

Remark 18.1 The term $\mathcal{G}^1 = 0$ if and only if $(\mathbf{f}, g, \boldsymbol{\psi}_0) = 0$.

18.5.2 BDIE System (N2)

From the equations (18.10), (18.11) and (18.12) we obtain

$$p + \mathcal{R}^\bullet \mathbf{v} + \Pi^d \boldsymbol{\varphi} = G_0 \quad \text{in } \Omega, \tag{18.15a}$$

$$\mathbf{v} + \boldsymbol{\mathcal{R}}\mathbf{v} + \mathbf{W}\boldsymbol{\varphi} = \mathbf{G} \quad \text{in } \Omega, \tag{18.15b}$$

$$\gamma^+ \boldsymbol{\mathcal{R}}\mathbf{v} + \frac{1}{2}\boldsymbol{\varphi} + \boldsymbol{\mathcal{W}}\boldsymbol{\varphi} = \gamma^+ \mathbf{G} \quad \text{on } \partial\Omega, \tag{18.15c}$$

Note that BDIE system (18.15a)–(18.15c) can be split into the BDIE system (**N2**), of 2 vector Eqs. (18.15b) and (18.15c) for 2 vector unknowns, $\mathbf{v}$ and $\boldsymbol{\varphi}$, and the scalar Eq. (18.15a) that can be used after solving the system to obtain the pressure, p. The system (**N2**) given by Eqs. (18.15a)–(18.15c) can be written using matrix notation as $\mathcal{N}^2 \mathcal{X} = \mathcal{G}^2$, where $\mathcal{X}$ represents the vector containing the unknowns of the system $\mathcal{X} = (p, \mathbf{v}, \boldsymbol{\varphi}) \in L^2(\Omega) \times \mathbf{H}^1(\Omega) \times \mathbf{H}^{\frac{1}{2}}(\partial\Omega)$. The matrix operator $\mathcal{N}^2$ is defined by

$$\mathcal{N}^2 = \begin{bmatrix} I & \mathcal{R}^\bullet & \Pi^d \\ 0 & I + \boldsymbol{\mathcal{R}} & \mathbf{W} \\ 0 & \gamma^+ \boldsymbol{\mathcal{R}} & \frac{1}{2}\mathbf{I} + \boldsymbol{\mathcal{W}} \end{bmatrix}, \qquad \mathcal{G}^2 = \begin{bmatrix} G_0 \\ \mathbf{G} \\ \gamma^+ \mathbf{G} \end{bmatrix}$$

Remark 18.2 Let $\boldsymbol{\Psi}_0 \in \mathbf{H}^{-\frac{1}{2}}(\partial\Omega)$. The term $\mathcal{G}^2 = 0$ if and only if $(\mathbf{f}, g, \boldsymbol{\psi}_0) = 0$.

In the following theorem, we shall see the equivalence of the original Neumann boundary value problem to the BDIE systems.

18.6 Equivalence and Invertibility Theorems

Theorem 18.3 (Equivalence Theorem) *Let* $\boldsymbol{f} \in \boldsymbol{L}^2(\Omega)$, $g \in L^2(\Omega)$ *and* $\boldsymbol{\psi}_0 \in \boldsymbol{H}^{-\frac{1}{2}}(\partial\Omega)$. *We have that*

(i) If some $(p, \boldsymbol{v}) \in L^2(\Omega) \times \boldsymbol{H}^1_{\mathcal{R}}(\Omega)$ *solves the Neumann BVP (18.3a)–(18.3c), then* $(p, \boldsymbol{v}, \boldsymbol{\varphi})$ *with*

$$\boldsymbol{\varphi} = \gamma^+ \boldsymbol{v} \in \boldsymbol{H}^{\frac{1}{2}}(\partial\Omega) \tag{18.16}$$

solves BDIE system (N1) and (N2).

(ii) If $(p, \boldsymbol{v}, \boldsymbol{\varphi}) \in L^2(\Omega) \times \boldsymbol{H}^1_{\mathcal{R}}(\Omega) \times \boldsymbol{H}^{\frac{1}{2}}(\partial\Omega)$ *solves the BDIE system (N1), then* $(p, \boldsymbol{v})$ *solves the BDIE system (N2) and the Neumann BVP (18.3a)–(18.3c) and the function* $\boldsymbol{\varphi}$ *satisfies* (18.16).

(iii) If $(p, \boldsymbol{v}, \boldsymbol{\varphi}) \in L^2(\Omega) \times \boldsymbol{H}^1_{\mathcal{R}}(\Omega) \times \boldsymbol{H}^{\frac{1}{2}}(\partial\Omega)$ *solves the BDIE system (N2), then* $(p, \boldsymbol{v})$ *solves the BDIE system (N1) and the Neumann BVP (18.3a)–(18.3c) and the function* $\boldsymbol{\varphi}$ *satisfies* (18.16).

(iv) The BDIE systems (N1) and (N2) are uniquely solvable in $L^2(\Omega) \times \boldsymbol{H}^1_{\mathcal{R}}(\Omega) \times \boldsymbol{H}^{\frac{1}{2}}(\partial\Omega)$.

Proof

(i) Let $(p, \mathbf{v}) \in L^2(\Omega) \times \mathbf{H}^1_{\mathcal{R}}(\Omega)$ be a solution of the BVP. Since $\mathbf{f} \in \mathbf{L}^2(\Omega)$, then $(p, \mathbf{v}) \in \mathbf{H}^{1,0}_{\mathcal{R}}(\Omega; \boldsymbol{\mathcal{A}})$. Let us define the function $\boldsymbol{\psi}$ by (18.16). Taking into account the Green identities (18.10)–(18.13), we immediately obtain that (p, v, ψ) solve system (N1) and (N2).

(ii) Let $(p, \mathbf{v}, \boldsymbol{\varphi}) \in L^2(\Omega) \times \mathbf{H}^1_{\mathcal{R}}(\Omega) \times \mathbf{H}^{\frac{1}{2}}(\partial\Omega)$ solves BDIE system (18.14a)–(18.14c). If we take the traction of (18.14a) and (18.14b) and subtract (18.14c) from it, we arrive at $\boldsymbol{\psi}_0 = \mathbf{T}^+(p, \mathbf{v})$ on $\partial\Omega$. Thus, the Neumann condition is satisfied. Also, we note that if $(p, \mathbf{v}) \in L^2(\Omega) \times \mathbf{H}^1_{\mathcal{R}}(\Omega)$ then $\boldsymbol{\mathcal{A}}(p, \mathbf{v}) = \mathbf{f} \in \mathbf{L}^2(\Omega)$. Due to relations (18.14a) and (18.14b) the hypotheses of the Lemma 18.1 are satisfied. As a result, we obtain that $(p, \mathbf{v})$ is a solution of $\boldsymbol{\mathcal{A}}(p, \mathbf{v}) = \mathbf{f}$ satisfying

$$\mathbf{V}(\boldsymbol{\psi}_0 - \mathbf{T}^+(p, \mathbf{v})) - \mathbf{W}(\boldsymbol{\varphi} - \gamma^+\mathbf{v}) = \mathbf{0}. \tag{18.17}$$

Now, inserting $\boldsymbol{\psi}_0 = \mathbf{T}^+(p, \mathbf{v})$ in (18.17), we have $\mathbf{W}(\boldsymbol{\varphi} - \gamma^+\mathbf{v}) = \mathbf{0}, \mathbf{y} \in \Omega$. Lemma 18.2(ii) implies $\boldsymbol{\varphi} = \gamma^+\mathbf{v}$. Therefore, satisfy (18.16).

(iii) Let $(p, \mathbf{v}, \boldsymbol{\varphi}) \in L^2(\Omega) \times \mathbf{H}^1(\Omega) \times \mathbf{H}^{\frac{1}{2}}(\partial\Omega)$ solve BDIE system (**N2**). If we take the trace of (18.15b) and subtract (18.15c) from it, we arrive at $\boldsymbol{\varphi} = \gamma^+\mathbf{v}$ on $\partial\Omega$. Then, inserting $\boldsymbol{\varphi} = \gamma^+\mathbf{v}$ in (18.17) gives $\mathbf{V}(\boldsymbol{\psi}_0 - \mathbf{T}^+(p, \mathbf{v})) = \mathbf{0}$, Lemma 18.2 (i) then implies $\boldsymbol{\psi}_0 = \mathbf{T}^+(p, \mathbf{v})$ on $\partial\Omega)$. Hence, the Neumann condition is satisfied.

(iv) The uniquely solvability of the BDIEs (18.14a)–(18.14c) follows from the uniquely solvable of the BVP as in Theorem 18.1.

□

References

1. Ayele, T.G., Dagnaw, M.A: Boundary-domain integral equation systems to the Dirichlet and Neumann problems for compressible Stokes equations with variable viscosity in 2D, Math. Methods Appl. Sci. **44**, 9876–9898 (2021)
2. Chkadua, O., Mikhailov, S.E., Natroshvili, D.: Analysis of direct boundary domain integral equations for a mixed BVP with variable coefficent I: equivalence and Invertibility. J. Integral Equ. Appl. **21**, 499–543 (2009)
3. Hsiao, G.C., Kress, R.: On an integral equation for the two- dimensional exterior Stokes problem, Appl. Numer. Math. **1**, 77–93 (1985)
4. Kohr, M., Wendland, W.L.: Variational boundary integral equations for the Stokes system. Applicable Anal. **85**, 1343-1372 (2006)
5. Ladyzhenskaya, O.A.: The Mathematical Theory of Viscous Incompressible Flow. Gordon and Breach, New York (1969)
6. McLean, W.: Strongly Elliptic Systems and Boundary Integral Equations. Cambridge University Press, Cambridge (2000)
7. Mikhailov, S.E.: Traces, extensions and co-normal derivatives for elliptic systems on Lipschitz domains. J. Math. Anal. Appl. **378**, 324–342 (2011)
8. Mikhailov, S.E., Portillo, C.F.: BDIE system to the mixed BVP for the Stokes equations with variable viscosity. In: Constanda, C., Kirsh, A. (eds.) Integral Methods in Science and Engineering: Theoretical and Computational Advances, pp. 401–412. Springer, Boston (2015)
9. Mikhailov, S.E., Portillo, C.F.: Analysis of boundary-domain integral equations to the mixed BVP for a compressible stokes system with variable viscosity. Commun. Pure Appl. Anal. **18**, 3059–3088 (2019)
10. Rjasanow, S., Steinbach, O.: The Fast Solution of Boundary Integral Equations. Springer, Berlin (2007)
11. Steinbach, O.: Numerical Approximation Methods for Elliptic Boundary Value Problems: Finite and Boundary Elements. Springer, Berlin (2007)

Chapter 19
Short Note on Generalised Bivariate Mittag-Leffler-Type Functions

Anvar Hasanov

Abstract In this short note, we aim to announce our joint result with Karimov and Hasanov (On Generalized Mittag-Leffler-Type Functions of Two Variables. Preprint, 2023. https://www.authorea.com/doi/pdf/10.22541/au.167575527.77144915) on generalised bivariate Mittag-Leffler-type functions $D_1(x, y), \ldots, D_5(x, y)$. Special attention we paid to functions $E_1(x, y), \ldots, E_{10}(x, y)$ as limiting cases of the functions $D_1(x, y), \ldots, D_5(x, y)$. Following Horn's method, we determine all possible cases of the convergence region of the function $D_1(x, y)$. Further, for the function $D_1(x, y)$ integral representations of the Euler-type have been presented. The one-dimensional and two-dimensional Laplace transforms of the function are also defined. We have constructed a system of partial differential equations that is linked to the function $D_1(x, y)$.

19.1 Introduction

The Mittag-Leffler function has gained importance and popularity through its applications [1]. Namely, it appears as a solution of fractional differential equations and integral equations of fractional order. In addition, the Mittag-Leffler function plays an important role in various fields of applied mathematics and engineering sciences, such as chemistry, biology, statistics, thermodynamics, mechanics, quantum physics, computer science, and signal processing[2]. In addition, the Mittag-Leffler function of many variables arises when solving some boundary value problems involving fractional Volterra type integrodifferential equations [3], initial-boundary value problems for a generalised polynomial diffusion equation with fractional time [4], and also initial-boundary value problems for time-fractional diffusion equations with positive constant coefficients [5].

A. Hasanov (✉)
V.I.Romanovskiy Institute of Mathematics, Tashkent, Uzbekistan
e-mail: anvarhasanov@mail.ru

M. Ruzhansky, K. Van Bockstal (eds.), *Extended Abstracts 2021/2022*, Research Perspectives Ghent Analysis and PDE Center 2,
https://doi.org/10.1007/978-3-031-42539-4_19

In [6], the Mittag-Leffler type function E_1 of two variables is introduced and studied, which in a particular case includes several Mittag-Leffler type functions of one variable. All possible cases are determined by the region of convergence. The system of hypergeometric equations is determined, which satisfies the function E_1, Euler type integral representations and the Mellin-Barnes contour integral, as well as the Laplace integral transformation is given

$$E_1\left(\begin{matrix}\gamma_1,\alpha_1;\gamma_2,\beta_1;\\ \delta_1,\alpha_2,\beta_2;\delta_2,\alpha_3;\delta_3,\beta_3;\end{matrix}\left|\begin{matrix}x\\ y\end{matrix}\right.\right)=$$
$$\sum_{m,n=0}^{\infty}\frac{(\gamma_1)_{\alpha_1 m}(\gamma_2)_{\beta_1 n}}{\Gamma\left(\delta_1+\alpha_2 m+\beta_2 n\right)}\frac{x^m}{\Gamma\left(\delta_2+\alpha_3 m\right)}\frac{y^n}{\Gamma\left(\delta_3+\beta_3 n\right)},$$
$$\gamma_1,\gamma_2,\delta_1,\delta_2,\delta_3,x,y\in\mathbb{C},\quad \min\{\alpha_1,\alpha_2,\alpha_3,\beta_1,\beta_2,\beta_3\}>0,$$

In that paper, another two-variable Mittag-Leffler type function was also introduced, but not studied:

$$E_2\left(\begin{matrix}\gamma_1,\alpha_1,\beta_1;\gamma_2,\alpha_2;\\ \delta_1,\alpha_3,\beta_2;\delta_2,\alpha_4;\delta_3,\beta_3;\end{matrix}\left|\begin{matrix}x\\ y\end{matrix}\right.\right)=$$
$$\sum_{m,n=0}^{\infty}\frac{(\gamma_1)_{\alpha_1 m+\beta_1 n}(\gamma_2)_{\alpha_2 m}}{\Gamma\left(\delta_1+\alpha_3 m+\beta_2 n\right)}\frac{x^m}{\Gamma\left(\delta_2+\alpha_4 m\right)}\frac{y^n}{\Gamma\left(\delta_3+\beta_3 n\right)},$$
$$\gamma_1,\gamma_2,\delta_1,\delta_2,\delta_3,x,y\in\mathbb{C},\quad \min\{\alpha_1,\alpha_2,\alpha_3,\alpha_4,\beta_1,\beta_2,\beta_3\}>0.$$

Another interesting special function was studied in [7]. The paper [8] considered the equations

$$f\left(x\right)=\begin{cases}{}_C D_{0t}^{\alpha}u\left(x,t\right)-u_{xx}\left(x,t\right), & t>0,\\ {}_C D_{t0}^{\beta}u\left(x,t\right)-u_{xx}\left(x,t\right), & t<0,\end{cases}$$

in a domain $\Omega=\{(x,t):0<x<1,-p<t<q\}$ where $\alpha,\beta,p,q\in\mathbb{R}^{+}$, $0<\alpha<1,\ 1<\beta<2$. For equation (13), the boundary value problem is considered and the solution to this problem is expressed by the functions E_1.

A boundary value problem in a domain $\Omega=\{(x,t):0<x<1,\ 0<t<T\}$ for the diffusion equation with a fractional time derivative is considered [9]

$${}^{PC}D_{0t}^{\alpha,\beta,\gamma,\delta}u\left(t,x\right)-u_{xx}\left(t,x\right)=f\left(t,x\right),\ \alpha,\beta,\gamma,\delta\in\mathbb{C},\ \ \mathrm{Re}\alpha>0,$$

where

$${}^{PC}D_{0t}^{\alpha,\beta,\gamma,\delta}y\left(t\right)={}^{P}I_{0t}^{\alpha,m-\beta,-\gamma,\delta}\frac{d^m}{dt^m}y\left(t\right),\ m-1<\mathrm{Re}\beta<m,\ m\in\mathbb{N},$$
$${}^{P}I_{0t}^{\alpha,\beta,\gamma,\delta}y\left(t\right)=\int_0^t\left(t-\xi\right)^{\beta-1}E_{\alpha,\beta}^{\gamma}\left[\delta(t-\xi)^{\alpha}\right]y\left(\xi\right)d\xi.$$

The solution to the problem is expressed by functions E_2.

19.2 Definitions of Some Functions of the Mittag-Leffler-Type

Having carefully studied the definitions of generalised hypergeometric functions and the Mittag-Leffler type functions, we understand the similarity of these functions with the Horn functions [10]. Given this situation, we define the following functions. Note that the parameters introduced in the functions satisfy the conditions $\gamma_i, \delta_i, x, y \in \mathbb{C}$, and $\alpha_i, \beta_i \in \mathbb{R}, \quad \min\{\alpha_i, \beta_i\} > 0$:

$$D_1\left(\begin{matrix}\gamma_1, \alpha_1, \beta_1; \gamma_2, \alpha_2; \gamma_3, \beta_2; \\ \delta_1, \alpha_3, \beta_3; \delta_2, \alpha_4; \delta_3, \beta_4;\end{matrix}\middle|\begin{matrix}x\\ y\end{matrix}\right) = \sum_{m,n=0}^{\infty} \frac{(\gamma_1)_{\alpha_1 m+\beta_1 n}(\gamma_2)_{\alpha_2 m}(\gamma_3)_{\beta_2 n}}{\Gamma(\delta_1+\alpha_3 m+\beta_3 n)} \frac{x^m}{\Gamma(\delta_2+\alpha_4 m)} \frac{y^n}{\Gamma(\delta_3+\beta_4 n)},$$

$$D_2\left(\begin{matrix}\gamma_1, \alpha_1, \beta_1; \gamma_2, \alpha_2; \gamma_3, \beta_2; \\ \delta_1, \alpha_3; \delta_2, \beta_3; \delta_3, \alpha_4; \delta_4, \beta_4;\end{matrix}\middle|\begin{matrix}x\\ y\end{matrix}\right) = \sum_{m,n=0}^{\infty} \frac{(\gamma_1)_{\alpha_1 m+\beta_1 n}(\gamma_2)_{\alpha_2 m}(\gamma_3)_{\beta_2 n}}{\Gamma(\delta_1+\alpha_3 m)\,\Gamma(\delta_2+\beta_3 n)} \frac{x^m}{\Gamma(\delta_3+\alpha_4 m)} \frac{y^n}{\Gamma(\delta_4+\beta_4 n)},$$

$$D_3\left(\begin{matrix}\gamma_1, \alpha_1; \gamma_2, \beta_1; \gamma_3, \alpha_2; \gamma_4, \beta_2; \\ \delta_1, \alpha_3, \beta_3; \delta_2, \alpha_4; \delta_3, \beta_4;\end{matrix}\middle|\begin{matrix}x\\ y\end{matrix}\right) = \sum_{m,n=0}^{\infty} \frac{(\gamma_1)_{\alpha_1 m}(\gamma_2)_{\beta_1 n}(\gamma_3)_{\alpha_2 m}(\gamma_4)_{\beta_2 n}}{\Gamma(\delta_1+\alpha_3 m+\beta_3 n)} \frac{x^m}{\Gamma(\delta_2+\alpha_4 m)} \frac{y^n}{\Gamma(\delta_3+\beta_4 n)},$$

$$D_4\left(\begin{matrix}\gamma_1, \alpha_1, \beta_1; \gamma_2, \alpha_2, \beta_2; \\ \delta_1, \alpha_3; \delta_2, \beta_3; \delta_3, \alpha_4; \delta_4, \beta_4;\end{matrix}\middle|\begin{matrix}x\\ y\end{matrix}\right) = \sum_{m,n=0}^{\infty} \frac{(\gamma_1)_{\alpha_1 m+\beta_1 n}(\gamma_2)_{\alpha_2 m+\beta_2 n}}{\Gamma(\delta_1+\alpha_3 m)\,\Gamma(\delta_2+\beta_3 n)} \frac{x^m}{\Gamma(\delta_3+\alpha_4 m)} \frac{y^n}{\Gamma(\delta_4+\beta_4 n)},$$

$$D_5\left(\begin{matrix}\gamma_1, \alpha_1, \beta_1; \gamma_2, \alpha_2, \beta_2; \\ \delta_1, \alpha_3, \beta_3; \delta_2, \alpha_4; \delta_3, \beta_4;\end{matrix}\middle|\begin{matrix}x\\ y\end{matrix}\right) = \sum_{m,n=0}^{\infty} \frac{(\gamma_1)_{\alpha_1 m+\beta_1 n}(\gamma_2)_{\alpha_2 m+\beta_2 n}}{\Gamma(\delta_1+\alpha_3 m+\beta_3 n)} \frac{x^m}{\Gamma(\delta_2+\alpha_4 m)} \frac{y^n}{\Gamma(\delta_3+\beta_4 n)},$$

$$E_3\left(\begin{matrix}\gamma_1,\alpha_1,\beta_1;\gamma_2,\alpha_2;\\ \delta_1,\alpha_3;\delta_2,\beta_2;\delta_3,\alpha_4;\delta_4,\beta_3;\end{matrix}\middle|\begin{matrix}x\\ y\end{matrix}\right)$$
$$=\sum_{m,n=0}^{\infty}\frac{(\gamma_1)_{\alpha_1 m+\beta_1 n}(\gamma_2)_{\alpha_2 m}}{\Gamma(\delta_1+\alpha_3 m)\Gamma(\delta_2+\beta_2 n)}\frac{x^m}{\Gamma(\delta_3+\alpha_4 m)}\frac{y^n}{\Gamma(\delta_4+\beta_3 n)},$$

$$E_4\left(\begin{matrix}\gamma_1,\alpha_1,\beta_1;\\ \delta_1,\alpha_2;\delta_2,\beta_2;\delta_3,\alpha_3;\delta_4,\beta_3;\end{matrix}\middle|\begin{matrix}x\\ y\end{matrix}\right)$$
$$=\sum_{m,n=0}^{\infty}\frac{(\gamma_1)_{\alpha_1 m+\beta_1 n}}{\Gamma(\delta_1+\alpha_2 m)\Gamma(\delta_2+\beta_2 n)}\frac{x^m}{\Gamma(\delta_3+\alpha_3 m)}\frac{y^n}{\Gamma(\delta_4+\beta_3 n)},$$

$$E_5\left(\begin{matrix}\gamma_1,\alpha_1;\\ \delta_1,\alpha_2,\beta_1;\delta_2,\alpha_3;\delta_3,\beta_2;\end{matrix}\middle|\begin{matrix}x\\ y\end{matrix}\right)$$
$$=\sum_{m,n=0}^{\infty}\frac{(\gamma_1)_{\alpha_1 m}}{\Gamma(\delta_1+\alpha_2 m+\beta_1 n)}\frac{x^m}{\Gamma(\delta_2+\alpha_3 m)}\frac{y^n}{\Gamma(\delta_3+\beta_2 n)},$$

$$E_6\left(\begin{matrix}\gamma_1,\alpha_1;\gamma_2,\beta_1;\gamma_3,\alpha_2;\\ \delta_1,\alpha_3,\beta_2;\delta_2,\alpha_4;\delta_3,\beta_3;\end{matrix}\middle|\begin{matrix}x\\ y\end{matrix}\right)$$
$$=\sum_{m,n=0}^{\infty}\frac{(\gamma_1)_{\alpha_1 m}(\gamma_2)_{\beta_1 n}(\gamma_3)_{\alpha_2 m}}{\Gamma(\delta_1+\alpha_3 m+\beta_2 n)}\frac{x^m}{\Gamma(\delta_2+\alpha_4 m)}\frac{y^n}{\Gamma(\delta_3+\beta_3 n)},$$

$$E_7\left(\begin{matrix}\gamma_1,\alpha_1;\gamma_2,\alpha_2;\\ \delta_1,\alpha_3,\beta_1;\delta_2,\alpha_4;\delta_3,\beta_2;\end{matrix}\middle|\begin{matrix}x\\ y\end{matrix}\right)$$
$$=\sum_{m,n=0}^{\infty}\frac{(\gamma_1)_{\alpha_1 m}(\gamma_2)_{\alpha_2 m}}{\Gamma(\delta_1+\alpha_3 m+\beta_1 n)}\frac{x^m}{\Gamma(\delta_2+\alpha_4 m)}\frac{y^n}{\Gamma(\delta_3+\beta_2 n)},$$

$$E_8\left(\begin{matrix}\gamma_1,\alpha_1,\beta_1;\\ \delta_1,\alpha_2,\beta_2;\delta_2,\alpha_3;\delta_3,\beta_3;\end{matrix}\middle|\begin{matrix}x\\ y\end{matrix}\right)$$
$$=\sum_{m,n=0}^{\infty}\frac{(\gamma_1)_{\alpha_1 m+\beta_1 n}}{\Gamma(\delta_1+\alpha_2 m+\beta_2 n)}\frac{x^m}{\Gamma(\delta_2+\alpha_3 m)}\frac{y^n}{\Gamma(\delta_3+\beta_3 n)},$$

$$E_9\left(\begin{matrix}-;\\ \delta_1,\alpha_1,\beta_1;\delta_2,\alpha_2;\delta_3,\beta_2;\end{matrix}\middle|\begin{matrix}x\\ y\end{matrix}\right)$$
$$=\sum_{m,n=0}^{\infty}\frac{1}{\Gamma(\delta_1+\alpha_1 m+\beta_1 n)}\frac{x^m}{\Gamma(\delta_2+\alpha_2 m)}\frac{y^n}{\Gamma(\delta_3+\beta_2 n)},$$

$$E_{10}\left(\begin{matrix}\gamma_1,\alpha_1,\beta_1;\gamma_2,\alpha_2;\\ \delta_1,\alpha_3;\delta_2,\beta_2;\end{matrix}\middle|\begin{matrix}x\\ y\end{matrix}\right)$$
$$=\sum_{m,n=0}^{\infty}(\gamma_1)_{\alpha_1 m+\beta_1 n}(\gamma_2)_{\alpha_2 m}\frac{x^m}{\Gamma(\delta_1+\alpha_3 m)}\frac{y^n}{\Gamma(\delta_2+\beta_2 n)}.$$

Note that the introduced generalised Mittag-Leffler functions in particular values of the parameters coincide with the known hypergeometric functions [10].

19.3 Determining the Region of the Convergence of a Function D_1

Following Horn [10], we determine the region of convergence of the introduced hypergeometric function D_1.

Definition 19.1 Let us call positive values r, s the associated radii of convergence of the double series

$$\sum_{m,n=0}^{\infty} A(m,n)x^m y^n, \tag{19.1}$$

if it converges absolutely at $|x| < r$, $|y| < s$ and diverges at $|x| > r$, $|y| > s$.

Let us also assume that the $\max(r) = R$, $\max(s) = S$. Points (r, s) lie on the curve C, which is located entirely in the rectangle $0 < r < R$, $0 < s < S$. This curve divides the rectangle into two parts; the part containing the point $r = s = 0$ is a two-dimensional image of the region of convergence of the double power series. Studying the convergence of the series (19.1), Horn introduced the functions

$$\Phi(\mu,\nu)=\lim_{t\to\infty} f(\mu t,\nu t),\quad \Psi(\mu,\nu)=\lim_{t\to\infty} g(\mu t,\nu t),$$

where

$$f(m,n)=\frac{A(m+1,n)}{A(m,n)},\quad g(m,n)=\frac{A(m,n+1)}{A(m,n)}, \tag{19.2}$$

and showed that $R = |\Phi(1,0)|^{-1}$, $S = |\Psi(0,1)|^{-1}$ and that C has a parametric representation $r = |\Phi(\mu,\nu)|^{-1}$, $s = |\Psi(\mu,\nu)|^{-1}$, $\mu,\ \nu > 0$.

Consider the function (19.2). It follows from the definition of the function D_1

$$f(\mu t, \nu t) = \frac{\Gamma(\gamma_1+\alpha_1+\alpha_1\mu t+\beta_1\nu t)\,\Gamma(\gamma_2+\alpha_2+\alpha_2\mu t)}{\Gamma(\gamma_1+\alpha_1\mu t+\beta_1\nu t)\,\Gamma(\gamma_2+\alpha_2\mu t)} \times\frac{\Gamma(\delta_1+\alpha_3\mu t+\beta_3\nu t)\,\Gamma(\delta_2+\alpha_4\mu t)}{\Gamma(\delta_1+\alpha_3+\alpha_3\mu t+\beta_3\nu t)\,\Gamma(\delta_2+\alpha_4+\alpha_4\mu t)},$$
$$g(\mu t, \nu t) = \frac{\Gamma(\gamma_1+\beta_1+\alpha_1\mu t+\beta_1\nu t)\,\Gamma(\gamma_3+\beta_2+\beta_2\nu t)}{\Gamma(\gamma_1+\alpha_1\mu t+\beta_1\nu t)\,\Gamma(\gamma_3+\beta_2\nu t)} \times\frac{\Gamma(\delta_1+\alpha_3\mu t+\beta_3\nu t)\,\Gamma(\delta_3+\beta_4\nu t)}{\Gamma(\delta_1+\beta_3+\alpha_3\mu t+\beta_3\nu t)\,\Gamma(\delta_3+\beta_4+\beta_4\nu t)}.$$

Due to the asymptotics of the Gamma function for large arguments [11]

$$\frac{\Gamma(z+\alpha)}{\Gamma(z+\beta)} \sim z^{\alpha-\beta}\left[1+\frac{(\alpha-\beta)(\alpha+\beta-1)}{2z}+O\left(z^{-2}\right)\right], \quad |\arg(z)| \le \pi,$$

we have

$$\begin{gathered} f(\mu t, \nu t) \sim \frac{1}{E} t^{-\Delta}, \quad \Delta = \alpha_3+\alpha_4-\alpha_1-\alpha_2, \\ E = \mu^{\alpha_4-\alpha_2}\frac{(\alpha_3\mu+\beta_3\nu)^{\alpha_3}(\alpha_4)^{\alpha_4}}{(\alpha_1\mu+\beta_1\nu)^{\alpha_1}(\alpha_2)^{\alpha_2}}, \quad G = \frac{(\alpha_3)^{\alpha_3}(\alpha_4)^{\alpha_4}}{(\alpha_1)^{\alpha_1}(\alpha_2)^{\alpha_2}}, \end{gathered} \tag{19.3}$$

Similarly, we define

$$\begin{gathered} g(\mu t, \nu t) \sim \frac{1}{E'}\cdot t^{-\Delta'}, \quad \Delta' = \beta_3+\beta_4-\beta_1-\beta_2, \\ E' = \nu^{\beta_4-\beta_2}\frac{(\alpha_3\mu+\beta_3\nu)^{\beta_3}(\beta_4)^{\beta_4}}{(\alpha_1\mu+\beta_1\nu)^{\beta_1}(\beta_2)^{\beta_2}}, \quad G' = \frac{(\beta_3)^{\beta_3}(\beta_4)^{\beta_4}}{(\beta_1)^{\beta_1}(\beta_2)^{\beta_2}}. \end{gathered} \tag{19.4}$$

Now, we consider some cases:

Case 1. Let $\Delta > 0$, $\Delta' > 0$. Then from (19.3) and (19.4) it follows

$$\Phi(\mu, \nu) = \lim_{t\to\infty} f(\mu t, \nu t) = 0, \quad \Psi(\mu, \nu) = \lim_{t\to\infty} g(\mu t, \nu t) = 0.$$

Positive numbers r and s are large numbers. The series converges for any value of the argument.

Case 2. Let $\Delta = 0$, $\Delta' = 0$. Then from (19.3) and (19.4) it follows

$$\Phi(\mu, \nu) = \frac{1}{E}, \qquad \Psi(\mu, \nu) = \frac{1}{E'},$$

which immediately led us to the parametric representation of the curve C in the plane (r, s) in the form $r = G$, $s = G'$. Therefore, the series converges absolutely for the values of $|x| < \rho$ and $|y| < \rho'$ where $\rho = \min\limits_{\mu,\nu>0}(E)$, $\rho' = \min\limits_{\mu,\nu>0}(E')$.

Case 3. Let $\Delta < 0$, $\Delta' < 0$. The series diverges $r = s = 0$. The series converges only at the point $x = y = 0$.

Case 4. Let $\Delta = 0$, $\Delta' > 0$. Then the series converges absolutely in the region $|x| < \rho$ and $|y| < \infty$.

Case 5. Let $\Delta > 0$, $\Delta' = 0$. Then the series converges absolutely in the region $|x| < \infty$ and $|y| < \rho'$.

19.4 Integral Representations of a Function D_1

For a generalised hypergeometric function of the Mittag-Leffler type D_1, the following integral representations of the Euler type are valid

$$D_1\left(\begin{matrix}\gamma_1,\alpha_1,\beta_1;\gamma_2,\alpha_2;\gamma_3,\beta_2;\\ \delta_1,\alpha_3,\beta_3;\delta_2,\alpha_4;\delta_3,\beta_4;\end{matrix}\left|\begin{matrix}x\\ y\end{matrix}\right.\right)=\frac{\Gamma(\mu)}{\Gamma(\gamma_2)\Gamma(\mu-\gamma_2)}$$
$$\times\int\limits_0^1\xi^{\gamma_2-1}(1-\xi)^{\mu-\gamma_2-1}D_1\left(\begin{matrix}\gamma_1,\alpha_1,\beta_1;\mu,\alpha_2;\gamma_3,\beta_2;\\ \delta_1,\alpha_3,\beta_3;\delta_2,\alpha_4;\delta_3,\beta_4;\end{matrix}\left|\begin{matrix}x\xi^{\alpha_2}\\ y\end{matrix}\right.\right)d\xi,$$
$$\operatorname{Re}\mu>\operatorname{Re}\gamma_2>0,$$

$$D_1\left(\begin{matrix}\gamma_1,\alpha_1,\beta_1;\gamma_2,\alpha_2;\gamma_3,\beta_2;\\ \delta_1,\alpha_3,\beta_3;\delta_2,\alpha_4;\delta_3,\beta_4;\end{matrix}\left|\begin{matrix}x\\ y\end{matrix}\right.\right)=\frac{\Gamma(\mu)}{\Gamma(\gamma_3)\Gamma(\mu-\gamma_3)}$$
$$\times\int\limits_0^1\xi^{\gamma_3-1}(1-\xi)^{\mu-\gamma_3-1}D_1\left(\begin{matrix}\gamma_1,\alpha_1,\beta_1;\gamma_2,\alpha_2;\mu,\beta_2;\\ \delta_1,\alpha_3,\beta_3;\delta_2,\alpha_4;\delta_3,\beta_4;\end{matrix}\left|\begin{matrix}x\\ y\xi^{\beta_2}\end{matrix}\right.\right)d\xi,$$
$$\operatorname{Re}\mu>\operatorname{Re}\gamma_3>0,$$

$$D_1\left(\begin{matrix}\gamma_1,\alpha_1,\beta_1;\gamma_2,\alpha_2;\gamma_3,\beta_2;\\ \delta_1,\alpha_3,\beta_3;\delta_2,\alpha_4;\delta_3,\beta_4;\end{matrix}\left|\begin{matrix}x\\ y\end{matrix}\right.\right)=\frac{\Gamma(\mu_1)\Gamma(\mu_2)}{\Gamma(\gamma_2)\Gamma(\gamma_3)\Gamma(\mu_1-\gamma_2)\Gamma(\mu_2-\gamma_3)}$$
$$\times\int\limits_0^1\int\limits_0^1\xi^{\gamma_2-1}\eta^{\gamma_3-1}(1-\xi)^{\mu_1-\gamma_2-1}(1-\eta)^{\mu_2-\gamma_3-1}$$
$$\times D_1\left(\begin{matrix}\gamma_1,\alpha_1,\beta_1;\mu_1,\alpha_2;\mu_2,\beta_2;\\ \delta_1,\alpha_3,\beta_3;\delta_2,\alpha_4;\delta_3,\beta_4;\end{matrix}\left|\begin{matrix}x\xi^{\alpha_2}\\ y\eta^{\beta_2}\end{matrix}\right.\right)$$
$$\times d\xi d\eta,\ \operatorname{Re}\mu_1>\operatorname{Re}\gamma_2>0,\ \operatorname{Re}\mu_2>\operatorname{Re}\gamma_3>0,$$

$$\int_0^1\int_0^1 \xi^{\delta_2-1}\eta^{\delta_3-1}(1-\xi)^{\sigma_1-1}(1-\eta)^{\sigma_2-1} D_1 \times \left(\begin{matrix}\gamma_1,\alpha_1,\beta_1;\gamma_2,\alpha_2;\gamma_3,\beta_2;\\ \delta_1,\alpha_3,\beta_3;\delta_2,\alpha_4;\delta_3,\beta_4;\end{matrix}\middle|\begin{matrix}x\xi^{\alpha_4}\\ y\eta^{\beta_4}\end{matrix}\right) d\xi d\eta$$
$$= \Gamma(\sigma_1)\Gamma(\sigma_2) D_1 \times \left(\begin{matrix}\gamma_1,\alpha_1,\beta_1;\gamma_2,\alpha_2;\gamma_3,\beta_2;\\ \delta_1,\alpha_3,\beta_3;\delta_2+\sigma_1,\alpha_4;\delta_3+\sigma_2,\beta_4;\end{matrix}\middle|\begin{matrix}x\\ y\end{matrix}\right),\ \operatorname{Re}\sigma_1>0,\ \operatorname{Re}\sigma_2>0,$$

$$D_1\left(\begin{matrix}\gamma_1,\alpha_1,\beta_1;\gamma_2,\alpha_2;\gamma_3,\beta_2;\\ \delta_1,\alpha_3,\beta_3;\delta_2,\alpha_4;\delta_3,\beta_4;\end{matrix}\middle|\begin{matrix}x\\ y\end{matrix}\right) = \frac{\Gamma(\mu)}{\Gamma(\gamma_1)\Gamma(\mu-\gamma_1)} \times \int_0^1 \xi^{\gamma_1-1}(1-\xi)^{\mu-\gamma_1-1} D_1 \times \left(\begin{matrix}\mu,\alpha_1,\beta_1;\gamma_2,\alpha_2;\gamma_3,\beta_2;\\ \delta_1,\alpha_3,\beta_3;\delta_2,\alpha_4;\delta_3,\beta_4;\end{matrix}\middle|\begin{matrix}x\xi^{\alpha_1}\\ y\xi^{\beta_1}\end{matrix}\right) d\xi,\ \operatorname{Re}\mu>\operatorname{Re}\gamma_1>0,$$

$$D_1\left(\begin{matrix}\gamma_1,\alpha_1,\beta_1;\gamma_2,\alpha_2;\gamma_3,\beta_2;\\ \mu_1+\mu_2,\alpha_3,\beta_3;\delta_2,\alpha_4;\delta_3,\beta_4;\end{matrix}\middle|\begin{matrix}x\\ y\end{matrix}\right) = \int_0^1 \xi^{\mu_1-1}(1-\xi)^{\mu_2-1} D_2\left(\begin{matrix}\gamma_1,\alpha_1,\beta_1;\gamma_2,\alpha_2;\gamma_3,\beta_2;\\ \mu_1,\alpha_3;\mu_2,\beta_3;\delta_2,\alpha_4;\delta_3,\beta_4;\end{matrix}\middle|\begin{matrix}x\xi^{\alpha_3}\\ y(1-\xi)^{\beta_3}\end{matrix}\right) d\xi,$$
$\operatorname{Re}\mu_1>0, \operatorname{Re}\mu_2>0,$

19.5 Integral Laplace Transforms

Let L_1 and L_2 denote the one-dimensional and two-dimensional Laplace transforms:

$$L_1\{f(t);p\} = \int_0^\infty f(t)e^{-pt}dt,\ \operatorname{Re}p>0, \tag{19.5}$$

$$L_2\{f(t_1,t_2);p,q\} = \int_0^\infty\int_0^\infty f(t_1,t_2)e^{-t_1p-t_2q}dt_1dt_2,\ \operatorname{Re}p>0,\ \operatorname{Re}q>0. \tag{19.6}$$

The following Laplace transformations are valid

$$L_1\left\{t^{\delta_1-1}D_1\left(\begin{matrix}\gamma_1,\alpha_1,\beta_1;\gamma_2,\alpha_2;\gamma_3,\beta_2;\\ \delta_1,\alpha_3,\beta_3;\delta_2,\alpha_4;\delta_3,\beta_4;\end{matrix}\left|\begin{matrix}xt^{\alpha_3}\\ yt^{\beta_3}\end{matrix}\right.\right):p\right\}$$
$$=\frac{1}{p^{\delta_1}}E_{11}\left(\begin{matrix}\gamma_1,\alpha_1,\beta_1;\gamma_2,\alpha_2;\gamma_3,\beta_2;\\ \delta_2,\alpha_4;\delta_3,\beta_4;\end{matrix}\left|\begin{matrix}\frac{x}{p^{\alpha_3}}\\ \frac{y}{p^{\beta_3}}\end{matrix}\right.\right),\ \mathrm{Re}\,p>0, \tag{19.7}$$

$$L_1\left\{t^{\rho-1}E_1\left(\begin{matrix}\gamma_1,\alpha_1;\gamma_2,\beta_1;\\ \delta_1,\alpha_2,\beta_2;\delta_2,\alpha_3;\delta_3,\beta_3;\end{matrix}\left|\begin{matrix}xt^{\mu_1}\\ yt^{\mu_2}\end{matrix}\right.\right);p\right\}$$
$$=\frac{\Gamma(\rho)}{p^{\rho}}D_1\left(\begin{matrix}\rho,\mu_1,\mu_2;\gamma_1,\alpha_1;\gamma_2,\beta_1;\\ \delta_1,\alpha_2,\beta_2;\delta_2,\alpha_3;\delta_3,\beta_3;\end{matrix}\left|\begin{matrix}\frac{x}{p^{\mu_1}}\\ \frac{y}{p^{\mu_2}}\end{matrix}\right.\right), \tag{19.8}$$
$$\mathrm{Re}\,\rho>0,\ \mathrm{Re}\,\mu_1>0,\ \mathrm{Re}\,\mu_2>0,$$

$$L_2\left\{t_1^{\rho_1-1}t_2^{\rho_2-1}E_8\left(\begin{matrix}\gamma_1,\alpha_1,\beta_1;\\ \delta_1,\alpha_2,\beta_2;\delta_2,\alpha_3;\delta_3,\beta_3;\end{matrix}\left|\begin{matrix}xt_1^{\mu_1}\\ yt_2^{\mu_2}\end{matrix}\right.\right);p,q\right\}=$$
$$=\frac{\Gamma(\rho_1)\Gamma(\rho_2)}{p^{\rho_1}q^{\rho_2}}D_1\left(\begin{matrix}\gamma_1,\alpha_1,\beta_1;\rho_1,\mu_1;\rho_2,\mu_2;\\ \delta_1,\alpha_2,\beta_2;\delta_2,\alpha_3;\delta_3,\beta_3;\end{matrix}\left|\begin{matrix}x\\ y\end{matrix}\right.\right),\quad \mathrm{Re}\,\rho_1>0,\quad \mathrm{Re}\,\rho_2>0, \tag{19.9}$$

$$L_2\left\{x^{\delta_1-1}y^{\delta_2-1}D_2\left(\begin{matrix}\gamma_1,\alpha_1,\beta_1;\gamma_2,\alpha_2;\gamma_3,\beta_2;\\ \delta_1,\alpha_3;\delta_2,\beta_3;\delta_3,\alpha_4;\delta_4,\beta_4;\end{matrix}\left|\begin{matrix}x^{\alpha_3}\\ y^{\beta_3}\end{matrix}\right.\right);p,q\right\}$$
$$=\frac{1}{p^{\delta_1}q^{\delta_2}}E_{11}\left(\begin{matrix}\gamma_1,\alpha_1,\beta_1;\gamma_2,\alpha_2;\gamma_3,\beta_2;\\ \delta_3,\alpha_4;\delta_4,\beta_4;\end{matrix}\left|\begin{matrix}\frac{1}{p^{\alpha_3}}\\ \frac{1}{q^{\beta_3}}\end{matrix}\right.\right),\ \mathrm{Re}\,p>0,\ \mathrm{Re}\,q>0, \tag{19.10}$$

$$E_{11}\left(\begin{matrix}\gamma_1,\alpha_1,\beta_1;\gamma_2,\alpha_2;\gamma_3,\beta_2;\\ \delta_1,\alpha_3;\delta_2,\beta_3;\end{matrix}\left|\begin{matrix}x\\ y\end{matrix}\right.\right)$$
$$=\sum_{m,n=0}^{\infty}(\gamma_1)_{\alpha_1 m+\beta_1 n}(\gamma_2)_{\alpha_2 m}(\gamma_3)_{\beta_2 n}\frac{x^m}{\Gamma(\delta_1+\alpha_3 m)}\frac{y^n}{\Gamma(\delta_2+\beta_3 n)}. \tag{19.11}$$

Equalities (19.5)–(19.11) are verified by direct calculations.

19.6 System of Partial Differential Equations

In this section, we present the result on which system of PDEs will satisfy the bivariate Mittag-Leffler-type function D_1. The proof of this statement will be given with details in the full publication.

Theorem 19.1 *Let* $\theta \equiv x(\partial/\partial x)$, $\phi \equiv y(\partial/\partial y)$, *then for* $\gamma_1, \gamma_2, \gamma_3, \delta_1, \delta_2, \delta_3, x, y \in \mathbb{C}$, $\alpha_j, \beta_j \in \mathbb{N}$, $(j = 1, 2, 3, 4)$ *the generalised bivariate Mittag-Leffler-type function* D_1 *satisfies the following system of partial differential equations:*

$$\begin{aligned}
&\prod_{i=1}^{\alpha_3} (\delta_1 + \alpha_3 - i + \alpha_3\theta + \beta_3\,\phi) \prod_{i=1}^{\alpha_4} (\delta_2 + \alpha_4 - i + \alpha_4\theta)\, x^{-1} \\
&\quad - \prod_{i=1}^{\alpha_1} (\gamma_1 + \alpha_1 - i + \alpha_1\theta + \beta_1\phi) \prod_{i=1}^{\alpha_2} (\gamma_2 + \alpha_2 - i + \alpha_2\theta)\, D_1(x, y) = 0, \\
&\prod_{i=1}^{\beta_3} (\delta_1 + \beta_3 - i + \alpha_3\theta + \beta_3\,\phi) \prod_{i=1}^{\beta_4} (\delta_2 + \beta_4 - i + \beta_4\,\phi)\, y^{-1} \\
&\quad - \prod_{i=1}^{\beta_1} (\gamma_1 + \beta_1 - i + \alpha_1\theta + \beta_1\,\phi) \prod_{i=1}^{\beta_2} (\gamma_2 + \beta_2 - i + \beta_2\,\phi)\, D_1(x, y) = 0.
\end{aligned}$$

Acknowledgments The research was initiated during the author's visit to the Intercontinental Research Centre "Analysis and PDE" (Ghent University, Belgium), supported by the FWO Odysseus 1 grant G.0H94.18N: Analysis and partial differential equations and by the Methusalem programme of the Ghent University Special Research Fund (BOF) (Grant number 01M01021).

References

1. Mittag-Leffler, G.M.: Sur la nouvelle fonction $E_\alpha(z)$. C R Acad. Sci Paris. **137**, 554–558 (1903)
2. Dzherbashian, M.M.: Integral Transforms and Representation of Functions in the Complex Domain. Nauka, Moscow (1966) [in Russian]
3. Samko, S., Kilbas, A., Marichev, O.: Fractional Integrals and Derivatives: Theory and Applications. Gordon and Breach, Switzerland (1993)
4. Luchko, Y.: Initial boundary value problems for the generalized multi-term time-fractional diffusion equation. J. Math. Anal. Appl. **374**, 538–548 (2011)
5. Li, Z., Liu, Y., Yamamoto, M.: Initial-boundary value problems for multi-term time-fractional diffusion equations with positive constant coefficients. Appl. Math. Comput. **257**, 381–397 (2015)
6. Garg, M., Manohar, P., Kalla, S.L.: A Mittag-Leffler type function of two variables. Integral Transform Spec. Funct. **24** (11), 934–944 (2013)
7. Maged, G.B.-S., Hasanov, A., Ruzhansky, M.: Some properties relating to the Mittag-Leffler function of two variables. Integral Transform Spec. Funct. **33** (5), 400–418 (2022)
8. Karimov, E., Al-Salti, N., Kerbal, S.: An inverse source non-local problem for a mixed type equation with a Caputo fractional differential operator. East Asian J. Appl. Math. **7** (2), 417–438 (2017)
9. Karimov, E., Hasanov, A.: A boundary-value problem for time-fractional diffusion equation involving regularized Prabhakar fractional derivative. In: Abstracts of the Conference. "Operator Algebras, Non-Associative Structures and Related Problem", pp. 178–180. Tashkent (2022)
10. Erdelyi, A., Magnus, W., Oberhettinger, F., Tricomi, F.G.: Higher Transcendental Functions, vol. 1. McGraw-Hill. New York (1953)
11. Luke, Y.L.: The Special Functions and Their Approximations, vol. I. Academic Press, New York-London (1969)

Chapter 20
Inverse Problems for Time-Fractional Mixed Equation Involving the Caputo Fractional Derivative

Erkinjon Karimov, Niyaz Tokmagambetov, and Shokhzodbek Khasanov

Abstract The key object of the present short note is the time-fractional mixed equation consisting of the sub-diffusion and fractional wave equations. Different inverse problems, including inverse source, inverse-initial and inverse-terminal problems for the above-mentioned mixed equation have been targeted for their unique solvability.

20.1 Introduction and Formulation of a Problem

It is well known that inverse problems for partial differential equations have numerous applications in real-life problems [1]. There are many types of inverse problems such as inverse-coefficient problem [2], inverse-initial problem [3], inverse-boundary problem [4], inverse-source problem [5] and etc. for partial differential equations of integer and fractional order. There are many inverse problems that have been studied for diffusion and wave equations, or their fractional versions. A comparatively less studied class of inverse problems is inverse problems for mixed equations, which consist of diffusion and wave equations. We mention, for instance, the works [6–8], where inverse problems for mixed equations were studied for unique solvability.

E. Karimov (✉)
Fergana State University, Fergana, Uzbekistan

N. Tokmagambetov
Institute of Mathematics and Mathematical Modeling, Almaty, Kazakhstan

Centre de Recerca Matemática, Edifici C, Bellaterra, Spain
e-mail: tokmagambetov@math.kz; tokmagambetov@crm.cat

S. Khasanov
Namangan Engineering Construction Institute, Namangan, Uzbekistan

M. Ruzhansky, K. Van Bockstal (eds.), *Extended Abstracts 2021/2022*,
Research Perspectives Ghent Analysis and PDE Center 2,
https://doi.org/10.1007/978-3-031-42539-4_20

Our main object is the following mixed equation

$$r(t)g(t,x)=\begin{cases} {}^C D_{0t}^{\alpha}u(t,x)-u_{xx}(t,x), & (t,x)\in\Omega^+,\ t>0,\\ {}^C D_{t0}^{\beta}u(t,x)-u_{xx}(t,x), & (t,x)\in\Omega^-,\ t<0, \end{cases} \tag{20.1}$$

where
$\Omega^+=\{(t,x):0<x<1,\ 0<t<q\},\Omega^-=\{(t,x):0<x<1,\ -p<t<0\}$, $p,q\in\mathbb{R}^+$, $\alpha,\beta\in\mathbb{R}$ such that $0<\alpha\le 1$, $1<\beta\le 2$, $g(t,x)$ is a given function, but $r(t)$, in general, is unknown function,

$${}^C D_{0t}^{\alpha}g(t,x)=\begin{cases} \dfrac{1}{\Gamma(1-\alpha)}\displaystyle\int_0^t \frac{g'_z(z,x)}{(t-z)^{\alpha}}dz, & 0<\alpha<1,\\ \dfrac{dg(t,x)}{dt}, & \alpha=1, \end{cases}$$

$${}^C D_{t0}^{\beta}g(t,x)=\begin{cases} \dfrac{1}{\Gamma(2-\beta)}\displaystyle\int_t^0 \frac{g''_z(z,x)}{(t-z)^{\beta-1}}dz, & 0<\alpha<1,\\ \dfrac{d^2g(t,x)}{dt^2}, & \beta=2 \end{cases}$$

are the left and right-sided Caputo fractional derivatives, correspondingly [9]. We will consider this equation in the rectangular domain $\Omega=\Omega^+\cup\Omega^-\cup\{t=0\}$.

Below, we formulate the time-dependent inverse source problem for mixed Eq. (20.1).

Problem 20.1 To find a pair of functions $\{u(t,x),r(t)\}$, which satisfies Eq. (20.1) in Ω and the following conditions:

- $u(t,x)\in C(\bar{\Omega})\cap C_x^2(\Omega^+\cup\Omega^+)$, ${}^C D_{0t}^{\alpha}u(t,x)\in C(\Omega^+)$, ${}^C D_{t0}^{\beta}u(t,x)\in C(\Omega^-)$;
- $u_x(t,0)=u_x(t,1),\ \ u(t,1)=0,\ \ -p\le t\le q$;
- $\lim\limits_{t\to+0}{}^C D_{0t}^{\alpha}u(t,x)=\lim\limits_{t\to-0}u_t(t,x),\ 0\le x\le 1$;
- $u(-p,x)=\psi(x),\ \ 0\le x\le 1$;
- $\int\limits_0^1 u(t,x)dx=N_2(t),\ t\in[-p;0];\quad \int\limits_0^1 u(t,x)dx=N_1(t),\ t\in[0,q]$.

Here, $N_j(t)$ $(j=1,2)$ are given functions.

20.2 Short Description of the Investigation

Using non-local conditions on space-variable and two over-determination conditions, we will find unknown function $r(t)$ as follows:

$$r(t) = \frac{{}^{C}D_{0t}^{\alpha} N_1(t)}{\int_0^1 g(t,x)dx}, \quad t \in [0;q], \quad r(t) = \frac{{}^{C}D_{t0}^{\beta} N_2(t)}{\int_0^1 g(t,x)dx}, \quad t \in [-p;0].$$

It is obvious that if $\int_0^1 g(t,x)dx \neq 0$, $t \in [-p;q], x \in [0;1]$, then one can uniquely find $r(t)$. That is the first condition on the given function $g(t,x)$.

We will search solution to the problem ISP as follows:

$$u(t,x) = 2A_0(t)(1-x) + \sum_{n=1}^{\infty} 4A_{1,n}(t)(1-x)\cos 2\pi nx + \sum_{n=1}^{\infty} 4A_{2,n}(t)\sin 2\pi nx, \quad t \geq 0, \tag{20.2}$$

$$u(t,x) = 2B_0(t)(1-x) + \sum_{n=1}^{\infty} 4B_{1,n}(t)(1-x)\cos 2\pi nx + \sum_{n=1}^{\infty} 4B_{2,n}(t)\sin 2\pi nx, \quad t \leq 0, \tag{20.3}$$

$$g(t,x) = 2g_0(t)(1-x) + \sum_{n=1}^{\infty} 4g_{1,n}(t)(1-x)\cos 2\pi nx + \sum_{n=1}^{\infty} 4g_{2,n}(t)\sin 2\pi nx, \tag{20.4}$$

where $A_0(t),\ B_0(t),\ A_{i,n}(t),\ B_{i,n}(t)\ (i=1,2)$ are unknown functions,

$$g_0(t) = \int_0^1 g(t,x)\cdot 1\,dx, \quad g_{1,n}(t) = \int_0^1 g(t,x)\cos 2\pi nx\,dx,$$

$$g_{2,n}(t) = \int_0^1 g(t,x)x\sin 2\pi nx\,dx.$$

The key motivation for such a form came from the method of separation of variables, which leads us to the non-self-adjoint problem. As a result of this, we need to expand functions into the bi-orthogonal series, presented above.

Substituting (20.2)–(20.4) into (20.1), we will get the following equations:

$$ {}^C D_{0t}^{\alpha} A_0(t) = r(t) g_0(t),\ t \geq 0, $$

$$ {}^C D_{0t}^{\alpha} A_{1,n}(t) + (2\pi n)^2 A_{1,n}(t) = r(t) g_{1,n}(t),\ t \geq 0, $$

$$ {}^C D_{0t}^{\alpha} A_{2,n}(t) + (2\pi n)^2 A_{2,n}(t) = 4\pi n A_{1,n}(t) + r(t) g_{2,n}(t),\ t \geq 0, $$

$$ {}^C D_{t0}^{\beta} B_0(t) = r(t) g_0(t),\ t \leq 0, $$

$$ {}^C D_{t0}^{\beta} B_{1,n}(t) + (2\pi n)^2 B_{1,n}(t) = r(t) g_{1,n}(t),\ t \leq 0, $$

$$ {}^C D_{t0}^{\beta} B_{2,n}(t) + (2\pi n)^2 B_{2,n}(t) = 4\pi n B_{1,n}(t) + r(t) g_{2,n}(t),\ t \leq 0. $$

General solutions of these equations can be obtained by Laplace transform:

$$ A_0(t) = A_0(0) + \frac{1}{\Gamma(\alpha)} \int_0^t (t-\tau)^{\alpha-1} r(\tau) g_0(\tau) d\tau, $$

$$ \begin{aligned} A_{1,n}(t) = {} & A_{1,n}(0) E_{\alpha,1}\left(-(2\pi n)^2 t^{\alpha}\right) \\ & + \int_0^t (t-\tau)^{\alpha-1} E_{\alpha,\alpha}\left(-(2\pi n)^2 (t-\tau)^{\alpha}\right) r(\tau) g_{1,n}(\tau) d\tau, \end{aligned} $$

$$ \begin{aligned} A_{2,n}(t) = {} & A_{2,n}(0) E_{\alpha,1}\left(-(2\pi n)^2 t^{\alpha}\right) \\ & + 4\pi n A_{1,n}(0) t^{\alpha} E_{\alpha,\alpha+1}^{2}\left(-(2\pi n)^2 t^{\alpha}\right) \\ & + 4\pi n \int_0^t (t-\tau)^{2\alpha-1} E_{\alpha,2\alpha}^{2}\left(-(2\pi n)^2 (t-\tau)^{\alpha}\right) r(\tau) g_{1,n}(\tau) d\tau \\ & + \int_0^t (t-\tau)^{\alpha-1} E_{\alpha,\alpha}^{2}\left(-(2\pi n)^2 (t-\tau)^{\alpha}\right) r(\tau) g_{2,n}(\tau) d\tau, \end{aligned} $$

$$ B_0(t) = B_0(0) + B'_0(0)\, t + \frac{1}{\Gamma(\beta)} \int_t^0 (\tau - t)^{\beta-1} r(\tau)\, g_0(\tau) d\tau, $$

$$B_{1,n}(t) = B_{1,n}(0)\, E_{\beta,1}\left(-(2\pi n)^2(-t)^\beta\right) + B'_{1,n}(0)\, t E_{\beta,2}\left(-(2\pi n)^2(-t)^\beta\right)$$

$$-\int_t^0 (\tau - t)^{\beta-1} E_{\beta,\beta}\left(-(2\pi n)^2(\tau - t)^\beta\right) r(\tau) g_{1,n}(\tau) d\tau,$$

$$B_{2,n}(t) = B_{2,n}(0)\, E_{\beta,1}\left(-(2\pi n)^2(-t)^\beta\right) - B'_{2,n}(0)\, t E_{\beta,2}\left(-(2\pi n)^2(-t)^\beta\right)$$

$$+ 4\pi n B_{1,n}(0)\, (-t)^\beta E^2_{\beta,\beta+1}\left(-(2\pi n)^2(-t)^\beta\right)$$

$$- 4\pi n B'_{1,n}(0)\, (-t)^{\beta+1} E^2_{\beta,\beta+2}\left(-(2\pi n)^2(-t)^\beta\right)$$

$$+ 4\pi n \int_t^0 (\tau - t)^{2\beta-1} E^2_{\beta,2\beta}\left(-(2\pi n)^2(\tau - t)^\beta\right) r(\tau) g_{1,n}(\tau) d\tau$$

$$+ \int_t^0 (\tau - t)^{\beta-1} E_{\beta,\beta}\left(-(2\pi n)^2(\tau - t)^\beta\right) r(\tau) g_{2,n}(\tau) d\tau.$$

In order to find unknown constants $A_0(0)$, $A_{1,n}(0)$, $A_{2,n}(0)$, $B_0(0)$, $B'_0(0)$, $B_{1,n}(0)$, $B'_{1,n}(0)$, $B_{2,n}(0)$, $B'_{2,n}(0)$, we will use conditions in time and glueing conditions on the line $t = 0$. In the process, we should impose the following condition

$$E_{\beta,1}\left(-(2\pi n)^2(p)^\beta\right) + (2\pi n)^2 p E_{\beta,2}\left(-(2\pi n)^2(p)^\beta\right) \neq 0.$$

That might be accepted as a geometrical restriction since there is a lower bound p of the considered rectangular domain. Nevertheless, we have to show that the above given presentation is not zero for large enough ns. This can be done by the fact that $\lim\limits_{z\to\infty} E_{\beta,1}(z) = 0$ $\lim\limits_{z\to\infty} z E_{\beta,2}(z) = -\frac{1}{\Gamma(2-\beta)}$.

The uniform convergence of infinite series will be proved using appropriate estimation of the Mittag-Leffler function and imposing certain conditions on the given data.

Another interesting problem is to find not only a solution, an unknown source, but also unknown data on time (it might be the initial or final time). The tool of investigation will remain the same but with specific evaluations.

Acknowledgments This research is funded by the Science Committee of the Ministry of Science and Higher Education of the Republic of Kazakhstan (Grant No. BR20281002). NT is also supported by the Beatriu de Pinós programme and by AGAUR (Generalitat de Catalunya) grant 2021 SGR 00087.

References

1. Isakov, V.: Inverse problems for partial differential equations. Applied Mathematical Sciences, vol. 127. Springer, New York (1998)
2. Brizitskii, R.V., Saritskaya, Z.Y.: Inverse coefficient problems for a non-linear convection-diffusion-reaction equation. Izvestiya Math. **82**, 14–30 (2018)
3. Masood, K., Messaoudi, S., Zaman, F.D.: Initial inverse problem in heat equation with Bessel operator. Int. J. Heat Mass Transf. **45**(14), 2959–2965 (2002)
4. Babaei, A.: Solving a time-fractional inverse heat conduction problem with an unknown nonlinear boundary condition. J. Math. Model. **7**, 85–106 (2019)
5. Ruzhansky, M., Tokmagambetov, N., Torebek, B.: Inverse source problems for positive operators. I: hypoelliptic diffusion and subdiffusion equations. J. Inverse ill-Posed Probl. **27**, 891–911 (2019)
6. Karimov, E.T., Kerbal, S., Al-Salti, N.: Inverse source problem for multi-term fractional mixed type equation. In: Ruzhansky, M. et al. (eds.) Advances in Real and Complex Analysis with Applications. Trends in Mathematics, pp. 289–301. Springer International Publishing, New York (2017)
7. Ashurov, R.R., Zununnov, R.T.: Inverse problem for determining the order of the fractional derivative in mixed-type equations. Lobachevskii J. Math. **42**, 2714–2729 (2021)
8. Durdiev, D.K.: Determining the coefficient of a mixed parabolic-hyperbolic equation with noncharacteristic type change line. Differ. Equ. **58**, 1618–1629 (2022)
9. Kilbas, A.A., Srivastava, H.M., Trujillo, J.J.: Theory and Applications of Fractional Differential Equations. Elsevier. Amsterdam (2006)

Chapter 21
Time Dependent Inverse Source Problems for Integrodifferential Kelvin-Voigt System

Khonatbek Khompysh and Aidos G. Shakir

Abstract In this work, time-dependent inverse source problems for an integrodifferential Kelvin-Voigt (Navier-Stokes-Voigt) system governing a flow of incompressible non-Newtonian fluids with viscoelastic properties is considered. The inverse problems considered consist of determining a time-dependent intensity of the density of external forces, along with the velocity field and the pressure of the fluids. The system is also supplemented with an initial and a boundary condition, and with two types of integral overdetermination conditions with respect to the space domain. Under suitable conditions on the data, the global-in-time existence and uniqueness of weak and strong solutions to the considered inverse problem were established.

21.1 Introduction

The study of inverse problems of Newtonian hydrodynamics, in particular for the Navier-Stokes equations as well as related systems of heat convection and magnetohydrodynamics, have been studied by many authors who have proposed their various approaches and methodologies. One of the main works is the monograph [19] of Prilepko and et al., which included their previous works on the theory of the Navier-Stokes equations since 1989. They first reduce the original inverse problems to an equivalent operator equation of the second kind, and then use the fixed point principle. As another approach to solving inverse problems of hydrodynamics, one can consider the works of Abylkairov [1, 2], who used the method of successive approximations. Other approaches use Carleman estimates; for example, we refer the readers to [5, 6, 8, 9, 15], and the references therein. However, all of the above approaches are applicable only in the case where the corresponding direct problems are uniquely solvable and their solutions have

K. Khompysh (✉) · A. G. Shakir
Al-Farabi Kazakh National University, Almaty, Kazakhstan
e-mail: konat_k@mail.ru

M. Ruzhansky, K. Van Bockstal (eds.), *Extended Abstracts 2021/2022*,
Research Perspectives Ghent Analysis and PDE Center 2,
https://doi.org/10.1007/978-3-031-42539-4_21

additional smoothness. But inverse problems for non-Newtonian hydrodynamics are not sufficiently studied from a mathematical point of view, see for instance [3, 4, 7, 12, 13], and they have important applications from a physical point of view. Thus, in this paper, we study the unique solvability of some inverse source problems for a system of integrodifferential Kelvin-Voigt equations describing the motion of incompressible non-Newtonian fluids, which take into account viscoelastic properties.

21.1.1 Statement of the Problem

Let Ω be a bounded domain in $\mathbb{R}^d$, $d \geq 2$, with a smooth boundary $\partial\Omega$, and $Q_T = \Omega \times (0, T)$ is a cylinder with fixed positive finite constant $0 < T < \infty$, and $\Gamma_T = \partial\Omega \times [0, T]$. Let us consider the following inverse problem of determining a time-dependent source function $f(t)$, a velocity field $\mathbf{v}(x, t)$, and a pressure $\pi(x, t)$, which satisfy the system of integrodifferential Kelvin-Voigt equations governing flows of an incompressible viscoelastic fluid

$$\mathbf{v}_t - \kappa\Delta\mathbf{v}_t - \nu\Delta\mathbf{v} - \int_0^t K(t-s)\Delta\mathbf{v}(\mathbf{x}, s)ds - \nabla\pi = f(t)\mathbf{g}(\mathbf{x}, t) \text{ in } Q_T, \tag{21.1}$$

$$\mathbf{div}\, \mathbf{v}(\mathbf{x}, t) = 0 \text{ in } Q_T, \tag{21.2}$$

the initial condition

$$\mathbf{v}(\mathbf{x}, 0) = \mathbf{v}_0(\mathbf{x}) \text{ in } \Omega, \tag{21.3}$$

the boundary condition

$$\mathbf{v}(\mathbf{x}, t) = 0 \quad \text{on} \quad \Gamma_T, \tag{21.4}$$

and the integral overdetermination condition

$$\int_\Omega \mathbf{v}\boldsymbol{\omega} d\mathbf{x} = e(t), \ t \in [0, T]. \tag{21.5}$$

Here and below, the bold letters denote vector-valued functions and spaces, and $\mathbf{v}(\mathbf{x}, t) = (v_1, v_2, \ldots, v_d)$ is a vector of velocity, $\pi(\mathbf{x}, t)$ is a pressure, and the positive constants ν and κ are the kinematic viscosity and the relaxation coefficient of the fluid, respectively. The vector-function $f(t)\mathbf{g}(\mathbf{x}, t) := \mathbf{F}(\mathbf{x}, t)$ is a density of external force with unknown intensity $f(t)$, and $\mathbf{v}_0(\mathbf{x})$, $\boldsymbol{\omega}(\mathbf{x})$, $\mathbf{g}(\mathbf{x}, t)$, $e(t)$, and $K(t)$ are given functions. The system of Eqs. (21.1)–(21.2) is also called the

Kelvin-Voigt system (Navier-Stokes-Voigt) with memory [17, 20] or the Oskolkov system [11] and it models the motion of an incompressible non-Newtonian fluid with viscoelastic properties. For the details of the physical background and its mathematical modelling, we refer the readers to [10, 18, 20]. Inverse problems of recovering a source $\mathbf{f}(\mathbf{x})$ depending on the space variable from the system (21.1)–(21.4) supplemented with an additional observation were studied in [13], and in [3], in the case $K(t) = 0$.

21.2 Preliminaries: An Equivalent Nonlocal Direct Problem

Throughout the paper, for the function spaces here we use a standard notation from the Navier-Stokes theory, for example in [16], and the values of C denote positive constants different from each other depending only on data of the problem.

Definition 21.1 The pair $(\mathbf{v}(x,t), f(t))$ is a weak solution to the inverse problem (21.1)–(21.5) if:

1. $\mathbf{v} \in L^\infty(0,T;\mathbf{H}^1(\Omega))$, $\mathbf{v}_t \in L^2(0,T;\mathbf{H}^1(\Omega))$, $f(t) \in L^2[0,T]$;
2. $\mathbf{v}(0) = \mathbf{v}_0$ a.e. in Ω;
3. For every $\boldsymbol{\varphi} \in \mathbf{H}^1(\Omega)$ and for a.a. $t \in (0,T)$ holds

$$\begin{aligned}&\frac{d}{dt}\left((\mathbf{v},\boldsymbol{\varphi})_{2,\Omega} + \kappa\,(\nabla\mathbf{v},\nabla\boldsymbol{\varphi})_{2,\Omega}\right) + \nu\,(\nabla\mathbf{v},\nabla\boldsymbol{\varphi})_{2,\Omega}\\ &\quad + \int_0^t K(t-s)\,(\nabla\mathbf{v},\nabla\boldsymbol{\varphi})_{2,\Omega}\,ds = f(t)\,(\mathbf{g},\boldsymbol{\varphi})_{2,\Omega}\,,\end{aligned} \tag{21.6}$$

where $\|\cdot\|_{2,\Omega}$ and $(\cdot,\cdot)_{2,\Omega}$ denote the norm and inner product in $\mathbf{L}^2(\Omega)$ space.

A weak solution is a strong solution if

$$\mathbf{v} \in L^\infty(0,T;\mathbf{H}^1(\Omega)\cap\mathbf{H}^2(\Omega)), \quad \mathbf{v}_t \in L^2(0,T;\mathbf{H}^1(\Omega)\cap\mathbf{H}^2(\Omega)),$$

and each equation holds in the distribution sense in the their corresponding domain.

Assume that data of the problem satisfy the following conditions

$$\mathbf{v}_0(\mathbf{x}) \in \mathbf{H}^1(\Omega); \tag{21.7}$$

$$\exists k_0 \in \mathbb{R}^+ : \ |g_0(t)| = \left|(\mathbf{g}(\mathbf{x},t),\boldsymbol{\omega}(\mathbf{x}))_{2,\Omega}\right| \ge k_0 > 0, \ \forall t \ge 0; \tag{21.8}$$

$$\mathbf{g}(\mathbf{x},t) \in L^\infty(0,T;\mathbf{L}^2(\Omega)); \tag{21.9}$$

$$\omega(\mathbf{x}) \in \mathbf{H}^1(\Omega), \ e(t) \in W_2^1([0,T]); \tag{21.10}$$

$$(\mathbf{v}_0, \boldsymbol{\omega})_{2,\Omega} = e(0); \tag{21.11}$$

$$K(t) \in L^2([0,T]) \Leftrightarrow \|K(t)\|_{L^2([0,T])} \equiv K_0 < \infty. \tag{21.12}$$

Arguing as in [4], first we reduce the original inverse problem (21.1)–(21.5) to an equivalent nonlocal direct problem. To do this, multiply the Eq. (21.1) by $\boldsymbol{\omega}(\mathbf{x})$ and integrate over Ω. Integrating by parts and using (21.5), (21.8), we get

$$\begin{aligned} f(t) =& \frac{1}{g_0(t)} \Big(e'(t) + \kappa\,(\nabla \mathbf{v}_t, \nabla\boldsymbol{\omega})_{2,\Omega} + \nu\,(\nabla \mathbf{v}, \nabla\boldsymbol{\omega})_{2,\Omega} \\ &+ \int_0^t K(t-s)\,(\nabla \mathbf{v}, \nabla\boldsymbol{\omega})_{2,\Omega}\, ds\Big) := F(t,\mathbf{v}). \end{aligned} \tag{21.13}$$

Substituting (21.13) into the system (21.1)–(21.2), we obtain the following system of nonlocal equations for unknown functions $\mathbf{v}$ and π

$$\begin{aligned} \mathbf{v}_t - \kappa\Delta\mathbf{v}_t - \nu\Delta\mathbf{v} - \int_0^t K(t-s)\Delta\mathbf{v}(\mathbf{x},s)ds - \nabla\pi &= F(t,\mathbf{v})\mathbf{g}(\mathbf{x},t), \\ \mathbf{div}\,\mathbf{v} &= 0, \end{aligned} \tag{21.14}$$

where $F(t,\mathbf{v})$ is defined by (21.13). The following lemma is valid.

Lemma 21.1 *Let the conditions (21.8)–(21.11) be fulfilled. Then the solvability of the inverse problem (21.1)–(21.5) is equivalent to the nonlocal direct problem (21.14), (21.3)–(21.4), i.e. if $(\mathbf{v}, p, f)$ is a solution of the inverse problem (21.1)–(21.5), then the pair $(\mathbf{v}, p)$ is the solution of the nonlocal direct problem (21.14), (21.3)–(21.4), and vice versa, if the pair $(\mathbf{v}, p)$ is a solution of the nonlocal direct problem (21.14), (21.3)–(21.4), then this pair $(\mathbf{v}, p)$ together with the function $f(t)$ defined by* (21.13) *gives a solution of the inverse problem (21.1)–(21.5).*

Proof We omit the proof because it is very similar to the analogical lemmas in [4]. □

21.3 Existence of a Weak Solution to Inverse Problem (21.1)–(21.5)

By Lemma 21.1, in order to prove the existence of solutions of inverse problem (21.1)–(21.5) it is sufficient to prove the existence of solutions of the nonlocal direct problem (21.14) and (21.3)–(21.4).

Theorem 21.1 *Let the conditions (21.7)–(21.12) be fulfilled. Assume also that there exists a positive constant m such that*

$$\frac{\kappa}{k_0^2} \sup_{t\in[0,T]} \|\mathbf{g}(t)\|_{2,\Omega}^2 \|\boldsymbol{\omega}\|_{\mathbf{H}^1(\Omega)}^2 \leq m < 2. \tag{21.15}$$

Then the nonlocal direct problem (21.14), (21.3)–(21.4) has a global (in time) unique weak solution $\mathbf{v}(\mathbf{x}, t)$ *in* Q_T. *Moreover, the weak solution satisfies the estimate*

$$\|\mathbf{v}(t)\|_{L^\infty(0,T;\mathbf{H}(\Omega)\cap\mathbf{H}^1(\Omega))} + \|\mathbf{v}_t\|_{L^2(0,T;\mathbf{H}(\Omega)\cap\mathbf{H}^1(\Omega))} \leq C < \infty, \tag{21.16}$$

Therefore, by Lemma 21.1, the corresponding inverse problem (21.1)–(21.5) has a global unique weak solution $(\mathbf{v}(\mathbf{x}, t), f(t))$ *and in addition to* (21.16) *holds*

$$\|f(t)\|_{L^2([0,T])} \leq C < \infty, \tag{21.17}$$

where C is a constant depending on data of the problem.

Proof Let $\{\varphi_k\}_{k\in\mathbb{N}}$ be an orthonormal family in $\mathbf{L}^2(\Omega)$ formed by functions of $\mathbf{H}$ whose linear combinations are dense in $\mathbf{H}^1(\Omega)$. Given $n \in \mathbb{N}$, let us consider the n-dimensional space $\mathbf{X}^n$ spanned by φ_k, $k = 1, \ldots, n$, respectively. For each $n \in \mathbb{N}$, we search for approximate solutions to the problem (21.1)–(21.4) and (21.13) in the form

$$\mathbf{v}^n(x,t) = \sum_{j=1}^{n} c_j^n(t)\varphi_j(x), \quad \varphi_j \in \mathbf{X}^n,$$

where unknown coefficient $c_j^n(t)$, $j = 1, \ldots, n$ are defined as solutions to the following Cauchy problem for the system of ordinary integrodifferential equations

$$\frac{d}{dt}\left(\left(\mathbf{v}^n, \varphi_k\right)_{2,\Omega} + \kappa\left(\nabla\mathbf{v}^n, \nabla\varphi_k\right)_{2,\Omega}\right) + \nu\left(\nabla\mathbf{v}^n, \nabla\varphi_k\right)_{2,\Omega} =$$
$$-\int_0^t K(t-s)\left(\nabla\mathbf{v}^n, \nabla\varphi_k\right)_{2,\Omega} ds + F^n(\mathbf{v}^n, t)\left(\mathbf{g}, \varphi_k\right)_{2,\Omega}, \; k = 1, 2, \ldots, n, \tag{21.18}$$

supplemented with the initial condition

$$\mathbf{v}^n(0) = \mathbf{v}_0^n \quad \text{in } \Omega, \tag{21.19}$$

where $F^n(t, \mathbf{v}^n) = F(t, \mathbf{v}^n)$ and $\mathbf{v}_0^n = \sum_{j=1}^n \left(\mathbf{v}_0, \boldsymbol{\varphi}_j\right)_{2,\Omega} \boldsymbol{\varphi}_j$ is a sequence in $\mathbf{L}^2(\Omega) \cap \mathbf{H}^1(\Omega)$ such that

$$\mathbf{v}_0^n \to \mathbf{v}_0(x) \text{ strong as } n \to \infty \text{ in } \mathbf{L}^2(\Omega) \cap \mathbf{H}^1(\Omega).$$

According to a general theory of ordinary differential equations, the system (21.18)–(21.19) has a solution $c_j^n(t)$ in $[0, t_0]$. By a priori estimates, which we shall establish below, the solution can be extended to $[0, T_0] \subset [0, T]$, where $[0, T_0]$ is a maximal time interval, such that the priori estimates are true.

Let us multiply k-th equation of (21.18) by $c_k^n(t)$ and $\frac{d c_k^n}{dt}$ and sum up the results from $k = 1$ until $k = n$, and integrate over Ω. Using the formula of integrating by parts, and adding results, we obtain

$$\begin{aligned} &\frac{1}{2}\frac{d}{dt}\left(\left\|\mathbf{v}^n\right\|_{2,\Omega}^2 + (\kappa+\nu)\left\|\mathbf{v}^n\right\|_{\mathbf{H}^1(\Omega)}^2\right) + \nu\left\|\mathbf{v}^n\right\|_{\mathbf{H}^1(\Omega)}^2 \\ &\qquad + \left\|\mathbf{v}_t^n(t)\right\|_{2,\Omega}^2 + \kappa\left\|\mathbf{v}_t^n(t)\right\|_{\mathbf{H}^1(\Omega)}^2 = J_1 + J_2, \end{aligned} \tag{21.20}$$

where

$$J_1 := F(t, \mathbf{v}^n)\left(\mathbf{g}, \mathbf{v}^n\right)_{2,\Omega} - \int_0^t K(t-\tau)\left(\nabla \mathbf{v}^n(t), \nabla \mathbf{v}^n(\tau)\right)_{2,\Omega} d\tau,$$

$$J_2 := F(t, \mathbf{v}^n)\left(\mathbf{g}, \mathbf{v}_t^n\right)_{2,\Omega} - \int_0^t K(t-\tau)\left(\nabla \mathbf{v}_t^n(t), \nabla \mathbf{v}^n(\tau)\right)_{2,\Omega} d\tau.$$

First, using Hölder and Young inequalities, we estimate $F^n(t, \mathbf{v}^n)$ as follows

$$\begin{aligned} \left|F^n(t, \mathbf{v}^n)\right| \le \frac{1}{k_0}\Bigg[&|e'(t)| + \kappa\left\|\mathbf{v}_t^n\right\|_{\mathbf{H}^1(\Omega)}\|\boldsymbol{\omega}\|_{\mathbf{H}^1(\Omega)} \\ &+\nu\left\|\mathbf{v}^n\right\|_{\mathbf{H}^1(\Omega)}\|\boldsymbol{\omega}\|_{\mathbf{H}^1(\Omega)} + \|\boldsymbol{\omega}\|_{\mathbf{H}^1(\Omega)} K_0\left(\int_0^t \left\|\mathbf{v}^n(\tau)\right\|_{\mathbf{H}^1(\Omega)}^2 d\tau\right)^{\frac{1}{2}}\Bigg]. \end{aligned} \tag{21.21}$$

Now, we use Hölder and Young inequalities and (21.21), to get estimates for J_1 and J_2:

$$\begin{aligned} |J_1| \le{}& \frac{3\varepsilon_1+\varepsilon_3}{2}\left\|\mathbf{v}^n\right\|_{2,\Omega}^2 + \frac{\|\mathbf{g}\|_{2,\Omega}^2}{2\varepsilon_1 k_0^2}|e'|^2 + \frac{\nu}{2}\|\mathbf{v}^n\|_{\mathbf{H}^1(\Omega)}^2 + \frac{K_0^2}{2\nu}\int_0^t \|\mathbf{v}^n(\tau)\|_{\mathbf{H}^1(\Omega)}^2 d\tau \\ &+ \frac{1}{2k_0^2}\|\mathbf{g}\|_{2,\Omega}^2\|\boldsymbol{\omega}\|_{\mathbf{H}^1(\Omega)}^2 \end{aligned}$$

$$\times\left(\frac{\kappa^2}{\varepsilon_3}\left\|\mathbf{v}_t^n\right\|^2_{\mathbf{H}^1(\Omega)}+\frac{\nu^2}{\varepsilon_1}\left\|\mathbf{v}^n\right\|^2_{\mathbf{H}^1(\Omega)}+\frac{K_0^2}{\varepsilon_1}\int_0^t\left\|\mathbf{v}^n(\tau)\right\|^2_{\mathbf{H}^1(\Omega)}d\tau\right),$$

$$|J_2|\le\frac{3\varepsilon_2+\varepsilon_3}{2}\left\|\mathbf{v}_t^n\right\|^2_{2,\Omega}+\frac{\|\mathbf{g}\|^2_{2,\Omega}}{2\varepsilon_2k_0^2}|e'|^2+\frac{\varepsilon_0}{2}\|\mathbf{v}_t^n\|^2_{\mathbf{H}^1(\Omega)}+\frac{K_0^2}{2\varepsilon_0}\int_0^t\|\mathbf{v}^n(\tau)\|^2_{\mathbf{H}^1(\Omega)}d\tau$$

$$+\frac{1}{2k_0^2}\|\mathbf{g}\|^2_{2,\Omega}\|\boldsymbol{\omega}\|^2_{\mathbf{V}^1(\Omega)}$$

$$\times\left[\frac{\kappa^2}{\varepsilon_3}\left\|\mathbf{v}_t^n\right\|^2_{\mathbf{H}^1(\Omega)}+\frac{\nu^2}{\varepsilon_2}\left\|\mathbf{v}^n\right\|^2_{\mathbf{H}^1(\Omega)}+\frac{K_0^2}{\varepsilon_2}\int_0^t\left\|\mathbf{v}^n(\tau)\right\|^2_{\mathbf{H}^1(\Omega)}d\tau\right].$$

Plugging these inequalities into (21.20) and adding the results, we get

$$\frac{d}{dt}\left(\left\|\mathbf{v}^n\right\|^2_{2,\Omega}+(\kappa+\nu)\left\|\mathbf{v}^n\right\|^2_{\mathbf{H}^1(\Omega)}\right)+\nu\left\|\mathbf{v}^n\right\|^2_{\mathbf{H}^1(\Omega)}+\alpha\left\|\mathbf{v}_t^n\right\|^2_{2,\Omega}+\beta\left\|\mathbf{v}_t^n\right\|^2_{\mathbf{H}^1(\Omega)}$$

$$\le C_1\left\|\mathbf{v}^n\right\|^2_{2,\Omega}+C_2\left\|\mathbf{v}^n\right\|^2_{\mathbf{H}^1(\Omega)}+C_3\int_0^t\|\mathbf{v}^n(\tau)\|^2_{\mathbf{H}^1(\Omega)}d\tau+C_4(t),\qquad(21.22)$$

where $C_i,\ i=1,2,3$ are positive constants depending on ε_i and data of the problem, and $C_4(t)\in L^1([0,T])$ is a given function, and

$$\alpha:=2\left(1-\frac{3\varepsilon_2}{2}-\frac{\varepsilon_3}{2}\right);\ \beta:=2\left(\kappa-\frac{\kappa^2}{\varepsilon_3k_0^2}\sup_{t\in[0,T]}\|\mathbf{g}(t)\|^2_{2,\Omega}\|\boldsymbol{\omega}\|^2_{\mathbf{H}^1(\Omega)}-\frac{\varepsilon_0}{2}\right).$$

Now, we choose $\varepsilon_i,\ i=0,1,2,3$ such that $\alpha,\beta>0$, and C_i to be finite. It is possible because (21.15), but $\varepsilon_0,\varepsilon_3$ cannot be chosen such that $m>2$, because $\varepsilon_3<2$ due to $\alpha>0$. Thus, choosing $\varepsilon_i,\ i=0,1,2,3$ with suitable values, and integrating (21.22) by s from 0 to t, we obtain the following integral inequality

$$\left\|\mathbf{v}^n\right\|^2_{2,\Omega}+\left\|\mathbf{v}^n\right\|^2_{\mathbf{H}^1(\Omega)}+\int_0^t\left(\left\|\mathbf{v}^n\right\|^2_{\mathbf{H}^1(\Omega)}+\left\|\mathbf{v}_t^n(t)\right\|^2_{\mathbf{H}^1(\Omega)}\right)ds$$

$$\le C_5\int_0^t\left(\left\|\mathbf{v}^n\right\|^2_{2,\Omega}+\left\|\mathbf{v}^n\right\|^2_{\mathbf{H}^1(\Omega)}\right)ds+C_6,\qquad(21.23)$$

where C_5 and C_6 positive constants independent of n. It follows from (21.23) that

$$\|\mathbf{v}^n\|^2_{L^\infty(0,T;\mathbf{H}^1(\Omega))} + \|\mathbf{v}^n\|^2_{L^2(0,T;\mathbf{H}^1(\Omega))} + \|\mathbf{v}^n_t\|^2_{L^2(0,T;\mathbf{H}^1(\Omega))} \le C.$$

The a priori estimate allows us to pass the limit in (21.18), and obtain a solution as a limit. The uniqueness of the solution follows from the estimates obtained above by the linearity of the system. □

21.4 Existence and Uniqueness of Strong Solution to (21.14) and (21.3)–(21.4)

In this section, we establish the global existence and uniqueness of strong solutions of (21.1)–(21.5).

Theorem 21.2 *Assume that all the conditions of Theorem 21.1 be fulfilled and*

$$\mathbf{v}_0(\mathbf{x}) \in \mathbf{H}^1(\Omega) \cap \mathbf{H}^2(\Omega)$$

holds. Then the problem (21.14), (21.3)–(21.4) has a global unique strong solution $\mathbf{v}(\mathbf{x}, t)$ *in* Q_T*, and in addition to* (21.16) *the following estimate holds true*

$$\|\mathbf{v}\|^2_{L^\infty(0,T;\mathbf{H}^1\cap\mathbf{H}^2(\Omega))} + \|\mathbf{v}_t\|^2_{L^2(0,T;\mathbf{H}^1\cap\mathbf{H}^2(\Omega))} \le C < \infty. \tag{21.24}$$

where C *is a positive constant depending on the data of the problem. Therefore, by Lemma 21.1, the corresponding inverse problem (21.1)–(21.5) has a global unique strong solution* $(\mathbf{v}(\mathbf{x}, t), f(t))$*.*

Proof We prove the existence of strong solutions to these problems by using the special basis, associated with the eigenfunctions of the Stokes operator [14]:

$$\mathbb{A}\varphi_k := -\mathbb{P}\Delta\varphi_k = \lambda_k\varphi_k, \quad \varphi_k(x) \in \mathbf{H}^1(\Omega) \cap \mathbf{H}^2(\Omega), \tag{21.25}$$

where $\mathbb{P} : \mathbf{L}^2(\Omega) \to \mathbf{H}(\Omega)$ is the Leray projector, and it is known [14] that the system $\{\varphi_k\}_{k\in\infty}$ of eigenfunctions of both spectral problem (21.25) is orthogonal in $\mathbf{H}$ and an orthonormal basis in the space $\mathbf{H}^1(\Omega) \cap \mathbf{W}^{2,2}(\Omega)$.

For strong solutions, the estimate (21.16) is still true. Thus, to complete the proof, it is sufficient to get estimates for $\Delta\mathbf{v}^n$ and $\Delta\mathbf{v}^n_t$. Let us rewrite (21.18) in the form

$$\frac{d}{dt}\left((\mathbf{v}^n, \varphi_k)_{2,\Omega} - \kappa\left(\Delta\mathbf{v}^n, \varphi_k\right)_{2,\Omega}\right) - \nu\left(\Delta\mathbf{v}^n, \varphi_k\right)_{2,\Omega}$$
$$= -\int_0^t K(t-s)\left(\Delta\mathbf{v}^n, \varphi_k\right)_{2,\Omega} ds + F^n(t, \mathbf{v}^n)\left(\mathbf{g}(\mathbf{x}, t), \varphi_k\right)_{2,\Omega}. \tag{21.26}$$

Now, multiply the k-th equation of (21.26) by $-\lambda_k \frac{dc_k^n(t)}{dt}$, and sum up with respect to k from 1 to n. Taking in account the Stokes operator $\mathbb{A}$, we have

$$\frac{\nu}{2}\frac{d}{dt}\left\|\mathbb{A}\mathbf{v}^n\right\|_{2,\Omega}^2 + \kappa\left\|\mathbb{A}\mathbf{v}_t^n\right\|_{2,\Omega}^2 = \left(\mathbf{v}_t^n, \mathbb{A}\mathbf{v}_t^n\right)_{2,\Omega}$$
$$-\int_0^t K(t-s)\left(\mathbb{A}\mathbf{v}^n(s), \mathbb{A}\mathbf{v}_t^n(t)\right)_{2,\Omega} ds + F^n(t,\mathbf{v}^n)\left(\mathbf{g}(\mathbf{x},t), \mathbb{A}\mathbf{v}_t^n\right)_{2,\Omega} \equiv I, \tag{21.27}$$

where $F^n(t, \mathbf{v}^n)$ is estimated by (21.21). Using the Hölder and Young inequalities and (21.21) together with (21.16), we estimate the right hand side of (21.27)

$$|I| \leq \frac{\kappa}{2}\left\|\mathbb{A}\mathbf{v}_t^n\right\|_{2,\Omega}^2 + \frac{2}{3\nu}\left(\left\|\mathbf{v}_t^n\right\|_{2,\Omega}^2 + \left|F^n\right|^2 \|\mathbf{g}(t)\|_{2,\Omega}^2 + K_0^2\int_0^t \left\|\mathbb{A}\mathbf{v}^n(s)\right\|_{2,\Omega}^2 ds\right).$$

Plugging this into (21.27) and integrating by s from 0 to $t \in [0, T]$, we have

$$\nu\left\|\mathbb{A}\mathbf{v}^n\right\|_{2,\Omega}^2 + \kappa\int_0^t \left\|\mathbb{A}\mathbf{v}_t^n\right\|_{2,\Omega}^2 ds \leq C_7 + \frac{4K_0^2 T}{3\nu}\int_0^t \left\|\mathbb{A}\mathbf{v}^n(s)\right\|_{2,\Omega}^2 ds, \tag{21.28}$$

where C_7 is positive constant independent of n. It follows from (21.28) that

$$\left\|\mathbf{v}^n\right\|_{L^\infty(0,T;\mathbf{H}^2(\Omega))}^2 + \left\|\mathbf{v}_t^n\right\|_{L^2(0,T;\mathbf{H}^2(\Omega))}^2 \leq C < \infty.$$

□

21.5 Some Special Cases of (21.1)–(21.5) Removing Restriction (21.15)

In turn, the condition (21.15) states that the problem (21.1)–(21.5) is locally solvable with respect to data of the problem. In the purpose of removing this restriction, in this section, we consider the inverse problem for the Eq. (21.1) with the special right-hand side $g(x,t) := \omega(x)$

$$\mathbf{v}_t - \kappa\Delta\mathbf{v}_t - \nu\Delta\mathbf{v} + \nabla\pi - \int\limits_0^t K(t-s)\Delta\mathbf{v}(s)ds = f(t)\boldsymbol{\omega}(\mathbf{x}), \quad (x,t) \in Q_T, \tag{21.29}$$

that supplemented with (21.2)–(21.5), where $\boldsymbol{\omega}(\mathbf{x})$ is the testing function in (21.5). For this inverse problem, i.e. for (21.29), (21.2)–(21.5), the following assertion is hold.

Theorem 21.3 *Assume that the conditions (21.7), (21.10), (21.11), and (21.12) are fulfilled. Then the inverse problem* (21.29)*, (21.2)–(21.5) has a global unique weak solution* $(\mathbf{v}(\mathbf{x},t), f(t))$*, and the estimates (21.16)–(21.17) hold true. If instead of (21.7) holds* $\mathbf{v}_0(x) \in \mathbf{H}^1(\Omega) \cap \mathbf{H}^2(\Omega)$*, then it has a global unique strong solution* $(\mathbf{v}(\mathbf{x},t), f(t))$*, and the estimates (21.16)–(21.17) and* (21.24) *are satisfied.*

Remark 21.1 All above results remain true without the assumption (21.15) if we consider the inverse problems (21.2)–(21.5) and (21.29), (21.2)–(21.5) with the overdetermination condition

$$\int_\Omega (\mathbf{v}\boldsymbol{\omega} + \kappa\nabla\mathbf{v}\nabla\boldsymbol{\omega})d\mathbf{x} = e(t), \quad t \in [0,T]$$

instead of (21.5). Moreover, similar results can also be established, but locally in time, for weak and strong solutions of inverse problems for nonlinear integrodifferential Eq. (21.1) with the convective term $(\mathbf{v}\cdot\nabla)\mathbf{v}$.

Acknowledgments This research work has been funded by Grant number AP09057950 the Ministry of Education and Science of the Republic of Kazakhstan (MES RK), Kazakhstan.

References

1. Abylkairov, U.U., Aitzhanov, S.E.: The inverse problem for a nonlinear system of Navier-Stokes equations with integral overdetermination. J. Math. Mech. Comput. Sci. **72**(1), 7–13 (2012) (in Kazakh)
2. Abylkairov, U.U., Aitzhanov, S.E.: Inverse problem for non-stationary system of magnetohydrodynamics. Boundary Value Probl. **1**, 1–17 (2015)
3. Abylkairov, U.U., Khompysh, K.: An inverse problem of identifying the coefficient in Kelvin-Voigt equations. Appl. Math. Sci. **9**(101–104), 5079–5088 (2015)
4. Antontsev, A.N., Khompysh, K.: An inverse problem for generalized Kelvin-Voigt equation with p-Laplacian and damping term. Inverse Probl. **37**, 085012 (2021)
5. Shebotarev, A.Y.: Determination of the right-hand side of the Navier-Stokes system of equations and inverse problems for the thermal convection equations. Comput. Math. Math. Phys. **51**(12), 2146–2154 (2011). https://doi.org/10.1134/s0965542511120098

6. Fan, J., Cristo, M.D., Jiang, Y., Nakamura, G.: Inverse viscosity problem for the Navier-Stokes equation. J. Math. Anal. Appl. **365**(2), 750–757 (2010)
7. Fedorov, V.E., Ivanova, N.D.: Inverse problem for Oskolkov's system of equations. Math. Methods Appl. Sci. **40**(17), 6123–6126 (2017)
8. Imanuvilov, O.Y., Yamamoto, M.: Global uniqueness in inverse boundary value problems for the Navier-Stokes equation and Lame system in two dimensions. Inverse Probl. **31**, 035004 (2015)
9. Jiang, Y., Fan, J., Nagayasu, S., Nakamura, G.: Local solvability of an inverse problem to the Navier-Stokes equation with memory term. Inverse Probl. **36**, 065007 (2020)
10. Joseph, D.D.: Fluid Dynamics of Viscoelastic Liquids. Springer-Verlag, New York (1990)
11. Karazeeva, N.A.: Solvability of initial boundary value problems for equations describing motions of linear viscoelastic fluids. J. Appl. Math. **1**, 59–80 (2005)
12. Khompysh, K., Kenzhebai, K.: An inverse problem for Kelvin-Voigt equation perturbed by isotropic diffusion and damping. Math. Methods Appl. Sci. **45**(7), 3817–3842 (2022)
13. Khompysh, K., Nugymanova, N.K.: Inverse problem for integrodifferential Kelvin-Voigt equation. J. Inverse Ill-Posed Probl. (2022). https://doi.org/10.1515/jiip-2020-0157
14. Ladyzhenskaya, O.A.: The Mathematical Theory of Viscous Incompressible Flow II. Nauka, Moscow (1970)
15. Li, X., Wang, J.N.: Determination of viscosity in the stationary Navier-Stokes equation. J. Differ. Equ. **242**, 24–39 (2007)
16. Lions, J.-L.: Quelques Methodes de Resolution des Problemes aux Limites non Liniaires. Dunod, Paris (1969)
17. Oskolkov, A.P.: Initial-boundary value problems for equations of motion of Kelvin–Voigt fluids and Oldroyd fluids. Proc. Steklov Inst. Math. **179**, 137–182 (1989)
18. Pavlovsky, V.A.: On the theoretical description of weak water solutions of polymers. Doklady Akademii Nauk SSSR **200**(4), 809–812 (1971)
19. Prilepko, A.I., Orlovsky, D.G., Vasin, I.A.: Method for Solving Inverse Problems in Mathematical Physics. Monograths and Textbooks in Pure and Applied Mathematics, vol 231, 709 p. Marcel Dekker, New York (2000)
20. Zvyagin, V.G., Turbin, M.V.: The study of initial-boundary value problems for mathematical models of the motion of Kelvin-Voigt fluids. J. Math. Sci. **168**, 157–308 (2010)

Chapter 22
A Nonlocal Initial Conditional Boundary Value Problem on Metric Graph

Jonibek R. Khujakulov

Abstract The main task of the present research is a nonlocal boundary value problem for a time-fractional differential equation involving the Hilfer fractional derivative on a metric star graph. The main goal of this work is to study the uniqueness and existence of a solution to the problem formulated. Using Fourier's method, we find a solution to the investigated problem in the form of the Fourier series. The uniqueness of the solution is proved by using a priori estimates of the solution.

22.1 Introduction

This research work deals with the problem of nonlocal initial conditions for a time-fractional differential equation (FDE) with Hilfer's operator on tree metric graphs. Notice that boundary value problems with nonlocal initial conditions were considered in [1] for reaction-diffusion equations, and in [2, 3] for degenerate parabolic equations. Ashurov and Fayziev investigated nonlocal problems in time for time-fractional subdiffusion equations in [4]. Moreover, a nonlocal initial problem for a second-order time-fractional and space-singular equation and a time-nonlocal boundary value problem for a time-fractional partial differential equation involving the bi-ordinal Hilfer fractional differential operators were respectively investigated by Karimov et al. [5], and Karimov and Toshtemirov [6].

J. R. Khujakulov (✉)
Termiz state university, Termez, Uzbekistan

M. Ruzhansky, K. Van Bockstal (eds.), *Extended Abstracts 2021/2022*,
Research Perspectives Ghent Analysis and PDE Center 2,
https://doi.org/10.1007/978-3-031-42539-4_22

22.2 Preliminaries

22.2.1 Fractional Derivatives and Integrals

Let us first present definitions of fractional-order integral and integrodifferential operators.

Definition 22.1 ([7]) The Riemann-Liouville (R-L) fractional integrals with order α of a function $f(t)$ is defined by

$$\left(I_{at}^{\alpha} f\right)(t) = \left(D_{at}^{-\alpha} f\right)(t) = \frac{1}{\Gamma(\alpha)} \int_a^t \frac{f(\tau)d\tau}{(t-\tau)^{1-\alpha}}, \quad (x > a, \quad \alpha > 0).$$

Definition 22.2 ([7]) The Caputo fractional derivative of order α is defined as follows:

$$\left({}_C D_{at}^{\alpha} f\right)(t) = \left(I_{at}^{n-\alpha} D^n f\right)(t) + \frac{1}{\Gamma(n-\alpha)} \int_a^t \frac{f^{(n)}(\tau)}{(t-\tau)^{\alpha-n+1}} d\tau,$$

where $D = d \setminus dt, n \in \mathbb{N}, t > a$ and $n = [\alpha] + 1$.

Definition 22.3 ([8]) Hilfer fractional derivative $D_{0t}^{\alpha,\mu}$ of order α and type μ with respect to t is defined by

$$\left(D_{0t}^{\alpha,\mu} u\right)(t) = I_{0t}^{\mu(n-\alpha)} D^n \left(I_{0t}^{(1-\mu)(n-\alpha)} u\right)(t), \quad n-1 < \alpha < n, 0 \le \mu \le 1, n \in \mathbb{N}.$$

whenever the right-hand side exists.

Definition 22.4 ([7]) A two-parameter Mittag-Leffler function is represented in the form

$$E_{\alpha,\beta}(z) = \sum_{n=0}^{\infty} \frac{z^n}{\Gamma(\alpha n + \beta)}, \quad \alpha > 0, \quad \beta, z \in \mathbb{C}.$$

Lemma 22.1 ([7]) *If* $\alpha < 2$, β *is an arbitrary real number,* μ *is such that* $\pi\alpha/2 < \mu < min\{\pi, \pi\alpha\}$ *and* C_1 *is a real constant, then*

$$\left|E_{\alpha,\beta}(z)\right| \le \frac{C_1}{1+|z|}, \qquad (\mu \le |arg(z)| \le \pi), |z| \ge 0.$$

22.2.2 Metric Graphs

Let a graph $\Gamma = V \bigcup E$ be a connected metric graph [9], where $E = \{B_k\}_{k=1}^{j}$ is the set of its edges, and $V = \{\nu_k\}_{k=1}^{n} \bigcup \{\gamma_k\}_{k=1}^{m_1}$ is the set of vertices. Let us determine the coordinates x_k on the edges of the graph using isometric mapping of these edges onto the intervals $(0, L_k)$, $k = 1, 2, ..., j$.

We will say that a vertex ν is in contact with an edge B_k if it is the end of this edge, and denote this as $B_k \sim \nu$. The number of elements of the set $\{b : b \sim \nu, b \in E\}$ is called vertex valency ν. If the valency of a vertex is equal to one, then it is called boundary. Let $\{\gamma_1, \gamma_2, \dots, \gamma_{m_1}\} = \partial\Gamma \subset V$ be the boundary vertices of the graph. In addition, without loss of generality, we will use x instead of x_k.

Definition 22.5 (Definition 1.3.6. in [9]) The space $L_2(\Gamma)$ on Γ consists of functions that are measurable and square-integrable on each edge B_k, $k = \overline{1, j}$ with the scalar product and the norm:

$$(f(x), g(x))_{L_2(\Gamma)} = \sum_k \int_0^{L_k} f^{(k)}(x) g^{(k)}(x) dx, \ \ \|f(x)\|^2_{L_2(\Gamma)} = \sum_k \|f(x)\|^2_{L_2(B_k)}.$$

In other words, $L_2(\Gamma)$ is the orthogonal direct sum of spaces $L_2(B_k)$. Further, for simplicity, next we will use $\int_\Gamma f(x)dx$ instead of $\sum\limits_k \int_0^{L_k} f^{(k)}(x)dx$.

22.3 Problem Formulation

On the each edges of the over-defined graph Γ, we consider fractional differential equations

$$D_{0t}^{\alpha,\mu} u^{(k)}(x,t) - u_{xx}^{(k)}(x,t) = f^{(k)}(x,t), \ \ (x,t) \in (B_k \times (0,T)), k = \overline{1, j}, \tag{22.1}$$

where $D_{0t}^{\alpha,\mu}$ is the Hilfer's operator, $l-1 < \alpha < l$, $l = \{1, 2\}$, $0 \le \mu \le 1$, and $f^{(k)}(x,t)$ are given functions.

We will study the following problem for Eq. (22.1) on Γ.

Problem 22.1 To find functions $u^{(k)}(x,t)$ in the domain $B_k \times (0,T)$ satisfying Eq. (22.1) with the following properties:

$$t^{l-\gamma} u(x,t) \in C(\Gamma \times [0,T]), t^{l-\gamma} u_x^{(k)}(x,t) \in C([0, L_k) \times [0,T]),$$

$$u_{xx}^{(k)}(x,t), D_{0t}^{\alpha,\mu} u^{(k)}(x,t) \in C((0, L_k) \times (0,T)), \gamma = \alpha + l\mu - \alpha\mu,$$

$$t^{2-\gamma} \frac{d}{dt} I_{0t}^{2-\gamma} u^{(k)}(x,t) \in C(\Gamma \times [0,T]), k = \overline{1, j}, \gamma = \alpha + 2\mu - \alpha\mu,$$

nonlocal and initial conditions:

$$
\begin{aligned}
&I_{0t}^{l-\gamma} u^{(k)}(x,t)\Big|_{t=0} = M I_{0t}^{l-\gamma} u^{(k)}(x,t)\Big|_{t=T}, \ \gamma = \alpha + l\mu - \alpha\mu \ \ x \in \overline{B_k}, \\
&\frac{d}{dt} I_{0t}^{l-\gamma} u^{(k)}(x,t)\Big|_{t=0} = \varphi^{(k)}(x), \ l = 2, \quad x \in B_k, k = \overline{1, j}.
\end{aligned}
\tag{22.2}
$$

At branching points (i.e., at internal vertices) of the graph, the solution must satisfy the following δ *type conditions* which are at a vertex ν as follows:

$$
\begin{cases}
u^{(k)}(x,t) \text{ are } \quad \text{continuous} \quad \text{at } \nu, \\
\sum\limits_{B_k \sim \nu} u_x^{(k)}(x,t)\Big|_{\nu} = 0,
\end{cases}
\tag{22.3}
$$

and boundary conditions:

$$
u^{(k)}(x,t)|_{\gamma_k} = 0, \qquad \gamma_k \in \partial\Gamma, t \in (0, T], \ k = \overline{1, m_1}. \tag{22.4}
$$

The functions $\varphi^{(k)}(x)$ are sufficiently smooth and satisfy conditions (22.3)–(22.4).

22.4 Uniqueness of Solution

We prove the following theorem to show that the solution of Eq. (22.1) satisfying the conditions (22.2)–(22.4) is unique.

Theorem 22.1 *If* $M^2 < \frac{1}{l-1+T^{\beta-1}E_{\alpha-l+1,\beta-l+1}(T^\alpha)}$ *for* $l-1 < \alpha < l$, $l = \{1, 2\}$, *the solution of problem (22.1)–(22.4) will be unique, where*

$$
\beta = \begin{cases} 1, & \text{if } l = 1, \\ \alpha, & \text{if } l = 2. \end{cases}
$$

Proof We introduce the following notation

$$
I_{0t}^{l-\gamma} u^{(k)}(x,t) = v^{(k)}(x,t), \quad k = \overline{1, j}. \tag{22.5}
$$

Then, Eq. (22.1) can be rewritten as follows

$$
{}_C D_{0t}^{\alpha} v^{(k)}(x,t) - v_{xx}^{(k)}(x,t) = f_l^{(k)}(x,t), \tag{22.6}
$$

where $f_l^{(k)}(x,t) = I_{0t}^{l-\gamma} f^{(k)}(x,t)$. Considering (22.2)–(22.4), and from (22.5) we deduce

$$\begin{aligned} v^{(k)}(x,t)|_{t=0} &= M v^{(k)}(x,t)|_{t=T}, \\ v_t^{(k)}(x,t)|_{t=0} &= \varphi^{(k)}(x), \ l=2, \ k=\overline{1,j}, \ x \in B_k \end{aligned} \tag{22.7}$$

and δ *type conditions* which are at a vertex v as follows:

$$\begin{cases} v^{(k)}(x,t) \text{ are continuous at } v, \\ \sum\limits_{B_k \sim v} v_x^{(k)}(x,t)\Big|_v = 0, \end{cases} \tag{22.8}$$

and boundary conditions

$$v^{(k)}(x,t)|_{\gamma_k} = 0, \qquad \gamma_k \in \partial\Gamma, \ t \in [0,T], \ k = \overline{1,m_1}. \tag{22.9}$$

First, we consider the case $0 < \alpha < 1$: Let us obtain an a priori estimate for the solution to problem (22.6)–(22.9). To this end, we take the inner products of both sides of Eq. (22.6) with the function $v^{(k)}(x,t)$, and obtain

$$\begin{aligned} \left({}_C D_{0t}^{\alpha} v(x,t), v(x,t)\right)_{L_2(\Gamma)} &- (v_{xx}(x,t), v(x,t))_{L_2(\Gamma)} \\ &= (f_1(x,t), v(x,t))_{L_2(\Gamma)}, \end{aligned} \tag{22.10}$$

where $f_1^{(k)}(x,t) = I_{0t}^{1-\gamma} f^{(k)}(x,t)$. We will use the notation $\|f\| = \|f\|_{L_2(\Gamma)}$ for convenience. Based on the inequality $v\,{}_C D_{ot}^{\alpha} v \geq \frac{1}{2}{}_C D_{ot}^{\alpha} v^2$ [10], we have

$$\int_0^{L_k} v^{(k)}(x,t)\,{}_C D_{0t}^{\alpha} v^{(k)}(x,t)dx \geq \frac{1}{2}\int_0^{L_k} {}_C D_{0t}^{\alpha} \left(v^{(k)}(x,t)\right)^2 dx. \tag{22.11}$$

Therefore,

$$\left({}_C D_{0t}^{\alpha} v(x,t), v(x,t)\right)_{L_2(\Gamma)} \geq \frac{1}{2}{}_C D_{0t}^{\alpha} \|v(x,t)\|^2. \tag{22.12}$$

After integrating by parts and using conditions (22.8)–(22.9) and from Definition 22.5, we get that

$$\begin{aligned} (v_{xx}(x,t), v(x,t))_{L_2(\Gamma)} &= \int_\Gamma v(x,t) v_{xx}(x,t)dx \\ &= -\int_\Gamma v_x^2(x,t)dx = -\|v_x(\cdot,t)\|^2. \end{aligned} \tag{22.13}$$

Now, we estimate

$$(f_1(x,t), v(x,t))_{L_2(\Gamma)} \leq 1/2 \left(\|v(x,t)\|^2 + \|f_1(x,t)\|^2 \right). \tag{22.14}$$

Taking (22.10) into account, from (22.12)–(22.14), we get that

$$_C D_{0t}^{\alpha} \|v(\cdot,t)\|^2 + 2\|v_x(\cdot,t)\|^2 \leq \left(\|v(\cdot,t)\|^2 + \|f_1(\cdot,t)\|^2 \right),$$

$$_C D_{0t}^{\alpha} \|v(\cdot,t)\|^2 \leq \|v(\cdot,t)\|^2 + \|f_1(\cdot,t)\|^2. \tag{22.15}$$

Based on the statement [10], and from (22.15) it follows that

$$\|v(\cdot,t)\|^2 \leq \|v(\cdot,0)\|^2 E_{\alpha,1}\left(t^{\alpha}\right) + \Gamma(\alpha) E_{\alpha,\alpha}\left(t^{\alpha}\right) I_{0t}^{\alpha} \|f_1(\cdot,t)\|^2, \tag{22.16}$$

$$\|v(\cdot,T)\|^2 \leq \|v(\cdot,0)\|^2 E_{\alpha,1}\left(T^{\alpha}\right) + N_1,$$

where $N_1 = \Gamma(\alpha) E_{\alpha,\alpha}\left(T^{\alpha}\right) I_{0t}^{\alpha} \|f_1(x,t)\|^2 |_{t=T}$. Employing the nonlocal condition (22.7), we have that

$$\|v(x,0)\|^2 \leq \frac{M^2 N_1}{1 - M^2 E_{\alpha,1}\left(T^{\alpha}\right)}. \tag{22.17}$$

Considering (22.16)–(22.17), we get that

$$\|v(x,t)\|^2 \leq \frac{\Gamma(\alpha) E_{\alpha,\alpha}\left(T^{\alpha}\right)}{1 - M^2 E_{\alpha,1}\left(T^{\alpha}\right)} I_{0t}^{\alpha} \|f_1(x,t)\|^2 |_{t=T}. \tag{22.18}$$

If $f^{(k)}(x,t) = 0$, then $v^{(k)}(x,t) \equiv 0$. Due to (22.5) we get $u^{(k)}(x,t) \equiv 0$. Therefore, we can say that the solution to problem (22.1)–(22.4) is unique if it exists for the case $l = 1$.

Now, we consider the case $1 < \alpha < 2$. We first derive an a priori estimate for the solution of the investigated problem. To this end, we take the inner products of both sides of Eq. (22.6) with the function $v_t^{(k)}(x,t)$, and obtain

$$\left(_C D_{0t}^{\alpha} v(x,t), v_t(x,t) \right)_{L_2(\Gamma)} - (v_{xx}(x,t), v_t(x,t))_{L_2(\Gamma)}$$
$$= (f_2(x,t), v_t(x,t))_{L_2(\Gamma)}. \tag{22.19}$$

where $f_2^{(k)}(x,t) = I_{0t}^{2-\gamma} f^{(k)}(x,t)$. Based on the inequality (22.11) and from the Definition 22.5, we have that

$$\left({}_C D_{0t}^{\alpha} v(x,t), v_t(x,t)\right)_{L_2(\Gamma)} = \int_\Gamma v_t(x,t) {}_C D_{0t}^{\alpha-1} v_t(x,t) dx \geq \frac{1}{2} {}_C D_{0t}^{\alpha-1} \|v_t(\cdot,t)\|^2,$$

$$-\left(v_{xx}(x,t), v_t(x,t)\right)_{L_2(\Gamma)} = -\int_\Gamma v_t(x,t) v_{xx}(x,t) dx = \frac{1}{2}\frac{\partial}{\partial t} \|v_x(x,t)\|^2.$$

Similar to (22.14), one can obtain that

$$\left(f_2(x,t), v_t(x,t)\right)_{L_2(\Gamma)} \leq \frac{1}{2}\|v_t(\cdot,t)\|^2 + \frac{1}{2}\|f_2(\cdot,t)\|^2.$$

Taking into account the performed transformations, from the identity (22.19), we deduce the inequality

$${}_C D_{0t}^{\alpha-1} \|v_t(\cdot,t)\|^2 + \frac{\partial}{\partial t}\|v_x(\cdot,t)\|^2 \leq \|v_t(\cdot,t)\|^2 + \|f_2(\cdot,t)\|^2.$$

By integrating this relation with respect to τ from 0 to t, we obtain the inequality

$$\begin{aligned} &I_{0t}^{2-\alpha}\|v_t(\cdot,t)\|^2 + \|v_x(\cdot,t)\|_0^2 \\ &\quad\leq \int_0^t \left(\|v_\tau(\cdot,\tau)\|^2 + \|f_2(\cdot,\tau)\|^2\right) d\tau + c_1\|\varphi\|^2 + \|v_x(\cdot,0)\|^2, \end{aligned} \tag{22.20}$$

where $c_1 = T^{2-\alpha}/\Gamma(3-\alpha)$. Based on (22.20), we deduce

$$\|v_x(\cdot,t)\|^2 \leq \int_0^t \left(\|v_\tau(\cdot,\tau)\|^2 + \|f(\cdot,\tau)\|^2\right) d\tau + c_1\|\varphi\|^2 + \|v_x(\cdot,0)\|^2, \tag{22.21}$$

and

$$\begin{aligned} I_{0t}^{2-\alpha}\|v_t(\cdot,t)\|^2 \leq \int_0^t \|v_\tau(\cdot,\tau)\|^2 d\tau + \int_0^t \|f_2(\cdot,\tau)\|^2 d\tau \\ + c_1\|\varphi\|^2 + \|v_x(\cdot,0)\|^2. \end{aligned}$$

Consequently, following [10], we arrive at

$$\int_0^t \|v_\tau(\cdot,\tau)\|^2 d\tau \le \Gamma(\alpha) E_{\alpha-1,\alpha-1}\left(T^\alpha\right) I_{0t}^\alpha \|f_2(\cdot,t)\|^2 \Big|_{t=T} + c_2 \|\varphi\|^2 + c_3 \|v_x(\cdot,0)\|^2, \quad (22.22)$$

where $c_2 = c_1 T^{\alpha-1} E_{\alpha-1,\alpha-1}(T^\alpha)$ and $c_3 = c_2/c_1$. If, in (22.21) we replace t with T, we have

$$\|v_x(\cdot,T)\|^2 \le N_2 + (c_3+1)\|v_x(\cdot,0)\|^2,$$

where

$$N_2 = \Gamma(\alpha) E_{\alpha-1,\alpha-1}\left(T^\alpha\right) I_{0t}^\alpha \|f_2(\cdot,t)\|^2 |_{t=T} + \int_0^T \|f_2(\cdot,\tau)\|^2 d\tau + (c_1+c_2)\|\varphi\|^2.$$

From the nonlocal condition (22.7), we get

$$\|v_x(\cdot,0)\|^2_{L_2(\Gamma)} \le \frac{M^2 N_2}{1-M^2(1+c_3)}.$$

We can write (22.22) in the following form:

$$\int_0^t \|v_\tau(\cdot,\tau)\|^2 d\tau \le \frac{1}{1-M^2(1+c_3)} \times \left[c_2\|\varphi\|^2 + c_3 M^2 \int_0^T \|f_2(\cdot,\tau)\|^2 d\tau + (1-M^2) I_{0t}^\alpha \|f_2(\cdot,t)\|^2 \Big|_{t=T}\right].$$

Using a standard scheme, one can prove that a solution to the problem (22.6)–(22.9) is unique. If $f^{(k)}(x,t)=0$ and $\varphi^{(k)}(x)\equiv 0$, then $v^{(k)}(x,t)\equiv 0$. Based on (22.5), we get $u^{(k)}(x,t)\equiv 0$. Therefore, we can say that the solution to problem (22.1)–(22.4) is unique, if it exists for the case $l=2$. □

22.5 Existence of Solution

We will use the method of separation of variables. We look for the solution of the homogeneous equation in the form:

$$u^{(k)}(x,t) = X^{(k)}(x) T(t),$$

and we will get

$$\frac{d^2}{dx^2}X^{(k)}(x)+\lambda^2X^{(k)}(x)=0,\ \ k=\overline{1,j}. \tag{22.23}$$

Moreover, from the conditions (22.3)–(22.4), we obtain

$$\begin{cases} X^{(k)}(x) \text{ are } \quad \text{continuous} \quad \text{at } \nu, \\ \sum\limits_{B_k\sim\nu}\frac{d}{dx}X^{(k)}(x)\Big|_{\nu}=0, \end{cases} \tag{22.24}$$

$$X^{(k)}(x)|_{\gamma_k}=0, \qquad \gamma_k\in\partial\Gamma,\ k=\overline{1,m_1}. \tag{22.25}$$

The general solution of Eq. (22.23) has the form:

$$X^{(k)}(x)=a_k\cos\lambda x+b_k\sin\lambda x;\ \ x\in B_k. \tag{22.26}$$

The spectral problem (22.23)–(22.25) in the case of general metric graphs is investigated in [9, 11, 12]. In this case, the graph is called "quantum" graph, and the operator $\frac{d^2}{dx^2}$, defined in each edge of the graph together with conditions (22.24)–(22.25), is called to be "edge-base" Laplacian [11].

Now, we formulate a theorem on the completeness of eigenfunctions of the "edge-based" graph Laplacian (or quantum graph) from [11].

Theorem 22.2 (See Proposition 3.2. in [11]) *Let Γ be a finite graph. Then, there exists eigenpairs (X_m,λ_m), $m=1,2,\dots$ for the edge based Laplacian, such that:*

(1) $0\le\lambda_1\le\lambda_2\le\cdots$,
(2) the X_m satisfy the Dirichlet condition,
(3) the X_m form a complete orthonormal basis for $L_2(\Gamma)$,
(4) $\lambda_m\to\infty$.

Theorem 22.3 *If $\varphi^{(k)}(x)\in C^1[0,L_k]$, $t^{l-\gamma}f^{(k)}(x,t)\in C(\Gamma\times[0,T])$, $t^{l-\gamma}\frac{\partial}{\partial x}f^{(k)}(x,t)\in C([0,L_k)\times[0,T])$ besides $\frac{d^2}{dx^2}\varphi^{(k)}(x)$ and $\frac{\partial^2}{\partial x^2}f^{(k)}(x,t)$ are absolute integrable functions in $(0,L_k)$ and $(B_k\times(0,T))$, respectively. Furthermore, conditions (22.3)–(22.4) are also valid for functions $f^{(k)}(x,t),\varphi^{(k)}(x)$, and $M\neq\frac{1}{E_{\alpha,1}(-\lambda_m^2T^\alpha)}$ holds, then the solution of the investigated problem exists.*

Proof Noting that $f^{(k)}(x,t)\in L_2(\Gamma)$ and expand $f(x,t)$ into the Fourier series in terms of eigenfunctions, i.e.

$$f(x,t)=\sum_{m=1}^{\infty}f_m(t)X_m(x), \tag{22.27}$$

where $f_m(t)$ is the coefficients of the Fourier series (22.27). Further, introducing a solution of Eq. (22.1) in the form

$$u(x,t)=\sum_{m=1}^{\infty}X_m(x)W_m(t). \tag{22.28}$$

Substituting (22.28) into Eq. (22.1), we obtain

$$\sum_{m=1}^{\infty}\left(D_{0t}^{\alpha,\mu}W_m(t)+\lambda_m^2W_m(t)-f_m(t)\right)X_m(x)=0.$$

Since the function $X_m(x)$ is an eigenfunction of the investigating problem, we obtain an inhomogeneous differential equation of fractional order

$$D_{0t}^{\alpha,\mu}W_m(t)+\lambda_m^2W_m(t)=f_m(t). \tag{22.29}$$

If, we consider the case $0<\alpha<1$, the general solution of the Eq. (22.29) has the form [13]:

$$W_m(t)=\int_0^t(t-\tau)^{\alpha-1}E_{\alpha,\alpha}\left[-\lambda_m^2(t-\tau)^{\alpha}\right]f_m(\tau)d\tau \\ +A_mt^{\gamma-1}E_{\alpha,\gamma}(-\lambda_m^2t^{\alpha}), \tag{22.30}$$

where $\gamma=\alpha+\mu-\alpha\mu$. Considering (22.28) and (22.30), we can write the general solution of Eq. (22.1) in the following form:

$$u^{(k)}(x,t)=\sum_{m=1}^{\infty}\left[\int_0^t(t-\tau)^{\alpha-1}E_{\alpha,\alpha}\left[-\lambda_m^2(t-z)^{\alpha}\right]f_m(\tau)d\tau \\ +A_mt^{\gamma-1}E_{\alpha,\gamma}(-\lambda_m^2t^{\alpha})\right]X_m^{(k)}(x).\ k=\overline{1,j}. \tag{22.31}$$

Applying the operator $I_{0t}^{1-\gamma}$ to the (22.31) and considering the Definition 22.1, we have

$$I_{0t}^{1-\gamma}u^{(k)}(x,t)=\sum_{m=1}^{\infty}A_mF_{1,m}(t)X_m^{(k)}(x)+\sum_{m=1}^{\infty}F_{2,m}(t)X_m^{(k)}(x).$$

$$F_{1,m}(t)=E_{\alpha,1}(-\lambda_m^2t^{\alpha}),$$

$$F_{2,m}(t)=\int_0^t(t-\tau)^{(\alpha-\gamma)}E_{\alpha,1+\alpha-\gamma}(-\lambda_m^2(t-\tau)^{\alpha})f_m(\tau)d\tau.$$

Based on the condition (22.2), we have

$$A_m = \frac{M}{1 - MF_{1,m}(T)} F_{2,m}(T), \quad M \neq \frac{1}{F_{1,m}(T)}$$

Finally, the solution to Eq. (22.1) satisfying the conditions (22.2)–(22.4) when $l = 1$, has the following form:

$$u^{(k)}(x,t) = \sum_{m=1}^{\infty} \Big[\frac{M}{1 - MF_{1,m}(T)} F_{2,m}(T) t^{\gamma-1} E_{\alpha,\gamma}(-\lambda_m^2 t^\alpha)$$
$$+ \int_0^t (t-\tau)^{\alpha-1} E_{\alpha,\alpha}\left(-\lambda_m^2 (t-\tau)^\alpha\right) f_m(\tau) d\tau \Big] X_m^{(k)}(x). \tag{22.32}$$

It is required to prove the uniform convergence of the infinite series, corresponding to the functions $u^{(k)}(x,t)$, $u_{xx}^{(k)}(x,t)$, $D_{0t}^{\alpha,\mu} u^{(k)}(x,t)$ in the domain $\Gamma \times [0, T]$. For further investigations, we use the following lemma:

Lemma 22.2 ([14]) *The following estimate holds true for (22.26):*

$$\left| X_m^{(k)}(x) \right| = \left| a_{m,k} \cos \lambda_m x + b_{m,k} \sin \lambda_m x \right| \leq \sqrt{2/L_k}.$$

From Lemma 22.1, one can get

$$|A_m| = \left| \frac{M}{1 - MF_{1,m}(T)} F_{2,m}(T) \right| \leq \left| \frac{M}{1 - MF_{1,m}(T)} \right| \left| F_{2,m}(T) \right| \leq M_1 / \lambda_m^2,$$

where $M_1 = const > 0$. From (22.27) and based on the conditions of the Theorem 22.2, we deduce that

$$|f_m(t)| = \left| \int_\Gamma f(x,t) X_m(x) dx \right| = \left| \frac{1}{\lambda_m^2} \int_\Gamma \frac{d^2}{dx^2} f(x,t) X_m(x) dx \right| \leq \frac{const}{\lambda_m^2}$$

and

$$D_{0t}^{\alpha,\mu} u^{(k)}(x,t) = \sum_{m=1}^{\infty} A_m \lambda_m^2 t^{\gamma-1} E_{\alpha,\gamma}\left(-\lambda_m^2 \tau^\alpha\right) X_m^{(k)}(x)$$
$$- \sum_{m=1}^{\infty} f_m(0) E_{\alpha,1}\left(-\lambda_m^2 \tau^\alpha\right) X_m^{(k)}(x)$$
$$- \sum_{m=1}^{\infty} X_m^{(k)}(x) \int_0^t E_{\alpha,1}(-\lambda_m^2 (t-\tau)^\alpha) f_m'(\tau) d\tau.$$

According to the asymptotes of $\lambda_m \sim c \cdot m$ (c is a constant number), see [12], we can conclude that the series of $D_{0t}^{\alpha,\mu} u^{(k)}(x, t)$ is uniformly convergent in the domain $\Gamma \times [0, T]$.

We proved the uniform convergence of above infinite series, corresponding to the solution and its appropriate derivatives, we can state that the solution (22.32) defines the regular solution of the problem.

The case $l = 2$ will be proved in a similar way. □

References

1. Pao, C.V.: Reaction diffusion equations with nonlocal boundary and nonlocal initial conditions. J. Math. Anal. Appl. **195**, 702–718 (1995)
2. Rassias, J.M., Karimov, E.T.: Boundary-value problems with nonlocal initial condition for parabolic equations with parameter. Eur. J. Pure Appl. Math. **6**, 948–957 (2010)
3. Rassias, J.M., Karimov, E.T.: Boundary-value problems with nonlocal initial condition for degenerate parabolic equations. Contemp. Anal. Appl. Math. **1**, 42–48 (2013)
4. Ashurov, R., Fayziev, Y.: On the nonlocal problems in time for time-fractional subdiffusion equations. Fractal Fractional **6**(1), 41 (2022)
5. Karimov, E., Mamchuev, M., Ruzhansky, M.: Nonlocal initial problem for second order time-fractional and space-singular equation. Hokkaido Math. J. **49**(2), 349-361 (2020)
6. Karimov, E., Toshtemirov, B.: On a time-nonlocal boundary value problem for time-fractional partial differential equation. Int. J. Appl. Math. **35**(3), 423–438 (2022)
7. Podlubny, I.: Fractional Differential Equations. Academic Press, San Diego (1999)
8. Hilfer, R.: Applications of Fractional Calculus in Physics. World Scientific, Singapore (2000)
9. Berkolaiko, G., Kuchment, P.: Introduction to Quantum Graphs. Mathematical Surveys and Monographs, vol 186. AMS, Providence (2013)
10. Alikhanov, A.A.: A priori estimate for solutions of boundary value problems for fractional-order equations. Differ. Equ. **46**(5), 660–666 (2010)
11. Friedman, J., Tillich, J.-P.: Wave equations for graphs and the edge-based Laplacian. Pac. J. Math. **216**(2) (2004)
12. Berkolaiko, G.: An elementary introduction to quantum graphs. In: Geometric and Computational Spectral Theory. Contemporary Mathematics, vol. 700. AMS, Providence (2017)
13. Kadirkulov, B.J., Jalilov, M.A.: On a nonlocal problem for fourth-order mixed type equation with the Hilfer operator. Bull. Inst. Math. **6**(1), 59–67 (2020)
14. Sobirov, Z.A., Abdullaev, O.K., Khujakulov, J.R.: Initial boundary value problem for a time fractional wave equation on a metric graph. Differ. Equ. Appl. **15**(1), 13–27 (2023)

Chapter 23
Second-Order Semiregular Non-Commutative Harmonic Oscillators: The Spectral Zeta Function

Marcello Malagutti

Abstract We study the properties of the spectral zeta function associated to a class of elliptic semiregular differential systems, including models of semiregular NCHOs, relevant to Quantum Optics. By "semiregular syste" we mean a pseudodifferential systems with a step $-j$ in the homogeneity of the jth-term in the asymptotic expansion of the symbol. More precisely, we state a theorem about the continuation of the spectral zeta function of a semiregular elliptic differential system $A^{\mathrm{w}} = (A^{\mathrm{w}})^* > 0$ as a function whose poles set has only one accumulation point that is the negative infinity.

23.1 Introduction

The spectral zeta function is one of the most important observables of the spectrum of an elliptic operator. For a complex Hilbert space H and a densely defined linear operator $P : H \to H$, we denote the set of the eigenvalues (repeated by multiplicity) of P by Spec P. When Spec P is discrete we can define the spectral zeta function of P as

$$\zeta_P(s) := \sum_{\lambda \in \operatorname{Spec} P} \lambda^{-s},$$

for any given complex number s for which it makes sense. In particular, if P is an elliptic, selfadjoint and positive global pseudodifferential operator of order $\mu > 0$ on $\mathbb{R}^n$, then $s \mapsto \zeta_P(s)$ is holomorphic for $\operatorname{Re} s > 2n/\mu$ since the defining series is absolutely convergent (see Corollary 4.4.4. in [18]). For instance, if we denote by $P = \frac{x^2 - \partial_x^2}{2}$ the harmonic oscillator defined as the maximal operator in $L^2(\mathbb{R})$, then

M. Malagutti (✉)
Department of Mathematics, University of Bologna, Bologna, Italy
e-mail: marcello.malagutti2@unibo.it

M. Ruzhansky, K. Van Bockstal (eds.), *Extended Abstracts 2021/2022*, Research Perspectives Ghent Analysis and PDE Center 2,
https://doi.org/10.1007/978-3-031-42539-4_23

Spec $P = \{k + 1/2;\ k \in \mathbb{Z}_+\}$ with multiplicity 1, and

$$\zeta_P(s) = \sum_{k\geq 0}(k + 1/2)^{-s} = (2^s - 1)\zeta(s),$$

where $\zeta(s)$ denotes the Riemann zeta function. Note that ζ_P is holomorphic for $\mathrm{Re}(s) > 1$, and has a meromorphic continuation to the whole complex plane. Furthermore, ζ_P has the only pole at $s = 1$, and we have $\zeta_P(s) = 0$ for $s = -2k$, $k \in \mathbb{Z}_+$ which are, thus, called trivial zeros. Moreover, the spectral zeta function entangles information about the spectrum of P in its analytical properties. For instance, the residues of the zeta function at its poles gives the coefficients of the Weyl Law for P by the Ikehara Tauberian theorem (see Section 14 of Shubin [26]. See also Proposition (IV.6) in [6] and the references in Ivrii [11]).

To show the relevance of the results in this paper it is important the distinction between regular and semiregular symbols: we call "semiregular" those symbols whose j-th term in the asymptotic expansion has order $\mu - j$. For a semiregular system of order μ we call semiprincipal symbol the term of degree $\mu - 1$ in the asymptotic expansion of its symbol and principal symbol the one of order μ.

In this work, we study the properties of the spectral zeta function associated with a positive elliptic semiregular positive partial differential systems with polynomial coefficients, including also models of semiregular NCHOs in the class of the Semiregular Metric Globally Elliptic Systems (SMGES) as those introduced in Section 3 of Malagutti and Parmeggiani [12]. In more details, we state a result about the continuation of the spectral zeta function $\zeta_{A^{\mathrm{w}}}$ which turns out to be a meromorphic function whose poles are real and accumulate at $-\infty$. Namely, we give the continuation as a linear combination of the meromorphic functions $s \mapsto \frac{1}{s-(n-j)+h/2}$, $j \geq 0$ and $h = 0, 1$, modulo a function that is holomorphic on a complex half-plan. Notice that indeed our extension can have poles in all the negative semi-integers, unlike the results in [10, 18] and [27] where the poles are all positive. The meromorphic continuation is obtained by following the approach of Theorem 7.2.1 in [18] where the parametrix approximation $U_A(t)$ of the heat-semigroup $e^{-tA^{\mathrm{w}}}$is used. More precisely, by the Mellin transform we can write $\zeta_{A^{\mathrm{w}}}$ as $s \mapsto \frac{1}{\Gamma(s)}\int_0^{+\infty} t^{s-1}\mathrm{Tr}\ e^{-tA^{\mathrm{w}}}\, dt$ for $\mathrm{Re} s > 2n/2 = n$ and, at this point, the asymptotic expansion $\sum_{j\geq 0} b_{-j}(t)$ (in the sense of Remark 6.1.5 at p. 83 of [18]) of $U_A(t)$ with $t \in \overline{\mathbb{R}}_+$ becomes crucial. In fact, the approximation of $s \mapsto \frac{1}{\Gamma(s)}\int_0^{+\infty} t^{s-1}\mathrm{Tr}\ e^{-tA^{\mathrm{w}}}\, dt$ by $s \mapsto \frac{1}{\Gamma(s)}\int_0^{+\infty} t^{s-1}\mathrm{Tr}\ U_A(t)\, dt$ leads to the study of integrals of the form

$$(2\pi)^{-n}\int_{\mathbb{R}^{2n}} \chi(X)\mathsf{Tr}\left(b_{-2j-h}(t, X)\right) dX,\ j \in \mathbb{N},\ h = 0, 1, \tag{23.1}$$

where χ is a chosen excision function and Tr is the classical matrix trace. In fact the computation of (23.1) gives the coefficients of the linear combination of the aforementioned meromorphic functions. These coefficients contribute to determine the residues and zeros of the spectral zeta function. Now one needs to go through a Taylor expansion argument as the time variable $t \to 0+$ of the terms arising from the study of $\operatorname{Tr} e^{-tA^{\mathrm{w}}} - \sum_{j=0}^{\nu}\sum_{h=0}^{1} \operatorname{Tr} B_{-2j-h}(t)$ (where B_{-k} with principal symbol b_{-k}). (The behavior of $e^{-tA^{\mathrm{w}}}$ as $t \to +\infty$ does not affect the result.)

This is a delicate argument since the behaviour of the coefficients of the linear combination of the above meromorphic functions must be controlled as $t \to 0+$.

The plan of the paper is the following. First of all, in Sect. 23.2, an Excursus of the previous results is provided. Then we briefly recall in Sect. 23.3 the concept of NCHOs and of the class of semiregular symbols and its main properties. After that, the notation adopted will be introduced in Sect. 23.4 along with the parabolic ψdifferential calculus needed to define the heat-semigroup parametrix which will be shown in Sect. 23.5. Moreover, in Sect. 23.6, we will control the behaviour of the coefficients. Finally, we will state our theorem in Sect. 23.7. Actually, in Sect. 23.7 we will also obtain a meromorphic continuation for the Hurwitz spectral zeta function $\zeta_{A^{\mathrm{w}}+\tau I}$ for all $\tau \geq 0$.

23.2 Excursus of Previous Results

The notion of spectral zeta function was introduced for the first time for the Laplacian on a two-dimensional Euclidean domains Ω by Carleman [3] who studied the Dirichlet-type series

$$\sum_{\lambda_j \in \operatorname{Spec}\Delta} \frac{\phi_{\lambda_j}(x_1)\phi_{\lambda_j}(x_2)}{\lambda_j^s}, \quad x_1, x_2 \in \Omega \tag{23.2}$$

where ϕ_{λ_j} is the eigenfunction of Δ associated to the eigenvalue λ_j. Later, in the case of a bounded Euclidean domain V of arbitrary dimension N, Minakshisundaram [14] showed through a method different from Carleman's that (23.2) is an entire function of s with zeros at negative integers and that

$$\sum_{\lambda_j \in \operatorname{Spec}\Delta} \frac{\phi_{\lambda_j}(x_1)^2}{\lambda_j^s}$$

can be continued as a meromorphic function of s with a unique simple pole at $N/2$ and negative integer zeros. Next, the analytic continuation of the spectral zeta function was studied by Minakshisundaram and Pleijel [15] for the Laplacian on a general compact manifold by a method that is a generalization of Carleman's. Seeley [25] studied the spectral zeta function of an elliptic ψdo on a compact manifold without boundary through the trace of complex powers of ψdos, furthermore giving the value of the zeta function at 0.

Many different techniques have been used to obtain properties of the spectral zeta function. Duistermaat and Guillemin [4] (see also, [5] and the references in Hormander [9]) studied systematically the spectral zeta function of ψdos on compact boundaryless manifolds basing their approach on the construction of a parametrix for the wave equation. Robert [22] (see also Aramaki [1]) extended meromorphically the spectral zeta function of an elliptic ψdo on $\mathbb{R}^n$ to the whole complex plane with simple poles that he computed along with the corresponding residues. He generalized to the global setting the techniques by Seeley to construct the parametrix of the resolvent by complex powers.

Moreover, following the discussion by Parmeggiani [17, 18] and Parmeggiani and Wakayama [20, 21], we call second-order regular Non-Commutative Harmonic Oscillators (NCHOs) the class of the regular global partial differential systems of second-order with polynomial coefficients. From now on we will omit the expression "second-order" since all the NCHOs considered will be of second-order.

Ichinose and Wakayama [10] obtained a meromorphic continuation of the spectral zeta function of a subclass of regular NCHOs and determined some of its special values. In addition, they showed that such a spectral zeta function has only a simple pole at 1 and the sequence of its trivial zeros coincides with the one of the Riemann zeta function, the non-positive even integers. Their approach is based on the Mellin transform of the heat-semigroup of the operator in the approximation given by a parametrix which they computed directly, without using the one for the resolvent, obtaining its asymptotic expansion (see (15) and (16) in their paper). Later, Parmeggiani [18] generalized that approach to obtain the meromorphic continuation of the spectral zeta function of all the regular NCHOs. Nevertheless, while gaining in generality unfortunately his result did not explicitly locate the trivial zeros of the continuation of the spectral zeta function as could Ichinose and Wakayama.

Ichinose and Wakayama's and Parmeggiani's papers deal with regular systems. Regarding the semiregular systems, Sugiyama explored in [27] the Hurwitz-type spectral zeta function for the quantum Rabi model (describing the interaction of light and matter of a two-level atom coupled with a single quantized photon of the electromagnetic field, see the seminal papers [23] and [24] by Rabi, see also [2] by Braak).

In our study, we follow the construction of the zeta function provided by Ichinose and Wakayama, in analogy to the approach by Parmeggiani in Theorem 7.2.1 of [18].

23.3 Non-Commutative Harmonic Oscillators (NCHOs)

Let us introduce the Non-Commutative Harmonic Oscillators (NCHOs).

Definition 23.1 A Non-Commutative Harmonic Oscillator (NCHO) is the Weyl-quantization $a^{\mathrm{w}}(x, D)$ of an $N \times N$ system of the form

$$a(x, \xi) = a_2(x, \xi) + a_0, \quad (x, \xi) \in \mathbb{R}^n \times \mathbb{R}^n = T^*\mathbb{R}^n,$$

where $a_2(x, \xi)$is an $N \times N$ matrix whose entries are homogeneous polynomials of degree 2 in the (x, ξ) variables, and a_0 is a constant $N \times N$ matrix (introduced by A. Parmeggiani and M. Wakayama in [20, 21]. See also [18, 19]).

Therefore it can also be said that an NCHO comes from the Weyl-quantization of a matrix-valued quadratic form in (x, ξ) adding a constant matrix term.

Remark 23.1 Parmeggiani and Wakayama choose the name NCHO for two main reasons

(i) the fact that a scalar harmonic oscillator is a single quadratic form in (x, ξ);
(ii) the two levels of non-commutativity that we have to deal with when studying these systems: one due to the matrix-valued nature of the symbol of the system, and the other to the Weyl-quantization rule

$$x_k \xi_j \leftrightarrow (x_k D_{x_j} + D_{x_j} x_k)/2, \quad (\text{where } D = -i\partial),$$

reflected through symplectic geometry by the Poisson-bracket relations

$$\{\xi_j, x_k\} = \delta_{jk}, \quad 1 \leq j, k \leq n.$$

Definition 23.2 A NCHO $a^{\mathrm{w}}(x, D)$ is said to be elliptic when

$$a_2 \text{ is a } N \times N \text{ matrix and } \det a_2(x, \xi) \text{ behaves exactly like } (|x|^2 + |\xi|^2)^N$$

for $|(x, \xi)|$ large.

When a_2 and a_0 are Hermitian matrices, the operator $a^{\mathrm{w}}(x, D)$ is "formally self-adjoint" (i.e. symmetric on $\mathscr{S}(\mathbb{R}^n; \mathbb{C}^N)$). Moreover if in addition $a^{\mathrm{w}}(x, D)$ is positive elliptic (i.e. the matrix $a_2(x, \xi)$ is positive definite for $(x, \xi) \neq (0, 0)$), then it is self-adjoint as an unbounded operator in $L^2(\mathbb{R}^n; \mathbb{C}^N)$ with a discrete real spectrum.

23.4 Parabolic Calculus

In this section, similar to what is done by Parenti and Parmeggiani in [16] (see also Section 6.1 of [18]), we will introduce a class of symbols suitable for the construction of a pseudodifferential approximation of $e^{-tA^{\mathrm{w}}}$. Let us recall $\overline{\mathbb{R}}_+ = [0, +\infty)$.

Definition 23.3 Let $r \in \mathbb{R}$. By $S(\mu, r)$ we denote the set of all smooth maps $b : \overline{\mathbb{R}}_+ \times \mathbb{R}^n \times \mathbb{R}^n \longrightarrow \mathsf{M}_N$ satisfying the following estimates: for any given $\alpha \in \mathbb{Z}_+^{2n}$ and any given $p, j \in \mathbb{Z}_+$there exists $C > 0$ such that

$$\sup \left| t^p (\frac{d}{dt})^j \partial_X^\alpha b(t, X) \right| \leq C m(X)^{r-|\alpha|+(j-p)\mu}.$$

For $b \in S(\mu, r)$ we then consider the pseudodifferential operator

$$b^{\mathrm{w}}(t, x, D)u(x) = (2\pi)^{-n} \iint e^{i(x-y,\xi)} b(t, \frac{x+y}{2}, \xi) u(y) dy d\xi, \ u \in \mathscr{S}(\mathbb{R}^n; \mathbb{C}^N),$$

and we shall say that $B \in \mathrm{OP}S(\mu, r)$ if $B = b^{\mathrm{w}}(t, x, D) + R$, where R is smoothing. In this setting, a smoothing operator R is any continuous map

$$R : \mathscr{S}'(\mathbb{R}^n; \mathbb{C}^N) \longrightarrow \mathscr{S}(\overline{\mathbb{R}}_+; \mathscr{S}(\mathbb{R}^n; \mathbb{C}^N)).$$

Then we introduce the "classical operators": in this case the key is to take in account the correct homogeneity properties. The basic example to keep in mind is the matrix $e^{-ta_\mu(x,\xi)}$.

Definition 23.4 We say that the operator $B \in \mathrm{OP}S(\mu, r)$, $B = b^{\mathrm{w}} + R$ is classical, and write $B \in \mathrm{OP}S_{\mathrm{cl}}(\mu, r)$, if there exists a sequence of functions $b_{r-2j} = b_{r-2j}(t, X)$, $j \geq 0, t \geq 0$ and $X \neq 0$, such that:

1. One has the homogeneity

$$b_{r-2j}(t, \tau X) = \tau^{r-2j} b_{r-2j}(\tau^\mu t, X), \ \forall \tau > 0, \ \forall j \geq 0;$$

2. The function

$$\mathbb{R}^{2n} \setminus \{0\} \ni X \longmapsto b_{r-2j}(\cdot, X) \in \mathscr{S}(\overline{\mathbb{R}}_+; M_N),$$

   is smooth for all $j \geq 0$;
3. For all $\nu \geq 1$

$$b(t, X) - \sum_{j=0}^{\nu=1} \chi(X) b_{r-2j}(t, X) \in S(\mu, r - 2\nu), \tag{23.3}$$

   where χ is an excision function.

Remark 23.2 We call $b_r = \sigma_r(B)$ the principal symbol of B.

Remark 23.3 Semi-regular classical symbols are defined accordingly, considering also terms with odd degree of homogeneity in the expansion formula (23.3), and the class of pseudodifferential operators associated to them is denoted by $\mathrm{OP}S_{\mathrm{sreg}}(\mu, r)$.

23.5 Parametrix of the Heat-Semigroup

In this section, we will construct the parametrix of the heat-semigroup of a semiregular positive elliptic pseudodifferential operator.

Lemma 23.1 *Let $A = A^*$, with $A \sim \sum_{j\geq 0} a_{2-j} \in S_{\mathrm{sreg}}(m^2, g; \mathsf{M}_N)$, be an elliptic second-order system such that $A^{\mathrm{w}} > 0$. Then, there exists $U_A \in \mathrm{OP}S_{\mathrm{sreg}}(2,0)$ such that*

$$\frac{d}{dt}U_A + A^{\mathrm{w}}U_A : \mathscr{S}'(\mathbb{R}^n; \mathbb{C}^N) \to \mathscr{S}(\overline{\mathbb{R}}_+; \mathscr{S}(\mathbb{R}^n; \mathbb{C}^N))$$

is smoothing, and

$$U_A|_{t=0} - I_N : \mathscr{S}'(\mathbb{R}^n; \mathbb{C}^N) \to \mathscr{S}(\mathbb{R}^n; \mathbb{C}^N)$$

is smoothing. Moreover, the principal symbol of U_A is

$$\overline{\mathbb{R}}_+ \times \left(\mathbb{R}^{2n} \setminus \{0\}\right) \ni (t, X) \mapsto e^{-ta_2(X)}.$$

23.6 Vanishing Property

Let A^{w} be as in the previous section. In this section we state the technical proposition that we need to control the behaviour of the b_{-j} constructed in Lemma 23.1 as $t \to 0+$, that is, its vanishing property, for a class of positive and self-adjoint elliptic differential systems with symbol in $S_{\mathrm{sreg}}(m^2, g; \mathsf{M}_N)$. Hence, we will suppose the symbol of A^{w} to be $a_2 + a_1 + a_0$ where a_j is an $N \times N$ matrix-valued function on $\mathbb{R}^{2n}$ with homogeneous polynomial of degree j entries for all $j = 0, 1, 2$.

Proposition 23.1 *Let $A = a_2 + a_1 + a_0$ be an elliptic of second-order where a_j is an $N \times N$ matrix-valued function on $\mathbb{R}^{2n}$ with homogeneous polynomial of degree j entries for all $j = 0, 1, 2$, let $A^{\mathrm{w}} > 0$, and let U_A be the heat-semigroup $e^{-tA^{\mathrm{w}}}$ parametrix constructed by Lemma 23.1. Then, denoting by $\sum_{j\geq 0} B_{-j}$ the expansion of U_A and by b_{-j} the principal symbol of B_{-j}, we have for all $j \geq 0$ and $h = 0, 1$*

$$b_{-2j-h}(t, \omega) = O(t^{j+h}),\ t \to 0+,$$

and for all $\alpha, \beta \in \mathbb{Z}_+^n$, with $|\alpha| = 2k + 1$, $k \geq 0$ and $|\beta| \leq 1$ we have:

$$\partial_X^{\alpha+\beta} b_{-2j-h}(t, \omega) = O(t^{j+k+h|\beta|+1}),\ t \to 0+,$$

where the constants in $O(\cdot)$ do not depend on $\omega \in \mathbb{S}^{2n-1}$.

23.7 Meromorphic Continuation of $\zeta_{A^{\mathrm{w}}}$

Let A^{w} be as in the Sect. 23.5. In this section, we will use the parametrix approximation of the heat-semigroup constructed in Lemma 23.1 to state the result about the continuation of the spectral zeta function of the class of positive and self-adjoint elliptic operators A^{w} satisfying the hypotheses of Proposition 23.1. Namely, $\zeta_{A^{\mathrm{w}}}$ can be rewritten modulo a term holomorphic on a half-plane of $\mathbb{C}$ as a linear complex combination of meromorphic functions. Moreover, we will give explicit formulas for the coefficients of this linear combination (see [13]).

Theorem 23.1 *Let $A = a_2 + a_1 + a_0$ be an elliptic system of second-order where a_j is an $N \times N$ matrix-valued function on $\mathbb{R}^{2n}$ with homogeneous polynomial of degree j entries for all $j = 0, 1, 2$. Moreover, suppose $A^{\mathrm{w}} > 0$.*

There exist constants $c_{-2j-h,n}$ with $0 \le j \le n-1$, $h = 0, 1$, and constants $c_{-2j-1,n}$, C_{-2j} with $j \ge n$, such that, for any given integer $\nu \in \mathbb{Z}_+$ with $\nu \ge n$,

$$\zeta_{A^{\mathrm{w}}}(s) = \frac{1}{\Gamma(s)} \Bigg[\left(\sum_{h=0}^{1} \sum_{j=0}^{n-1} \frac{c_{-2j-h,n}}{s-(n-j)+h/2} \right) + \left(\sum_{j=n}^{\nu} \frac{c_{-2j-1,n}}{s-(n-j)+1/2} \right) + \left(\sum_{j=n}^{\nu} \frac{C_{-2j}}{s-(n-j)} \right) + H_\nu(s) \Bigg], \tag{23.4}$$

where $\Gamma(s)$ is the Euler gamma function, and H_ν is holomorphic in the region $\mathrm{Re}\, s > (n-\nu) - 1$. Consequently, the spectral zeta function $\zeta_{A^{\mathrm{w}}}$ is meromorphic in the whole complex plane $\mathbb{C}$ with at most simple poles at $s = n, n-1, n-2, \ldots, 1$ and $s = n - \frac{1}{2}, n - \frac{3}{2}, n - \frac{5}{2}, \ldots$. One has

$$c_{-2j-h,n} = (2\pi)^{-n} \int_0^{+\infty} \int_{\mathbb{S}^{2n-1}} \mathit{Tr}\Big(b_{-2j-h}(\rho^2, \omega)\Big)\, \rho^{2(n-j)-1-h}\, d\omega\, d\rho, \tag{23.5}$$

where $0 \le j \le n-1$, $h = 0, 1$ or $j \ge n$, $h = 1$. In (23.5) the b_{-2j-h} are the terms in the symbol of the parametrix $U_A \in \mathrm{OP}S_{\mathrm{sreg}}(2, 0)$ constructed in the proof of Lemma 23.1,

$$U_A \sim \sum_{j \ge 0} B_{-j}.$$

Remark 23.4 An interesting problem can be to use the asymptotics for resolvent expansions and trace regularisations by Hitrik and Polterovich [7, 8].

Theorem 23.1 has the following corollary for the Hurwitz-type spectral zeta function of A^{w}.

Corollary 23.1 *Let $A = a_2 + a_1 + a_0$ be an elliptic system of second-order where a_j is an $N \times N$ matrix-valued function on $\mathbb{R}^{2n}$ with homogeneous polynomial of degree j entries for all $j = 0, 1, 2$. Moreover, suppose $A^{\mathrm{w}} > 0$.*

For all $\tau > 0$ there exist constants $c_{-2j-h,n}(\tau)$ with $0 \le j \le n-1$, $h = 0, 1$, and constants $c_{-2j-1,n}(\tau)$, $C_{-2j}(\tau)$ with $j \ge n$, such that, for any given integer $\nu \in \mathbb{Z}_+$ with $\nu \ge n$,

$$\zeta_{A^{\mathrm{w}}+\tau I}(s) = \frac{1}{\Gamma(s)}\left[\left(\sum_{h=0}^{1}\sum_{j=0}^{n-1}\frac{c_{-2j-h,n}(\tau)}{s-(n-j)+h/2}\right) + \left(\sum_{j=n}^{\nu}\frac{c_{-2j-1,n}(\tau)}{s-(n-j)+1/2}\right) + \left(\sum_{j=n}^{\nu}\frac{C_{-2j}(\tau)}{s-(n-j)}\right) + H_\nu(\tau)(s)\right], \tag{23.6}$$

where $\Gamma(s)$ is the Euler gamma function, and $s \mapsto H_\nu(\tau)(s)$ is holomorphic in the region $\operatorname{Re} s > (n-\nu)-1$. Consequently, the spectral zeta function $\zeta_{A^{\mathrm{w}}+\tau I}$ is meromorphic in the whole complex plane $\mathbb{C}$ with at most simple poles at $s = n, n-1, n-2, \ldots, 1$ and $s = n-\frac{1}{2}, n-\frac{3}{2}, n-\frac{5}{2}, \ldots$. One has

$$c_{-2j-h,n}(\tau) = (2\pi)^{-n}\int_0^{+\infty}\int_{\mathbb{S}^{2n-1}} Tr\Big(b_{-2j-h}(\rho^2,\omega)\Big)\,\rho^{2(n-j)-1-h}\,d\omega d\rho$$
$$-\tau(2\pi)^{-n}\int_0^{+\infty}\int_{\mathbb{S}^{2n-1}}\int_0^{\rho^2} e^{-(\rho^2-t')a_2}\,Tr(b_{2-2j-h}(t',\omega))\rho^{2(n-j)-1-h}\,dt'd\omega d\rho, \tag{23.7}$$

where $0 \le j \le n-1$, $h = 0, 1$ or $j \ge n$, $h = 1$. In (23.7), the b_{-2j-h} are the terms in the symbol of the parametrix $U_A \in \mathrm{OP}S_{\mathrm{sreg}}(2, 0)$ constructed in the proof of Lemma 23.1,

$$U_A \sim \sum_{j\ge 0} B_{-j},$$

where we set $b_k \equiv 0$ for all $k = 1, 2$.

Acknowledgments I wish to thank Prof. Alberto Parmeggiani for the helpful discussions.

References

1. Aramaki, J.: Complex powers of vector valued operators and their application to asymptotic behavior of eigenvalues. J. Funct. Anal. **87**(2), 294–320 (1989)

2. Braak, D.: Analytic solutions of basic models in quantum optics. In: Anderssen, R. et al. (eds.) Applications + Practical Conceptualization + Mathematics = fruitful Innovation - Proceedings of the Forum of Mathematics for Industry 2014. Mathematics for Industry, vol. 11, pp. 75–92. Springer, Berlin (2015)
3. Carleman, T.: Propriétés asymptotiques des fonctions fondamentales des membranes vibrantes. Skand. Matent. Kongress (1934)
4. Duistermaat, J.J., Guillemin, V.W.: The spectrum of positive elliptic operators and periodic bicharacteristics. Invent Math **29**, 39–79 (1975) https://doi-org.ezproxy.unibo.it/10.1007/BF01405172
5. Guillemin, V.W.: 25 years of fourier integral operators. In: Brüning, J., Guillemin, V.W. (eds) Mathematics Past and Present Fourier Integral Operators. Springer, Berlin, Heidelberg (1994) https://doi.org/10.1007/978-3-662-03030-1_1
6. Helffer, B., Robert, D.: Comportement asymptotique précisé du spectre d'opérateurs globalement elliptiques dans $\mathbb{R}^n$. Séminaire Équations aux dérivées partielles (Polytechnique) , exp. no. 2, pp. 1–22 (1980–1981)
7. Hitrik, M., Polterovich, I.: Regularized traces and Taylor expansions for the heat semigroup. J. Lond. Math. Soc. **68**(2), 402–418 (2003)
8. Hitrik, M., Polterovich, I.: Resolvent expansions and trace regularizations for Schr¨odinger operators. Advances in Differential Equations and Mathematical Physics (Birmingham, AL, 2002). Contemporary Mathematics, vol. 327, pp. 161–173. American Mathematical Society, Providence (2003)
9. Hormander, L.: The spectral function of an elliptic operator. Acta Math. **121**, 193–218 (1968). https://doi.org/10.1007/BF02391913
10. Ichinose, T., Wakayama, M.: On the spectral zeta function for the non-commutative harmonic oscillator. Rep. Math. Phys. **59**(3), 421–432 (2007)
11. Ivrii, V.: 100 years of Weyl's law. Bull. Math. Sci. **6**, 379–452 (2016). https://doi.org/10.1007/s13373-016-0089-y
12. Malagutti, M., Parmeggiani, A.: Spectral Asymptotic Properties of Semiregular Non-Commutative Harmonic Oscillators. To appear in Communications in Math. Phys. (Preprint 2022)
13. Malagutti, M.: On the Spectral Zeta Function of Second Order Semiregular Non-Commutative Harmonic Oscillators. Bulletin des Sciences Mathématiques (2023–06) https://doi.org/10.1016/j.bulsci.2023.103286
14. Minakshisundaram, S.: A generalization of epstein zeta functions. Can. J. Math. **1**(4), 320–327 (1949). https://doi.org/10.4153/CJM-1949-029-3
15. Minakshisundaram, S., Pleijel, A.: Some properties of eigenfunctions of the Laplace operator on Riemannian manifolds. Can. J. Math. **1**(3), 242–256 (1949)
16. Parenti, C., Parmeggiani, A.: A Lyapunov Lemma for elliptic systems. Ann. Glob. Anal. Geom. **25**, 27–41 (2004)
17. Parmeggiani, A.: On the spectrum of certain noncommutative harmonic oscillators. Proceedings of the conference "Around Hyperbolic Problems - in memory of Stefano"; Ann. Univ. Ferrara Sez. VII Sci. Mat. **52**, 431–456 (2006)
18. Parmeggiani, A.: Spectral Theory of Non-Commutative Harmonic Oscillators: An Introduction. Lecture Notes in Mathematics, vol. 1992, xii+254 pp. Springer-Verlag, Berlin (2010) https://doi.org/10.1007/978-3-642-11922-4
19. Parmeggiani, A.: Non-commutative harmonic oscillators and related problems. Milan J. Math. **82**(2), 343–387 (2014) https://doi.org/10.1007/s00032-014-0220-z
20. Parmeggiani, A., Wakayama, M.: Oscillator representations and systems of ordinary differential equations. Proc. Natl. Acad. Sci. USA **98**, 26–30 (2001)
21. Parmeggiani, A., Wakayama, M.: Non-commutative harmonic oscillators-I,-II, corrigenda and remarks to I. Forum Math. **14**, 539–604, 669–690 (2002) ibid. **15**, 955–963 (2003)
22. Robert, D.: Propriétés spectrales d'opérateurs pseudo-différentiels. Commun. Partial Differ. Equ. **3**(9), 755–826 (1978). https://doi.org/10.1080/03605307808820077
23. Rabi, I.I.: On the process of space quantization. Phys. Rev. **49**, 324-328 (1936)

24. Rabi, I.I.: Space quantization in a gyrating magnetic field. Phys. Rev. **51**, 652–654 (1937)
25. Seeley, R.T.: Complex powers of an elliptic operator. Proceedings of Symposia in Pure Mathematics, vol. 10, pp. 288–307. American Mathematical Society, Providence (1968)
26. Shubin, M.A.: Pseudodifferential Operators and Spectral Theory, 2nd edn., xii+288 pp. Springer-Verlag, Berlin (2001)
27. Sugiyama, S.: Spectral zeta functions for the quantum Rabi models. Nagoya Math. J. **229**, 52–98 (2018)

Chapter 24
Global Well-Posedness with Loss of Regularity for a Class of Singular Hyperbolic Cauchy Problems

Rahul Raju Pattar and N. Uday Kiran

Abstract The goal of this paper is to summarize global well-posedness results for a class of strictly hyperbolic Cauchy problems with coefficients in $L^1([0, T]; C^\infty(\mathbb{R}^n))$ growing polynomially in x and singular in t. The problems we study are of the strictly hyperbolic type with respect to a generic weight and a metric on the phase space. The singular behaviour is captured by the blow-up of the first and second t-derivatives of the coefficients, which allows the coefficients to either blow up or oscillate near $t = 0$. A crucial step in arriving at the results is conjugation by a pseudodifferential operator of the form $e^{\nu(t)\Theta(x,D_x)}$, where $\Theta(x, D_x)$ explains the quantity of the loss by linking it to the metric on the phase space and the singular behaviour, while $\upsilon(t)$ gives a scale for the loss. Depending on the order of this operator, the solution experiences zero, arbitrarily small, finite or infinite loss in relation to the initial datum. We also show the anisotropic cone conditions in our setting.

24.1 Introduction

The hyperbolic Cauchy problems with irregular coefficients govern the propagation of waves through an inhomogeneous medium and have applications in geophysics and medical imaging (see, for instance, [5]). An important feature of such problems is the loss of derivatives in relation to the initial datum. It is now well established that this loss is due to an irregularity of the coefficients in the time variable (see, for instance, [3, 4]). Recently, many authors have also considered the interplay between irregularities in time and certain suitable bounds on the space (see, for instance, [1]).

R. R. Pattar (✉)
Tata Institute of Fundamental Research, Centre for Apllicable Mathematics, Bengaluru, India
e-mail: rahul22@tifrbng.res.in

N. Uday Kiran
Sri Sathya Sai Institute of Higher Learning, Puttaparthi, India
e-mail: nudaykiran@sssihl.edu.in

M. Ruzhansky, K. Van Bockstal (eds.), *Extended Abstracts 2021/2022*, Research Perspectives Ghent Analysis and PDE Center 2,
https://doi.org/10.1007/978-3-031-42539-4_24

We consider the Cauchy problem for the strictly hyperbolic equation:

$$\partial_t^2 u - \sum_{i,j=1}^{n} a_{i,j}(t,x)\partial_{x_i}\partial_{x_j}u + \sum_{j=1}^{n} b_j(t,x)\partial_{x_j}u + b_{n+1}(t,x)u = f(t,x), \tag{24.1}$$

with initial data $u(0,x) = f_1(x)$, $\partial_t u(0,x) = f_2(x)$, $(t,x) \in [0,T] \times \mathbb{R}^n$. The coefficients are polynomially growing in x and are singular (either blowing up or infinitely oscillating) in t. The growth in x is characterized by generic weights $\omega(x)$ and $\Phi(x)$, i.e., the coefficients are smooth in x and satisfy the estimates

$$\left|\partial_x^\beta a_{i,j}(\cdot,x)\right| \le C_\beta \omega(x)^2 \Phi(x)^{-|\beta|},$$

$$\left|\partial_x^\beta b_j(\cdot,x)\right| \le C_\beta \omega(x)\Phi(x)^{-|\beta|} \quad \text{and} \quad \left|\partial_x^\beta b_{n+1}(\cdot,x)\right| \le C_\beta \Phi(x)^{-|\beta|},$$

for some positive constant C_β and multi-index $\beta \in \mathbb{N}_0^n$. The functions $\omega(x)$ and $\Phi(x)$ are positive monotone increasing in $|x|$ such that $1 \le \omega(x) \lesssim \Phi(x) \lesssim \langle x\rangle = \left(1+|x|^2\right)^{1/2}$. These functions specify the structure of the differential equation in x variable. The properties of these functions are outlined in [10, Section 2]. An example of a function whose growth is characterized by these weights is given in Fig. 24.1.

The singular behaviour is characterised by either infinite oscillations or blow-up near $t = 0$ for the coefficients $a_{i,j}(t,\cdot)$. The following definitions of these behaviours are inspired from [6].

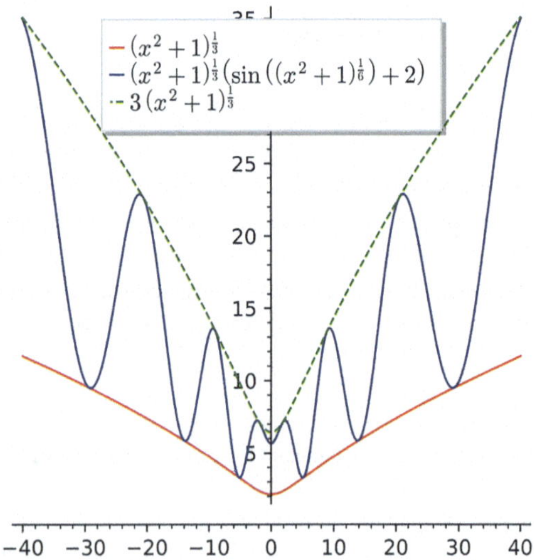

Fig. 24.1 In blue is a function with weight $\omega(x) = \langle x\rangle^{\frac{2}{3}}$ decaying purely sublinearly with $\Phi(x) = \langle x\rangle^{\frac{1}{3}}$

Definition 24.1 (Oscillatory Behavior) Let $c = c(t) \in L^\infty([0, T]) \cap C^2((0, T])$ satisfy the estimate

$$\left|\frac{d^j}{dt^j}c(t)\right| \lesssim \left(\frac{|\ln t|^{\tilde{\gamma}\mathbf{I}_q}}{t^q}\right)^j,$$

for $j = 1, 2$ and $q \geq 1$. The function $\mathbf{I}_q$ is such that $\mathbf{I}_q \equiv 1$ if $q = 1$ else $\mathbf{I}_q \equiv 0$. We say that the oscillating behaviour of the function $c(t)$ is

- very slow if $q = 1$ and $\tilde{\gamma} = 0$,
- slow if $q = 1$ and $\tilde{\gamma} \in (0, 1)$,
- fast if $q = \tilde{\gamma} = 1$,
- very fast if $q > 1$ or else $\tilde{\gamma} > 1$ when $q = 1$.

Definition 24.2 (Blow-Up Rate) Let $c = c(t) \in L^1((0, T]) \cap C^1((0, T])$ satisfy the estimates

$$|c(t)| \lesssim \frac{1}{t^p}|\ln t|^{\tilde{\gamma}\mathbf{I}_q},$$

$$|\partial_t c(t)| \lesssim \frac{1}{t^q}|\ln t|^{(\tilde{\gamma}-1)\mathbf{I}_q},$$

with $q \in [1, \infty)$, $p \in [0, 1)$, $p \leq q - 1$ and $\tilde{\gamma} > 0$. The function $\mathbf{I}_q$ is such that $\mathbf{I}_q \equiv 1$ if $q = 1$ else $\mathbf{I}_q \equiv 0$. We say that the blow-up rate of the function $c(t)$ is

- mild if $q = 1$, $p = 0$, $\tilde{\gamma} \in (0, 1)$,
- logarithmic if $q = 1$, $p = 0$, $\tilde{\gamma} = 1$,
- strong if $q = 1$, $p = 0$, $\tilde{\gamma} \in (1, \infty)$,
- very strong if $q > 1$, $p \in [0, 1)$. Examples of functions that are oscillating and blowing-up are given in Figs. 24.2 and 24.3, respectively.

In order to study the interplay of the singularity in t and unboundedness in x, we consider a class of metrics on the phase space of the form

$$g_{\Phi,k} = \frac{|dx|^2}{\Phi(x)^2} + \frac{|d\xi|^2}{\langle\xi\rangle_k^2} \tag{24.2}$$

where $\Phi(x)$ is as in (1.1.1) and $\langle\xi\rangle_k = \left(k^2 + |\xi|^2\right)^{1/2}$ for an appropriately chosen $k \geq 1$. Note that the functions $\omega(x)$ and $\Phi(x)$ are associated with the weight and the metric, respectively. An important geometric quantity associated with this metric is the Placnk function $h(x, \xi) = (\Phi(x)\langle\xi\rangle_k)^{-1}$. The properties of this metric and the corresponding symbolic calculus are discussed in [10, Section 3].

In the following sections, we summarise our new geometric approach where we use a metric on the phase space and the corresponding symbolic calculus to link the singular behaviour in t and the growth rate in x to the loss of regularity of the solution in relation to the initial datum.

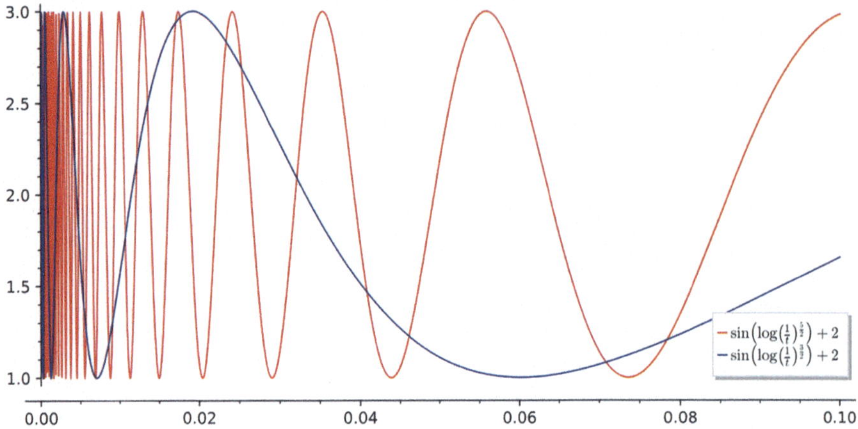

Fig. 24.2 Examples of slow $(2 + (\sin(\ln t))^{\frac{3}{2}})$ and fast $(2 + (\sin(\ln t))^{\frac{5}{2}})$ oscillating functions

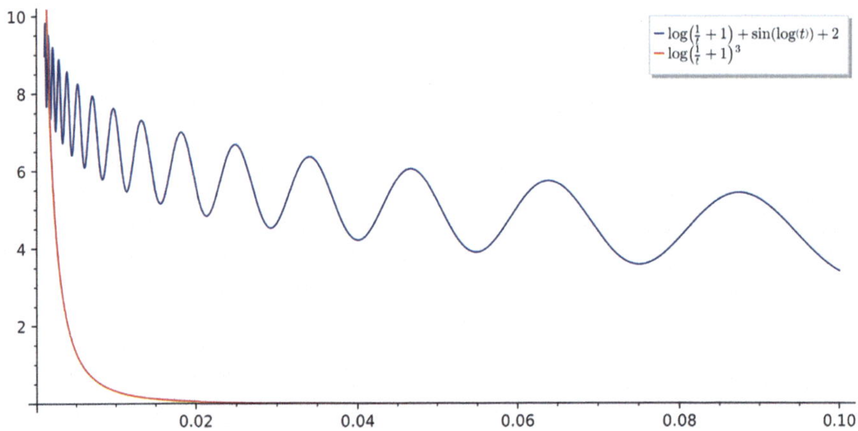

Fig. 24.3 Examples of logarithmically blowing up $(\ln(1 + 1/t) + \sin(\ln t) + 2)$ and Strongly blowing up$((\ln(1 + 1/t))^3)$ functions

24.2 Our Philosophy of Conjugation

It is well-known in the theory of hyperbolic operators (see [1, 3]) that the low-regularity in t needs to be compensated by a higher regularity in the x variable. For example, when the coefficients are Hölder continuous $(C^\alpha([0, T]), \alpha \in (0, 1))$ in t, one needs to compensate this irregularity in t with Gevrey regularity $(G^\sigma(\mathbb{R}^n), \sigma < 1/(1-\alpha))$ in x, as seen in [3]. This balancing operation is the key to understand the loss of regularity. As seen in the literature, conjugation brings this balance. Infact, conjugation links irregularity in t, growth with respect to x variable of the coefficients to the weight functions defining the solution spaces in a precise manner, see [3, Proposition 2.8] and [1, Proposition 3.1]. For example, when the coefficients are in $C^\alpha([0,T]; B^\infty(\mathbb{R}^n))$, the conjugating operator is of the form $e^{c_0\langle D_x\rangle^{1/\sigma}}$, $c_0 > 0$ while for the SG setting with coefficients in $C^\alpha([0,T]; C^\infty(\mathbb{R}^n))$, the conjugating operator used in the literature is of the form $e^{\nu(t)(\langle x\rangle^{1/\sigma}+\langle D_x\rangle^{1/\sigma})}$ for some function $\nu \in C([0,T]) \cap C^1((0,T])$.

In the global setting, solutions experience an infinite loss of both derivatives and decay when the coefficients display strong blow-up, very strong blow-up or very fast oscillations in t. In order to overcome the difficulty of tracking a precise loss in our context we introduce a class of parameter-dependent infinite order pseudodifferential operators of the form $e^{\nu(t)\Theta(x,D_x)}$ for the purpose of conjugation. These operators compensate, microlocally, the loss of regularity of the solutions. The operator $\Theta(x, D_x)$ is in general nonselfadjoint and it explains the compensation for the singularity in t and decides the quantity of the loss, while the monotone decreasing continuous function $\nu(t)$ gives a scale for the loss. Hence, we call the conjugating operator as the loss operator.

Our philosophy of conjugation is to design the loss operator that links the singular behaviour, the growth rate in x and the geometry of the phase space to the loss of regularity. In fact, the operator $\Theta(x, D_x)$ is such that $\Theta(x,\xi)$ is a function of $h(x,\xi) = (\Phi(x)\langle\xi\rangle_k)^{-1}$ which is the Planck function associated to the metric $g_{\Phi,k}$ in (24.2) and the quantity $\nu'(t)\Theta(x,\xi)$ majorises the symbols corresponding to the lower-order terms obtained after the application of a suitable diagonalisation technique to the first-order system corresponding to the Cauchy problem in (24.1). See [10, Theorem 4.1] and [7, Theorem 4.0.1] for more details. The appearance of the Planck function $h(x,\xi)$ in the definition of $\Theta(x,\xi)$ is justified as it controls the extent of localisation on the phase space. This philosophy of our conjugation is pictorially demonstrated in Fig. 24.4.

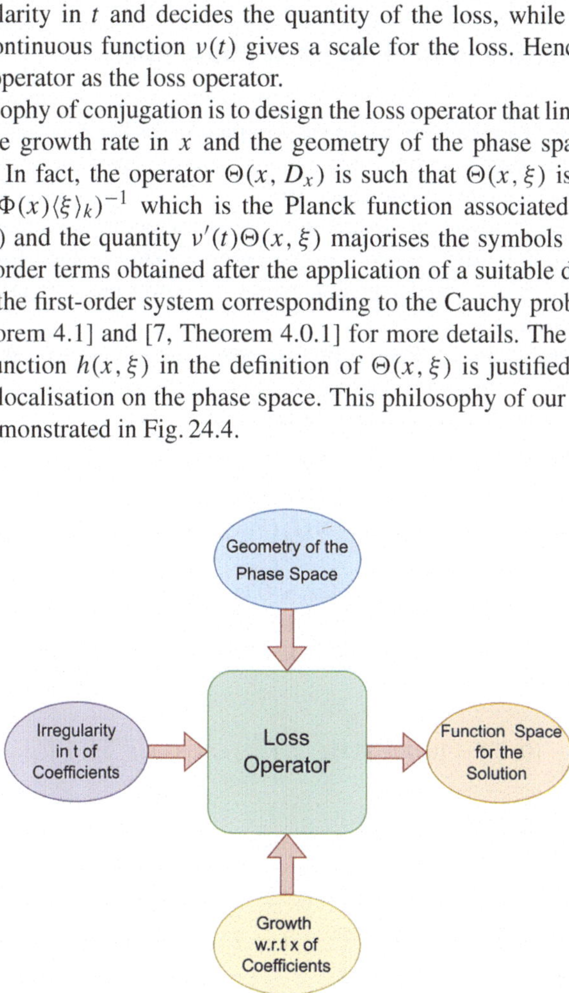

Fig. 24.4 Our philosophy of conjugation

One of our key observations is that the symbol of the operator arising after conjugation is governed by a metric $\tilde{g}_{\Phi,k}$ that is conformally equivalent to the initial metric $g_{\Phi,k}$. The metric $\tilde{g}_{\Phi,k}$ is of the form

$$\tilde{g}_{\Phi,k} = \Theta(x,\xi)^2 g_{\Phi,k}.$$

The operators with loss of regularity index are transformed to good operators by conjugation with the loss operator. This helps us to derive good a priori estimates of solutions in the Sobolev space associated with the loss operator by an application of sharp Gårding inequality followed by Grönwall inequality.

24.3 Subdivision of the Extended Phase Space

Before we perform a conjugation, we need to preprocess the operator with singular coefficients. This preprocessing step uses a careful amalgam of a localisation technique on the extended phase space ($[0,T] \times \mathbb{R}^n_x \times \mathbb{R}^n_\xi$) and a diagonalisation procedure already available in the literature [6] to handle the singularity. It helps in diagonalising the top-order terms in a first-order system obtained from a singular hyperbolic partial differential equation while the lower-order terms are handled through conjugation.

As the Planck function holds information about the extent of localisation on the phase space, our localisation technique that dictates the subdivision of the extended phase space is dependent on this critical information. For a fixed (x,ξ), we define the time splitting points $t^{(j)}_{x,\xi}$, $1 \le j \le m_0$, for some $m_0 \in \mathbb{N}$ as

$$t^{(j)}_{x,\xi} = NG_j(h(x,\xi)),$$

where N is the positive integer, $G_j(r)$, $r > 0$ is a function that depends on the order of singularity (see, for example, Section [10, Section 2.3]) and $h(x,\xi)$ is the Planck function associated with the metric $g_{\Phi,k}$. For a fixed (x,ξ) we split the time interval as

$$[0,T] = [0,t^{(1)}_{x,\xi}) \cup [t^{(1)}_{x,\xi}, t^{(2)}_{x,\xi}) \cup \cdots \cup [t^{(m_0)}_{x,\xi}, T].$$

We define the regions as below:

$$Z_j(N) = \{((t,x,\xi)) : t^{(j-1)}_{x,\xi} \le t < t^{(j)}_{x,\xi}\}, \quad j = 1,\ldots,m_0-1$$
$$Z_{m_0}(N) = \{((t,x,\xi)) : t^{(m_0)}_{x,\xi} \le t \le T\},$$

with $t^{(0)}_{x,\xi} = 0$. In all our cases $m_0 \le 2$.

The way we deploy these regions is that we first perform an excision of the irregular symbol (for example, see [10, Section 5.1]) so that the resulting symbol is smooth near $t = 0$ that is in $Z_1(N)$. The difference of these symbols is localised in $Z_1(N)$. Further, we deploy the diagonalisation procedure (for example, see [10, Section 5]) to restrict the singularities arising from t-derivatives to the appropriate regions. This kind of localisation of the singularities allows one to come up with a function $\nu(t)\Theta(x,\xi)$ that is used to define the loss operator (for example, see [10, Section 5.3]). For more details and interpretation, we refer the reader to [7–11].

24.4 Results

In the following we summarise the well-posedness results for singular hyperbolic Cauchy problems in a tabular fashion (See Tables 24.1 and 24.2). Recall the Definitions 24.1 and 24.2. Rows in bold are our results in the context. Figure 24.5 shows the sketch of proofs of our results.

24.5 Cone Condition

The L^1 integrability of the singularity plays a crucial role in arriving at the finite propagation speed. If the Cauchy data in (3.1) is such that $f \equiv 0$ and f_1, f_2 are supported in the ball $|x| \leq R$, then the solution to the Cauchy problem (24.1) is supported in the ball $|x| \leq R + \gamma_0 \omega(x)\tilde{\theta}(t)$ where $\tilde{\theta}$ is a monotone decreasing function such that

$$|\partial_x^\beta a_{i,j}(t,x)| \leq C_\beta \omega(x)^2 \tilde{\theta}(t), \text{ for } 1 \leq i, j \leq n.$$

The quantity $t\tilde{\theta}(t)$ is bounded in $[0,T]$. The constant γ_0 is such that the quantity $\gamma_0\omega(x)\tilde{\theta}(t)$ dominates the characteristic roots, i.e.,

$$\gamma_0 = \sup\left\{\sqrt{a(t,x,\xi)}\omega(x)^{-1}\tilde{\theta}(t)^{-1} : (t,x,\xi) \in [0,T] \times \mathbb{R}_x^n \times \mathbb{R}_\xi^n, |\xi| = 1\right\},$$

where $a(t,x,\xi) = \sum_{i,j=1}^n a_{i,j}(t,x)\xi_i\xi_j$. Note that the support of the solution increases as $|x|$ increases since $\omega(x)$ is monotone increasing function of $|x|$.

Let $K\left(x^0,t^0\right)$ denote the cone with the vertex $\left(x^0,t^0\right)$:

$$K\left(x^0,t^0\right) = \left\{(t,x) \in [0,T] \times \mathbb{R}^n : \left|x - x^0\right| \leq \gamma_0\omega(x)\tilde{\theta}\left(t^0 - t\right)\left(t^0 - t\right)\right\}.$$

Observe that the slope of the cone is anisotropic, that is, it varies with both x and t. See Fig. 24.6.

Table 24.1 Loss of regularity in case of oscillatory coefficients. Rows in bold correspond to the results

Order of Oscillations		Regularity in t of coefficients	Growth in x of coefficients		Loss of regularity index for solution	Ref.
q	$\tilde{\gamma}$		ω	Φ		
1	(0, 1)	$L^\infty([0,T]) \cap C^2((0,T])$	1	1	Zero to arbitrarily small	[12]
1	**(0,1)**	$\mathbf{C^2((0,T])}$	$\omega(x)$	$\Phi(x)$	**Zero to arbitrarily small**	**[10]**
1	1	$L^\infty([0,T]) \cap C^\infty((0,T])$	1	1	Finite	[6]
1	1	$L^\infty([0,T]) \cap C^\infty((0,T])$	$\langle x \rangle$	$\langle x \rangle$	Finite	[13]
1	$[1,\infty)$	$C^2((0,T])$	$\omega(x)$	$\Phi(x)$	**Finite to Infinite**	**[10]**
$(1,\infty)$	–	$C^1((0,T])$	1	1	Infinite	[2]
$\left(1,\frac{3}{2}\right)$	–	$C^1((0,T])$	$\omega(x)$	$\Phi(x)$	**Infinite**	**[7]**

Table 24.2 Loss of regularity in case of coefficients blowing-up at $t = 0$. Rows in bold correspond to our results

Rate of blow-up			Regularity in t of coefficients	Growth in x of coefficients		Loss of regularity index for solution	Ref.
p	q	$\tilde{\gamma}$		ω	Φ		
0	**1**	$(0, 1)$	$C^1((0, T])$	$\omega(x)$	$\Phi(x)$	**Arbitrarily small**	**[9]**
0	1	1	$C^1((0, T])$	–	–	Finite	[4]
0	1	1	$C^1((0, T])$	1	1	Finite	[2]
0	**1**	1	$C^1((0, T])$	$\omega(x)$	$\Phi(x)$	**Finite**	**[8]**
0	**1**	$(1, \infty)$	$C^2((0, T])$	$\omega(x)$	$\Phi(x)$	**Infinite**	**[10]**
$[0, 1)$	$(1, \infty)$	–	$C^1((0, T])$	–	–	Infinite	[4]
$\left(0, \frac{1}{2}\right)$	$\left(1, \frac{3}{2}\right)$	–	$C^1((0, T])$	$\omega(x)$	$\Phi(x)$	**Infinite**	**[11]**

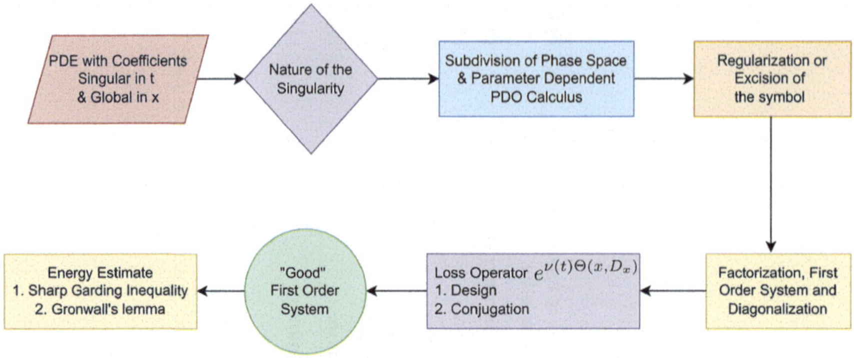

Fig. 24.5 Sketch of proofs of our well-posedness results

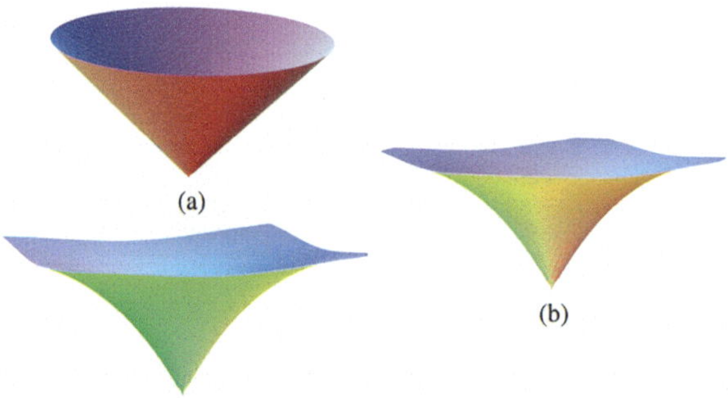

Fig. 24.6 The figures (**a**), (**b**) and (**c**) are the cones for the three cases of $(\omega(x), \tilde{\theta}(t)) : (1, 1)$, $(1, \ln(1 + 1/t))$ and $(\langle x \rangle^{\frac{1}{3}}, \ln(1 + 1/t))$, respectively

Theorem 24.1 *The Cauchy problem (3.1) has a cone dependence, that is, if*

$$f|_{K(x^0,t^0)} = 0, \quad f_i|_{K(x^0,t^0)\cap\{t=0\}} = 0, i = 1, 2$$

then

$$u|_{K(x^0,t^0)} = 0.$$

For the proof, we refer to [7–11].

Acknowledgments The works stated in this paper were carried out by R. R. Pattar during his doctoral studies under the mentorship of U. Kiran at Sri Sathya Sai Institute of Higher Learning. The authors dedicate this paper to the founder chancellor of the institute Bhagawan Sri Sathya Sai Baba on the occassion of his 97th birthday.

References

1. Ascanelli, A., Cappiello, M.: Hölder continuity in time for SG hyperbolic systems. J. Differ. Equ. **244**(8), 2091–2121 (2008)
2. Cicognani, M.: The Cauchy problem for strictly hyperbolic operators with non-absolutely continuous coefficients. Tsukuba J. Math. **27**, 1–12 (2003)
3. Cicognani, M., Lorenz, D.: Strictly hyperbolic equations with coefficients low-regular in time and smooth in space. J. Pseudo Differ. Oper. Appl. **9**, 643–675 (2018)
4. Colombini, F., Del Santo, D., Kinoshita, T.: Well-posedness of the Cauchy problem for a hyperbolic equation with non-Lipschitz coefficients. Ann. Scoula. Norm.-Sci. **1**(2), 327–358 (2002)
5. Hörmann, G., de Hoop, M.: Microlocal analysis and global solutions of some hyperbolic equations with discontinuous coefficients. Acta Appl. Math. **67**, 173–224 (2001)
6. Kubo, A., Reissig, M.: Construction of parametrix to strictly hyperbolic Cauchy problems with fast oscillations in nonLipschitz coefficients. Commun. Partial Differ. Equ. **28**, 1471–1502 (2003)
7. Pattar, R.R., Kiran, N.U.: Global well-posedness of a class of strictly hyperbolic Cauchy problems with coefficients non-absolutely continuous in time. Bull. Sci. Math. **171**, 103037 (2021)
8. Pattar, R.R., Kiran, N.U.: Strictly hyperbolic Cauchy problems on $\mathbb{R}^n$ with unbounded and singular coefficients. Annali dell' Universitá di Ferrara (2021)
9. Pattar, R.R., Kiran, N.U.: Strictly hyperbolic equations with coefficients sublogarithmic in time (2021). arXiv:2111.11701
10. Pattar, R.R., Kiran, N.U.: Energy estimates and global well-posedness for a broad class of strictly hyperbolic Cauchy problems with coefficients singular in time. J. Pseudo-Differ. Oper. Appl. **13**, 9 (2022)
11. Pattar, R.R., Kiran, N.U.: Global well-posedness of a class of singular hyperbolic Cauchy problems. Monatshefte für Mathematik (2022)
12. Reissig, M.: Hyperbolic equations with non-Lipschitz coefficients. Rend. Sem. Mat. Univ. Politec. Torino **61**, 135–181 (2003)
13. Uday Kiran, N., Coriasco, S., Battisti, U.: Hyperbolic operators with non-Lipschitz coefficients. Recent Advances in Theoretical & Computational Partial Differential Equations with Applications. Panjab University, Chandigarh (2016)

Chapter 25
On a Mixed Equation Involving Prabhakar Fractional Order Integral-Differential Operators

Erkinjon Karimov, Niyaz Tokmagambetov, and Muzaffar Toshpulatov

Abstract In this short note, we would like to give the main parts of the investigation of direct and inverse problems for mixed equations involving regularised Prabhakar fractional order integral-differential operator. Namely, we present the explicit solution of the Cauchy problem for an ordinary differential equation with the Prabhakar fractional derivative. We also have presented important statements on the bivariate Mittag-Leffler function. Namely, Euler-type integral representations and certain estimations for the bivariate Mittag-Leffler type function $E_2(x, y)$. The main targets are direct and inverse source problems for mixed equations involving regularised Prabhakar derivatives. We will present here the algorithm of the proof and a short description of the key steps.

25.1 Auxiliary Part: Cauchy Problem

To find a solution to the equation

$$ {}^{PC}D_{ax}^{\alpha,\beta,\gamma,\delta} y(x) - \lambda y(x) = f(x), \quad x > a, a \in \mathbb{R}, \tag{25.1}$$

E. Karimov
Fergana State University, Fergana, Uzbekistan

N. Tokmagambetov (✉)
Centre de Recerca Matemática, Edifici C, Bellaterra, Spain

Institute of Mathematics and Mathematical Modeling, Almaty, Kazakhstan
e-mail: tokmagambetov@crm.cat; tokmagambetov@math.kz

M. Toshpulatov
Andijan State University, Andijan, Uzbekistan

M. Ruzhansky, K. Van Bockstal (eds.), *Extended Abstracts 2021/2022*,
Research Perspectives Ghent Analysis and PDE Center 2,
https://doi.org/10.1007/978-3-031-42539-4_25

satisfying initial conditions

$$y^{(k)}(a) = A_k, \quad k = 0, 1, 2, \dots m - 1, \tag{25.2}$$

where $\alpha > 0$, $m - 1 < \beta \le m$, $m \in \mathbb{N}$, $\gamma, \lambda, \delta, A_k \in \mathbb{R}$, $f(x)$ is a given function,

$${}^{PC}D_{ax}^{\alpha,\beta,\gamma,\delta} y(x) = {}^{P}I_{ax}^{\alpha,m-\beta,-\gamma,\delta} \frac{d^m}{dx^m} f(x)$$

is a regularized Prabhakar derivative of order β [1],

$${}^{P}I_{ax}^{\alpha,\beta,\gamma,\delta} f(x) = \int_a^x (x-\xi)^{\beta-1} E_{\alpha,\beta}^{\gamma}(\delta(x-\xi)^{\alpha}) f(\xi) d\xi, \quad x > a \tag{25.3}$$

is the Prabhakar fractional integral of order β [1] and $E_{\alpha,\beta}^{\gamma}(z)$ is the Prabhakar function [2].

Lemma 25.1 *If $f(x) \in AC^m[a,\infty)$, then the problem has a unique solution represented as follows*

$$y(x) = \sum_{k=0}^{m-1} A_k \frac{(x-a)^k}{k!}$$
$$+ \sum_{k=0}^{m-1} A_k (x-a)^k (x-a)^{\beta} \lambda \Gamma(\gamma) \cdot E_2 \left(\begin{matrix} \gamma, \gamma, 1, 1, 0 \\ \beta+k+1, \beta, \alpha, \gamma, \gamma, 1, 1 \end{matrix} \middle| \begin{matrix} \lambda(x-\xi)^{\beta} \\ \delta(x-\xi)^{\alpha} \end{matrix} \right)$$
$$+ \Gamma(\gamma) \int_a^x (x-\xi)^{\beta-1} f(\xi) \cdot E_2 \left(\begin{matrix} \gamma, \gamma, 1, 1, 0 \\ \beta, \beta, \alpha, \gamma, \gamma, 1, 1 \end{matrix} \middle| \begin{matrix} \lambda(x-\xi)^{\beta} \\ \delta(x-\xi)^{\alpha} \end{matrix} \right) d\xi. \tag{25.4}$$

Here

$$E_2(x; y) = \sum_{i=0}^{\infty} \sum_{j=0}^{\infty} \frac{(\gamma_1)_{\alpha_1 m + \beta_1 n} (\gamma_2)_{\alpha_2 m}}{\Gamma(\delta_1 + \alpha_3 m + \beta_2 n)} \cdot \frac{x^m}{\Gamma(\delta_2 + \alpha_4 m)} \cdot \frac{y^n}{\Gamma(\delta_3 + \beta_3 n)} \tag{25.5}$$

is a bi-variate Mittag-Leffler function [3].

Proof We apply ${}^{P}I_{ax}^{\alpha,\beta,\gamma,\delta}$ on (25.1) and by using the formula

$${}^{P}I_{ax}^{\alpha,\beta,\gamma,\delta} ({}^{PC}D_{ax}^{\alpha,\beta,\gamma,\delta} y(x)) = y(x) - \sum_{k=0}^{m-1} y^{(k)}(a) \frac{(x-a)^k}{k!},$$

we obtain

$$y(x) - \sum_{k=0}^{m-1} y^{(k)}(a)\frac{(x-a)^k}{k!} - \lambda^P I_{ax}^{\alpha,\beta,\gamma,\delta} y(x) = \lambda^P I_{ax}^{\alpha,\beta,\gamma,\delta} f(x).$$

Using (25.2), we can rewrite it as follows

$$y(x) - \lambda \cdot {}^P I_{ax}^{\alpha,\beta,\gamma,\delta} y(x) = \tilde{f}(x), \tag{25.6}$$

where $\tilde{f}(x) = {}^P I_{ax}^{\alpha,\beta,\gamma,\delta} f(x) + \sum_{k=0}^{m-1} A_k \frac{(x-a)^k}{k!}$. Equation (25.6) is a Volterra integral equation of the second kind. We solve it by the method of successive approximation:

$$y_0 = \tilde{f}(x),$$

$$y_1 = \tilde{f}(x) + \lambda^P I_{ax}^{\alpha,\beta,\gamma,\delta} y_0,$$

$$\vdots$$

$$y_n = \tilde{f}(x) + \sum_{i=1}^{n} \lambda^{i\,P} I_{ax}^{\alpha,i\beta,i\gamma,\delta} \tilde{f}(x).$$

In the final step, we get that

$$y(x) = \lim_{n\to\infty} y_n = \tilde{f}(x) + \sum_{i=1}^{\infty} \lambda^{i\,P} I_{ax}^{\alpha,\beta i,\gamma i,\delta} \tilde{f}(x).$$

Considering the representation of $\tilde{f}(x)$, we get

$$y(x) = \sum_{k=0}^{m-1} A_k \frac{(x-a)^k}{k!} + {}^P I_{ax}^{\alpha,\beta,\gamma,\delta} f(x) + \sum_{i=1}^{\infty} \lambda^i \cdot {}^P I_{ax}^{\alpha,\beta i,\gamma i,\delta} \left(\sum_{k=0}^{m-1} A_k \frac{(x-a)^k}{k!} \right)$$
$$+ \sum_{i=1}^{\infty} \lambda^i \cdot {}^P I_{ax}^{\alpha,\beta i,\gamma i,\delta} \left({}^P I_{ax}^{\alpha,\beta i,\gamma i,\delta} f(x) \right).$$

Now, by using the formula

$${}^P I_{ax}^{\alpha,\beta_1,\gamma_1,\delta} \circ {}^P I_{ax}^{\alpha,\beta_2,\gamma_2,\delta} f = {}^P I_{ax}^{\alpha,\beta_1+\beta_2,\gamma_1+\gamma_2,\delta} f,$$

we deduce

$$y(x) = \sum_{k=0}^{m-1} A_k \frac{(x-a)^k}{k!} + \sum_{i=1}^{\infty} \lambda^i \cdot {}^P I_{ax}^{\alpha,\beta i,\gamma i,\delta} \left(\sum_{k=0}^{m-1} A_k \frac{(x-a)^k}{k!} \right)$$
$$+ \sum_{i=0}^{\infty} \lambda^i \cdot {}^P I_{ax}^{\alpha,\beta i+\beta,\gamma i+\gamma,\delta} f(x). \tag{25.7}$$

Now, we simplify the second summand of (25.7):

$$\sum_{i=1}^{\infty} \lambda^i {}^P I_{ax}^{\alpha,\beta i,\gamma i,\delta} \left(\sum_{k=0}^{m-1} A_k \frac{(x-a)^k}{k!} \right)$$
$$= \sum_{k=0}^{m-1} \frac{A_k}{k!} \sum_{i=1}^{\infty} \lambda^i \int_a^x (x-\xi)^{\beta i-1} E_{\alpha,\beta i}^{\gamma i}(\delta(x-\xi)^\alpha)(\xi-a)^k d\xi$$
$$= \sum_{k=0}^{m-1} \frac{A_k}{k!} \sum_{i=1}^{\infty} \lambda^i \sum_{j=0}^{\infty} \frac{\Gamma(\gamma i+j)}{\Gamma(\alpha j+\beta i)\Gamma(\gamma i)} \cdot \int_a^x (x-\xi)^{\beta i-1} (\delta(x-\xi)^\alpha)^j (\xi-a)^k d\xi$$
$$= \sum_{k=0}^{m-1} \sum_{i=1}^{\infty} A_k (x-a)^k (\lambda(x-a)^\beta)^i E_{\alpha,\beta i+k+1}^{\gamma i}(\delta(x-a)^\alpha).$$

Thus

$$\sum_{i=1}^{\infty} \lambda^i {}^P I_{ax}^{\alpha,\beta i,\gamma i,\delta} \left(\sum_{k=0}^{m-1} A_k \frac{(x-a)^k}{k!} \right)$$
$$= \sum_{i=1}^{\infty} (\lambda(x-a)^\beta)^i \sum_{k=0}^{m-1} A_k (x-a)^k E_{\alpha,\beta i+k+1}^{\gamma i}(\delta(x-a)^\alpha).$$

Now, we simplify the third summand of (25.7) using (25.3):

$$\sum_{i=0}^{\infty} \lambda^i {}^P I_{ax}^{\alpha,\beta i+\beta,\gamma i+\gamma,\delta} f(x)$$
$$= \int_a^x (x-\xi)^{\beta-1} \sum_{i=0}^{\infty} [\lambda(x-\xi)^\beta]^i \cdot E_{\alpha,\beta+\beta i}^{\gamma+\gamma i}(\delta(x-\xi)^\alpha) f(\xi) d\xi.$$

Thus, (25.7) is formed as follows:

$$y(x)=\sum_{k=0}^{m-1}A_k\frac{(x-a)^k}{k!}+\sum_{k=0}^{m-1}A_k(x-a)^k\sum_{i=1}^{\infty}(\lambda(x-a)^{\beta})^iE_{\alpha,\beta i+k+1}^{\gamma i}(\delta(x-a)^{\alpha})$$
$$+\int_a^x(x-\xi)^{\beta-1}f(\xi)\sum_{i=0}^{\infty}(\lambda(x-\xi)^{\beta})^iE_{\alpha,\beta+\beta i}^{\gamma+\gamma i}(\delta(x-\xi)^{\alpha})d\xi. \tag{25.8}$$

Next, we rewrite Eq. (25.8) by the function (25.5). For this, we do the following evaluations:

$$\sum_{i=0}^{\infty}[\lambda(x-a)^{\beta}]^{i+1}\sum_{j=0}^{\infty}\frac{(\gamma+\gamma i)_j(\delta(x-a)^{\alpha})^j}{j!\Gamma(\alpha j+\beta+\beta i+k+1)}$$
$$=\sum_{i=0}^{\infty}\sum_{j=0}^{\infty}\lambda(x-a)^{\beta}\cdot\frac{\Gamma(\gamma+\gamma i+j)}{j!\Gamma(\gamma+\gamma i)}\frac{[\lambda(x-a)^{\beta}]^i[\delta(x-a)^{\alpha}]^j}{\Gamma(\alpha j+\beta i+\beta+k+1)}$$
$$\Rightarrow\left\{\begin{array}{l}\gamma_1=\gamma,\alpha_1=\gamma,\beta_1=1,\gamma_2=1,\alpha_2=0\\ \alpha_3=\beta,\beta_2=\alpha,\delta_1=\beta+k+1\\ \delta_2=\gamma,\alpha_4=\gamma,\delta_3=1,\beta_3=1\end{array}\right\}$$
$$\Rightarrow(x-a)^{\beta}\Gamma(\gamma)\cdot E_2\left(\begin{array}{c}\gamma,\gamma,1,1,0\\ \beta+k+1,\beta,\alpha,\gamma,\gamma,1,1\end{array}\middle|\begin{array}{c}\lambda(x-a)^{\beta}\\ \delta(x-a)^{\alpha}\end{array}\right)$$

and

$$\sum_{i=0}^{\infty}[\lambda(x-\xi)^{\beta}]^i\sum_{j=0}^{\infty}\frac{\Gamma(\gamma+\gamma i+j)}{j!\Gamma(\gamma+\gamma i)}\cdot\frac{(\delta(x-\xi)^{\alpha})^j}{\Gamma(\alpha j+\beta+\beta i)}$$
$$\Rightarrow\left\{\begin{array}{l}\gamma_1=\gamma,\alpha_1=\gamma,\beta_1=1,\alpha_3=\beta,\beta_2=\alpha,\delta_1=\beta\\ \delta_2=\gamma,\alpha_4=\gamma,\delta_3=\beta_3=1,\gamma_2=1,\alpha_2=0\end{array}\right\}$$
$$\Rightarrow\Gamma(\gamma)E_2\left(\begin{array}{c}\gamma,\gamma,1,1,0\\ \beta,\beta,\alpha,\gamma,\gamma,1,1\end{array}\middle|\begin{array}{c}\lambda(x-\xi)^{\beta}\\ \delta(x-\xi)^{\alpha}\end{array}\right).$$

Therefore, (25.8) can be written as (25.4). Lemma 25.1 is proved. □

We note that by Mikusinsky's operational method, a similar problem was solved in [4], and the solution was represented by a double series. Similar research is done for the Hilfer-Prabhakar case [5]. The key properties of the Hilfer-Prabhakar derivative were studied in [6]. Below, we present, in particular cases, the explicit solutions, which will be used further. Here, the parameters $\alpha, \gamma, \delta, \lambda, A_1, A_2, B_0$ are supposed to be real numbers.

The problem

$$\begin{cases} {}^{PC}D_{0x}^{\alpha,\beta_1,\gamma,\delta} y(x) - \lambda y(x) = f(x), & 1 < \beta_1 \leq 2 \\ y(0) = A_1,\ y'(0) = A_2 \end{cases}$$

has an explicit solution given by

$$\begin{aligned} y(x) = A_1 + A_2 x + A_1 \lambda x^{\beta_1} \Gamma(\gamma) \cdot E_2 &\left(\begin{matrix} \gamma, \gamma, 1, 1, 0 \\ \beta_1 + 1, \beta_1, \alpha, \gamma, \gamma, 1, 1 \end{matrix} \middle| \begin{matrix} \lambda x^{\beta_1} \\ \delta x^{\alpha} \end{matrix} \right) \\ + A_2 x^{\beta_1 + 1} \lambda \Gamma(\gamma) \cdot E_2 &\left(\begin{matrix} \gamma, \gamma, 1, 1, 0 \\ \beta_1 + 2, \beta_1, \alpha, \gamma, \gamma, 1, 1 \end{matrix} \middle| \begin{matrix} \lambda x^{\beta_1} \\ \delta x^{\alpha} \end{matrix} \right) \\ + \Gamma(\gamma) \int_0^x (x - \xi)^{\beta_1 - 1} f(\xi) E_2 &\left(\begin{matrix} \gamma, \gamma, 1, 1, 0 \\ \beta_1, \beta_1, \alpha, \gamma, \gamma, 1, 1 \end{matrix} \middle| \begin{matrix} \lambda (x - \xi)^{\beta_1} \\ \delta (x - \xi)^{\alpha} \end{matrix} \right) d\xi. \end{aligned} \tag{25.9}$$

The Cauchy problem

$$\begin{cases} {}^{PC}D_{ax}^{\alpha,\beta_2,\gamma,\delta} g(x) - \lambda g(x) = q(x), & 0 < \beta_2 \leq 1 \\ y(a) = B_0 \end{cases}$$

has an explicit solution given by

$$\begin{aligned} g(x) = B_0 + B_0 (x - a)^{\beta_2} \cdot \lambda \cdot \Gamma(\gamma) \cdot E_2 &\left(\begin{matrix} \gamma, \gamma, 1, 1, 0 \\ \beta_2 + 1, \beta, \alpha, \gamma, \gamma, 1, 1 \end{matrix} \middle| \begin{matrix} \lambda (x - a)^{\beta_2} \\ \delta (x - a)^{\alpha} \end{matrix} \right) \\ + \Gamma(\gamma) \int_a^x (x - \xi)^{\beta_2 - 1} q(\xi) \cdot E_2 &\left(\begin{matrix} \gamma, \gamma, 1, 1, 0 \\ \beta_2, \beta_2, \alpha, \gamma, \gamma, 1, 1 \end{matrix} \middle| \begin{matrix} \lambda (x - \xi)^{\beta_2} \\ \delta (x - \xi)^{\alpha} \end{matrix} \right) d\xi. \end{aligned} \tag{25.10}$$

25.2 Direct and Inverse Problems

We note that boundary problems for partial differential equations involving Prabhakar derivative were investigated in several works [7–9]. Especially, in the work [10], the direct problem for the subdiffusion equation involving regularised Prabhakar derivative has been tested for unique solvability.

We consider the following mixed equation

$$\frac{1-\operatorname{sign}(t-a)}{2}\cdot {}^{PC}D_{ot}^{\alpha,\beta_1,\gamma,\delta}u(t,x)$$

$$+\frac{1+\operatorname{sign}(t-a)}{2}\cdot {}^{PC}D_{at}^{\alpha,\beta_2,\gamma,\delta}u(t,x)-u_{xx}(t,x)=f(t,x) \qquad (25.11)$$

in a domain $\Omega = \Omega_1 \cup \Omega_2 \cup J$, where $J = \{(t,x) : t = a, 0 < x < 1\}$, $\Omega_1 = \{(t,x) : 0 < x < 1, a < t < b\}$, $\Omega_2 = \{(t,x) : 0 < x < 1, 0 < t < a\}$, $a, b \in R^+$ such that $b > a$, $1 < \beta_1 \le 2$, $0 < \beta_2 \le 1$, $\alpha, \gamma, \delta \in \mathbb{R}$, $f(t,x)$ is a given function.

Problem 25.1 (Direct Problem) Find a regular solution of (25.11) in Ω satisfying the following conditions:

$$u(t,0) = u(t,1) = 0, \quad 0 \le t \le b, \qquad (25.12)$$

$$u(0,x) = \varphi(x), \quad 0 \le x \le 1, \qquad (25.13)$$

$$\lim_{t\to a+} {}^{PC}D_{at}^{\alpha,\beta_2,\gamma,\delta}u(t,x) = \lim_{t\to a-} u_t(t,x), \quad 0 < x < 1. \qquad (25.14)$$

Definition 25.1 Function $u(t,x)$ is called a regular solution of (25.11) in Ω, if $u(t,x) \in C(\bar{\Omega})$, $u_{xx}(t,x) \in C(\Omega_1 \cup \Omega_2)$, ${}^{PC}D_{0t}^{\alpha,\beta_1,\gamma,\delta}u(t,x) \in C(\Omega_2)$, ${}^{PC}D_{0t}^{\alpha,\beta_2,\gamma,\delta}u(t,x) \in C(\Omega_1)$ and satisfies (25.11) in Ω.

We note that the given function $\varphi(x)$ should satisfy the following matching conditions:

$$\varphi(0) = \varphi(1) = 0.$$

We search for a solution to the problem as follows:

$$u(t,x) = \sum_{k=0}^{\infty} {}_1T_k(t)\cdot \sin k\pi x, \quad 0 \le t \le a \qquad (25.15)$$

and

$$u(t,x) = \sum_{k=0}^{\infty} {}_2T_k(t)\cdot \sin k\pi x, \quad a \le t \le b. \qquad (25.16)$$

We substitute (25.15) and (25.16) into (25.11) and will get the following fractional differential equations:

$$ {}^{PC}D_{0t}^{\alpha,\beta_1,\gamma,\delta}\cdot {}_1T_k(t)-(k\pi)^2\cdot {}_1T_k(t)=f_k(t),\quad 0<t<a, \tag{25.17}$$

$$ {}^{PC}D_{at}^{\alpha,\beta_2,\gamma,\delta}\cdot {}_2T_k(t)-(k\pi)^2\cdot {}_2T_k(t)=f_k(t),\quad a<t<b, \tag{25.18}$$

where $f_k(t)=2\cdot\int_0^1 f(t,x)\sin k\pi x dx$ is the Fourier coefficients of the function $f(t,x)=\sum\limits_{k=0}^{\infty} f_k(t)\sin k\pi x$.

General solutions of (25.17) and (25.18) can be written as follows (see (25.9) and (25.10)):

$$\begin{aligned} {}_1T_k(t) &= {}_1A_k+{}_2A_k t-{}_1A_k\cdot(k\pi)^2\cdot t^{\beta_1}\cdot\Gamma(\gamma)\cdot E_2\\ &\quad\times\left(\begin{matrix}\gamma,\gamma,1,1,0\\ \beta_1+1,\beta_1,\alpha,\gamma,\gamma,1,1\end{matrix}\middle|\begin{matrix}-(k\pi)^2t^{\beta_1}\\ \delta t^{\alpha}\end{matrix}\right)\\ &\quad-{}_2A_k\cdot t^{\beta_1+1}\cdot(k\pi)^2\cdot E_2\left(\begin{matrix}\gamma,\gamma,1,1,0\\ \beta_1+2,\beta_1,\alpha,\gamma,\gamma,1,1\end{matrix}\middle|\begin{matrix}-(k\pi)^2t^{\beta_1}\\ \delta t^{\alpha}\end{matrix}\right)\\ &\quad+\Gamma(\gamma)\cdot\int_0^t(t-\xi)^{\beta_1-1}\cdot f_k(\xi)\cdot E_2\left(\begin{matrix}\gamma,\gamma,1,1,0\\ \beta_1,\beta_1,\alpha,\gamma,\gamma,1,1\end{matrix}\middle|\begin{matrix}-(k\pi)^2\cdot(t-\xi)^{\beta_1}\\ \delta(t-\xi)^{\alpha}\end{matrix}\right)d\xi,\end{aligned} \tag{25.19}$$

$$\begin{aligned} {}_2T_k(t) &= {}_1B_k\\ &\quad+{}_1B_k\cdot(k\pi)^2\cdot(t-a)^{\beta_2}\cdot\Gamma(\gamma)\cdot E_2\left(\begin{matrix}\gamma,\gamma,1,1,0\\ \beta_2+1,\beta_2,\alpha,\gamma,\gamma,1,1\end{matrix}\middle|\begin{matrix}-(k\pi)^2(t-a)^{\beta_2}\\ \delta(t-a)^{\alpha}\end{matrix}\right)\\ &\quad+\Gamma(\gamma)\cdot\int_a^t(t-\xi)^{\beta_2-1}\cdot f_k(\xi)\cdot E_2\left(\begin{matrix}\gamma,\gamma,1,1,0\\ \beta_2,\beta_2,\alpha,\gamma,\gamma,1,1\end{matrix}\middle|\begin{matrix}-(k\pi)^2(t-\xi)^{\beta_2}\\ \delta(t-\xi)^{\alpha}\end{matrix}\right)d\xi.\end{aligned} \tag{25.20}$$

Here ${}_1A_k$, ${}_2A_k$ and ${}_1B_k$ are unknown constants to be found. To find these unknown constants, we will use condition (25.13) and the glueing condition (25.14). In order to prove the uniform convergence of infinite series (25.15) and (25.16), we will need the appropriate estimation for the bivariate Mittag-Leffler function E_2. This can be done using the following statement.

Lemma 25.2 *If* $(\alpha_3 > \alpha_4, \alpha_3 > 0, \beta_2 > \beta_3, \beta_2 > 0)$, *the the following Euler-type integral representations holds true:*

$$E_2\left(\begin{matrix}\gamma_1, \alpha_1, \beta_1, \gamma_2, \alpha_2 | \, x \\ \delta_1, \alpha_3, \beta_2, \delta_2, \alpha_4, \delta_3, \beta_3 | \, y\end{matrix}\right) = \frac{1}{\Gamma(\gamma_1)} \cdot \frac{1}{\Gamma(\gamma_2)}$$

$$\cdot \int_0^1 \int_0^\infty \xi^{\frac{\delta_1}{2}-1} \cdot \eta^{\gamma_1 - 1} \cdot (1-\xi)^{\frac{\delta_1}{2}-1} \cdot e^{-\eta} e_{\alpha_3, -\alpha_4}^{\frac{\delta_1}{2}, \delta_2}(x\eta^{\alpha_1}\xi^{\alpha_3})$$

$$\cdot e_{\beta_2, -\beta_3}^{\frac{\delta_1}{2}, \delta_3}(y\eta^{\beta_1}(1-\xi)^{\beta_2}) d\xi d\eta,$$

where $e_{\alpha,\beta}^{\mu,\delta}(z) = \sum\limits_{k=0}^{\infty} \frac{z^n}{(\alpha k + \mu)(\delta - \beta k)}$, $\alpha > \beta, \alpha > 0$ *is the Wright-type function.*

Problem 25.2 (Space-Dependent Inverse Source Problem) To find a pair of functions $\{u(t, x); h(x)\}$ satisfying Eq. (25.11) in the domain Ω, together with conditions (25.12)–(25.14) and the following overdetermination condition

$$u(T, x) = \psi(x), 0 \leq x \leq 1,$$

where we suppose that $f(t, x) = h(x)g(t, x)$ with the given function $g(t, x)$. We will use the same tool as in Problem 25.1. The additional condition to the given data will appear due to the overdetermination condition.

Problem 25.3 (Time-Dependent Inverse Source Problem) To find a pair of functions $\{u(t, x); r(t)\}$ satisfying Eq. (25.11) in the domain Ω, together with conditions (25.13)–(25.14) and the following nonlocal conditions

$$u(t, 0) = u(t, 1), \ 0 \leq t \leq b, \ u_x(t, 1) = 0, \ 0 < t < b$$

and overdetermination conditions

$$\int_0^1 u(t, x) dx = F_1(t), 0 \leq t \leq a, \ \int_0^1 u(t, x) dx = F_2(t), a \leq t \leq b.$$

Here, we suppose that $f(t, x) = r(t)g(t, x)$ with the given function $g(t, x)$. We will use the same tool as in Problem 25.1.

Acknowledgments This research is funded by the Science Committee of the Ministry of Science and Higher Education of the Republic of Kazakhstan (Grant No. BR20281002). NT is also supported by the Beatriu de Pinós programme and by AGAUR (Generalitat de Catalunya) grant 2021 SGR 00087.

References

1. D'Ovidio, M., Polito, F.: Fractional diffusion-telegraph equations and their associated stochastic solutions. Theory Probab. Appl. **62**(4), 552–574 (2018). arXiv: 1307.1696 (2013)
2. Prabhakar, T.R.: A singular integral equation with a generalized Mittag-Leffler function in the kernel. Yokohama Math. J. **19**, 7–15 (1971)
3. Garg, M., Manohar, P., Kalla, S.L.: A Mittag- Leffler-type function of two variables. Integral Transf. Special Funct. **24**(11), 934–944 (2013)
4. Rani, N., Fernandez, A.: Solving prabhakar differential equations using Mikusinski's operational calculus. Comput. Appl. Math. **41**(107), 15 (2022) https://doi.org/10.1007/s40314-022-01794-6
5. Rani, N., Fernandez, A., Tomovski, Z.: An operational calculus approach to Hilfer-Prabhakar fractional derivatives. Banach J. Math. Anal. **17**(33) (2023). https://doi.org/10.1007/s43037-023-00258-1
6. Garra, R., et al.: Hilfer-Prabhakar derivatives and some applications. Appl. Math. Comput. **242**, 576–589 (2014)
7. Bokhari, A., et al.: Regularized Prabhakar derivative for partial differential equations. Comput. Methods Differ. Equ. **10**(3), 726–737 (2022)
8. Elhadedy, H., et al.: Exact solution for heat equation inside a sphere with heat absorption using the regularized Hilfer-Prabhakar derivative. J. Appl. Math. Comput. Mech. **21**(2), 27–37 (2022)
9. Povstenko, Y., Klekot, J.: Fractional heat conduction with heat absorption in a sphere under Dirichlet boundary condition. Comput. Appl. Math. **37**(4), 4475–4483 (2018)
10. Karimov, E.T., Hasanov, A.: On a boundary-value problem in a bounded domain for a time-fractional diffusion equation with the Prabhakar fractional derivative. Bulletin of the Karaganda University. Mathematics Series, **111**(3), 39–46 (2023). http://dx.doi.org/10.31489/2023M3/

Chapter 26
Inverse Problem of Determining a Time-Dependent Source in a Fractional Langevin-Type Partial Differential Equation

Bakhodirjon Toshtemirov

Abstract In the current paper, we are interested in studying the time-dependent inverse source problem for the space-degenerate fractional Langevin-type PDE involving a bi-ordinal Hilfer fractional derivative. Sufficient conditions for the given data were established for the existence and uniqueness of the solution. The technique to show the existence result is based on the uniform convergence of the series.

26.1 Introduction

The study of inverse source problems has been an important target in mathematical research due to its applications in science and engineering [1, 2]. Showing the global (in time) existence and uniqueness of a solution is a complicated task. Another goal in inverse source problems is the description of a constructive algorithm for finding a solution. We suggest the references [3] and [4] to provide readers with detailed information on some methods for solving inverse source problems devoted to determining the factor dependent on t in the source, such as analytical and numerical techniques.

Letting the source term have the form $f(x,t) = a(t)h(x)$, then the inverse problem consists of determining a source term $a(t)$ and a function $u(x,t)$ (the temperature distribution in the heat process), from the initial data $\psi(x)$ (the initial temperature in the heat process) and boundary conditions that come from regularity conditions. As an additional space measurement, we will use the integral condition taken over the whole domain.

B. Toshtemirov (✉)
Ghent University, Ghent, Belgium

Alasala College, Dammam, Saudi Arabia
e-mail: bakhodirjon.toshtemirov@ugent.be

M. Ruzhansky, K. Van Bockstal (eds.), *Extended Abstracts 2021/2022*, Research Perspectives Ghent Analysis and PDE Center 2,
https://doi.org/10.1007/978-3-031-42539-4_26

26.1.1 Mathematical Setting

Let $\gamma_1 < \gamma_2 - \delta_2$, $T > 0$ be an arbitrarily fixed time and $\Omega = \{(x,t) : -1 < x < 1,\ 0 < t \le T\}$.

The inverse source problem (ISP) here is to find a pair $\{u(x,t), a(t)\}$ functions for given $h(x)$, $\psi(x)$, $\varphi(x)$ such that

$$D_{0+}^{(\alpha_1,\beta_1)\mu_1}\left(D_{0+}^{(\alpha_2,\beta_2)\mu_2}u(x,t) - \frac{\partial}{\partial x}\left[(1-x^2)u_x(x,t)\right]\right) = h(x)a(t), \tag{26.1}$$

$$\lim_{t\to 0+} I_{0+}^{(1-\mu_2)(1-\beta_2)}u(x,t) = \psi(x),\ -1 \le x \le 1, \tag{26.2}$$

$$\lim_{t\to 0+} I_{0+}^{1-\gamma_1-\delta_2}u(x,t) = \varphi(x),\ \ -1 \le x \le 1, \tag{26.3}$$

where $h(x)$, $\psi(x)$ and $\varphi(x)$ are given functions, $D_{0+}^{\alpha_i,\beta_i}\mu_i$ is a bi-ordinal Hilfer fractional derivative defined by

$$D_{0t}^{(\alpha_i,\beta_i)\mu_i}y(x) := I_{0+}^{\mu_i(1-\alpha_i)}\frac{d}{dx}I_{0+}^{(1-\mu_i)(1-\beta_i)}y(x), \tag{26.4}$$

where $0 < \alpha_i, \beta_i < 1$, $0 \le \mu_i \le 1$, $\delta_i = \beta_i + \mu_i(\alpha_i - \beta_i)$, $\gamma_i = \beta_i + \mu_i(1-\beta_i)$, $i = \overline{1,2}$ and $I_{0+}^{\gamma}y(x)$ is the Riemann-Liouville integral operator of order γ of a function $y(x)$ [5]. Also, for some investigations of different problems with the bi-ordinal Hilfer derivative, we refer to [6, 7].

We provide the over-determination condition as a way to make the inverse problem uniquely solved:

$$\int_{-1}^{1} u(x,t)dx = E(t), \tag{26.5}$$

where $E(t) \in AC^2([0,T],\mathbb{R})$. We also consider the following regularity conditions for the solution of the inverse source problem (26.1–26.5)

$$t^{1-\delta_2-\gamma_1}u \in C(\overline{\Omega}),\quad t^{1-\delta_2-\gamma_1}u_x \in C(\overline{\Omega}),\quad t^{1-\delta_2-\gamma_1}a \in C[0,T],$$

$$D_{0+}^{(\alpha_1,\beta_1)\mu_1}D_{0+}^{(\alpha_2,\beta_2)\mu_2}u \in C(\Omega),\quad u_{xx} \in C(\Omega).$$

The direct problem related to the ISP was studied in [8] for Eq. (26.1) and here we recall some properties of Legendre polynomials which are defined by

$$P_k(x) = \frac{1}{2^k \cdot k!}\frac{d^k(x^2-1)^k}{dx^k}.$$

The Legendre polynomials (see W. Kaplan [9, p. 511]) form a complete orthogonal system in $[-1, 1]$ and any piecewise continuous function g can be expressed in the form of Fourier-Legendre series with respect to the system $\{P_k(x)\}$:

$$g(x) = \sum_{k=0}^{\infty} c_k P_k(x), \quad c_k = \frac{(g, P_k)}{\|P_k\|^2} = \frac{2k+1}{2} \int_{-1}^{1} g(x) P_k(x) dx.$$

In this paper, we are also concerned with studying the time-dependent inverse source problem for Eq. (26.1) and we take another more favourable condition in order to facilitate calculations instead of a nonlocal condition.

Next, we present an important property of the Mittag-Leffler-type function, which is used in the sequel.

Lemma 26.1 (See [5]) *Let* $\alpha < 2$, $\beta \in \mathbb{R}$ *and* $\frac{\pi\alpha}{2} < \mu < \min\{\pi, \pi\alpha\}$. *Then, the following estimate holds true*

$$|E_{\alpha,\beta}(z)| \leq \frac{M}{1+|z|}, \quad \mu \leq |argz| \leq \pi, \quad |z| \geq 0,$$

where M is a positive constant.

26.2 Main Results

First, we state our main result.

Theorem 26.1 *Let* $\gamma_1 < \gamma_2 - \delta_2$, $0 < \left(\int_{-1}^{1} h(x)dx \right)^{-1} < M$ *such that* $h^{(4)} \in L^2(-1, 1)$, $\psi'' \in L^2(-1, 1)$, $\varphi'' \in L^2(-1, 1)$, $E'' \in L^1(0, T)$, *then the unique solution of the inverse source problem (26.1–26.5) exists.*

We now discuss the uniqueness and existence of a solution in separate subsections.

26.2.1 Uniqueness of the Solution

Let there exist two solutions $u_1(x, t)$ and $u_2(x, t)$ of the main problem and consider the function $u(x, t) = u_1(x, t) - u_2(x, t)$ which is a solution of Eq. (26.1) in the homogeneous case with homogeneous initial conditions

$$\lim_{t\to 0+} I_{0+}^{(1-\mu_2)(1-\beta_2)} u(x,t) = 0, \ -1 \le x \le 1, \tag{26.6}$$

$$\lim_{t\to 0+} I_{0+}^{1-\gamma_1-\delta_2} u(x,t) = 0, \ -1 \le x \le 1, \tag{26.7}$$

Let us consider the following function

$$u_k(t) = \int_{-1}^{1} u(x,t) P_k(x) dx, \ k = 0, 1, 2, \ldots,. \tag{26.8}$$

Based on (26.8), we consider the function below

$$v_k(t) = \int_{\varepsilon-1}^{1-\varepsilon} u(x,t) P_k(x) dx, \ k = 0, 1, 2, \ldots,. \tag{26.9}$$

where ε is a very small positive number. Applying the operator $D_{0+}^{(\alpha_1,\beta_1)\mu_1} D_{0+}^{(\alpha_2,\beta_2)\mu_2}$ with respect to t to both sides of equality (26.9) and using the homogeneous equation corresponding with (26.1) yield that

$$\begin{aligned} D_{0+}^{(\alpha_1,\beta_1)\mu_1} D_{0+}^{(\alpha_2,\beta_2)\mu} v_k(t) &= \int_{\varepsilon-1}^{1-\varepsilon} D_{0+}^{(\alpha_1,\beta_1)\mu_1} D_{0+}^{(\alpha_2,\beta_2)\mu_2} u(x,t) P_k(x) dx \\ &= \int_{\varepsilon-1}^{1-\varepsilon} P_k(x) D_{0+}^{(\alpha_1,\beta_1)\mu_1} \frac{\partial}{\partial x}\left[(1-x^2)u_x(x,t)\right] dx. \end{aligned}$$

Integrating by parts twice the right side of the last equality and calculating the limit as $\varepsilon \to 0$ give that

$$D_{0+}^{(\alpha_1,\beta_1)\mu_1}\left[D_{0+}^{(\alpha_2,\beta_2)\mu_2} + \lambda_k\right] u(t) = 0.$$

Obviously, it can be shown that this equation with homogeneous conditions (26.6)–(26.7) has only a trivial solution $u_k(t) \equiv 0, \ t \in [0, T]$ and therefore, from (26.8) we get that

$$\int_{-1}^{1} u(x,t) P_k(x) dx = 0, \ k = 0, 1, 2, \ldots .$$

Therefore, using the completeness property of the system $\{P_k(x)\}$, it is deduced that $u(x,t) \equiv 0$ in Ω, which proves the uniqueness of a solution to the ISP. Similarly, the uniqueness result for $a(t)$ can be obtained by following the steps described above.

26.2.2 Existence Result

By applying the standard procedure of the Fourier method, we obtain the following representation for the solution of (26.1–26.5) for arbitrary $a \in C[0,T]$:

$$u(x,t) = \sum_{k=0}^{\infty} u_k(t) P_k(x), \tag{26.10}$$

where

$$\begin{aligned} u_k(t) &= \psi_k t^{\gamma_2 - 1} E_{\delta_2,\gamma_2}\left(-\lambda_k t^{\delta_2}\right) + \varphi_k t^{\delta_2+\gamma_1-1} E_{\delta_2,\delta_2+\gamma_1}(-\lambda_k t^{\delta_2}) \\ &+ h_k \int_0^t (t-s)^{\delta_2+\delta_1-1} E_{\delta_2,\delta_2+\delta_1}\left[-\lambda_k (t-s)^{\delta_2}\right] a(s) ds, \end{aligned} \tag{26.11}$$

with

$$\psi_k = \frac{2k+1}{2} \int_{-1}^{1} \psi(x) P_k(x) dx, \qquad \varphi_k = \frac{2k+1}{2} \int_{-1}^{1} \varphi(x) P_k(x) dx,$$

$$h_k = \frac{2k+1}{2} \int_{-1}^{1} h(x) P_k(x) dx, \quad k = 0, 1, 2, \dots .$$

Taking fractional derivative $D_{0+}^{(\alpha_2,\beta_2)\mu_2}$ and $D_{0+}^{(\alpha_1,\beta_1)\mu_1}$ under the integral sign of the over-determination condition (26.5) and in view of Eq. (26.1), we obtain

$$a(t) = \left(\int_{-1}^{1} h(x) dx \right)^{-1} \cdot D_{0+}^{(\alpha_1,\beta_1)\mu_1} D_{0+}^{(\alpha_2,\beta_2)\mu_2} E(t). \tag{26.12}$$

Substituting (26.12) into (26.11) and assuming that $D_{0+}^{(\alpha_2,\beta_2)\mu_2}E \in L^1(0,T)$, $E \in C[0,T]$ and $E' \in L^1(0,T)$, we get step by step that

$$u_k(t) = \psi_k t^{\gamma_2-1} E_{\delta_2,\gamma_2}\left(-\lambda_k t^{\delta_2}\right) + \varphi_k t^{\delta_2+\gamma_1-1} E_{\delta_2,\delta_2+\gamma_1}(-\lambda_k t^{\delta_2})$$
$$+ h_k h^* \left(E(t) - \lambda_k \int_0^t (t-s)^{\delta_2-1} E_{\delta_2,\delta_2}\left[-\lambda_k (t-s)^{\delta_2}\right] E(s)ds \right), \tag{26.13}$$

where $h^* := \left(\int_{-1}^{1} h(x)dx\right)^{-1}$.

Now, let us establish the necessary condition for $E(t)$ in order to show the existence of (26.12) and here we might take a stronger condition for $E(t)$.

By considering the definition of bi-ordinal Hilfer fractional derivative, we make sure that $I_{0+}^{1-\gamma_1}\left(D_{0+}^{(\alpha_2,\beta_2)\mu_2}E\right) \in AC[0,T]$ or $D_{0+}^{(\alpha_2,\beta_2)\mu_2}E \in AC[0,T]$. From the last condition, we derive that $E' \in C[0,T]$ and $E'' \in L^1(0,T)$. These conditions ensure that $a \in C[0,T]$.

Let (f,g) be the scalar product of the functions f and g in $L^2(-1,1)$. By means of the properties of Legendre polynomials, one can have

$$\psi_k = \frac{2k+1}{2} \int_{-1}^{1} \psi(x) P_k(x) dx$$
$$= \frac{2k+1}{2} \int_{-1}^{1} \psi(x) \frac{1}{2k+1} \left[P'_{k+1}(x) - P'_{k-1}(x)\right] dx.$$

Integration by parts yields

$$\psi_k = -\frac{1}{2} \int_{-1}^{1} \psi(x) \left[P_{k+1}(x) - P_{k-1}(x)\right] dx = -\frac{1}{2}\left[(\psi', P_{k+1}) - (\psi', P_{k-1})\right].$$

Considering the Schwartz inequality $|(f,g)| \le \|f\|\|g\|$, we estimate ψ_k:

$$|\psi_k| \le \frac{1}{2}\left|(\psi', P_{k+1})\right| + \frac{1}{2}\left|(\psi', P_{k-1})\right|$$
$$\le \frac{1}{2}\left[\|\psi'\| \cdot \|P_{k+1}\| + \|\psi'\| \cdot \|P_{k+1}\|\right]$$

$$
\begin{aligned}
&\leq \frac{1}{2}\|\psi'\|\left(\frac{\sqrt{2}}{(2k+3)^{\frac{1}{2}}}+\frac{\sqrt{2}}{(2k-1)^{\frac{1}{2}}}\right)\\
&\leq \frac{\|\psi'\|\sqrt{2}}{(2k-1)^{\frac{1}{2}}}.
\end{aligned}
$$

Repeating this process, we can obtain

$$
|\psi_k| \leq \frac{4\sqrt{2}}{(2k-3)^{\frac{3}{2}}}\|\psi''(\cdot)\|. \tag{26.14}
$$

Clearly, the estimate (26.14) is applicable to the function h:

$$
|h_k| \leq \frac{4\sqrt{2}}{(2k-3)^{3/2}}\|h''(\cdot)\|.
$$

Moreover, repeating the process twice (integration by parts) for the function h, one can obtain the following estimate

$$
|h_k| \leq \frac{6\sqrt{2}}{(2k-7)^{7/2}}\|h^{(4)}(\cdot)\|.
$$

Similarly, these estimates are also available for the function φ.

Now, by taking estimate for the series $t^{1-\delta_2-\gamma_1}u(x,t)$, for all $t \in [0,T]$, we get that

$$
\begin{aligned}
\left|t^{1-\delta_2-\gamma_1}u(x,t)\right| &\leq \sum_{k=0}^{\infty}|u_k(t)|\cdot|P_k(x)|\\
&\leq \sum_{k=0}^{\infty}|\psi_k|\left|t^{\gamma_2-\delta_2-\gamma_1}E_{\delta_2,\gamma_2}(-\lambda_k t^{\delta_2})\right|\\
&\quad+\sum_{k=0}^{\infty}|\varphi_k|\left|E_{\delta_2,\delta_2-\gamma_2+\gamma_1+1}(-\lambda_k t^{\delta_2})\right|\\
&\quad+\sum_{k=0}^{\infty}|h_k||h^*|\left(|E(t)|+|E(0)|\,\lambda_k\left|t^{\delta_2}E_{\delta_2,\delta_2+1}(-\lambda_k t^{\delta_2})\right|\right)\\
&\quad+\sum_{k=0}^{\infty}|h_k||h^*|\lambda_k\int_0^t\left|(t-s)^{\delta_2}E_{\delta_2,\delta_2+1}\right.\\
&\quad\times\left.[-\lambda_k(t-s)^{\delta_2}]\right|\left|E'(s)\right|ds
\end{aligned}
$$

$$
\begin{aligned}
&\leq \sum_{k=0}^{\infty} |\psi_k| \left(\frac{Mt^{\gamma_2 - 1}}{1 + \lambda_k t^{\delta_2}} \right) + \sum_{k=0}^{\infty} |\varphi_k| \left(\frac{Mt^{\delta_2 + \gamma_1 - 1}}{1 + \lambda_k t^{\delta_2}} \right) \\
&\quad + \sum_{k=0}^{\infty} |h_k||h^*| \left(|E(t)| + |E(0)| \lambda_k \left(\frac{Mt^{\delta_2}}{1 + \lambda_k t^{\delta_2}} \right) \right) \\
&\quad + \sum_{k=0}^{\infty} |h_k||h^*| \int_0^t \left(\frac{M\lambda_k (t-s)^{\delta_2}}{1 + \lambda_k (t-s)^{\delta_2}} \right) |E'(s)| \, ds \\
&\leq \|\psi''(\cdot)\| \sum_{k=0}^{\infty} \frac{4\sqrt{2}}{(2k-3)^{3/2}} \left(\frac{Mt^{\gamma_2 - 1}}{1 + \lambda_k t^{\delta_2}} \right) \\
&\quad + \|\varphi''(\cdot)\| \sum_{k=0}^{\infty} \frac{4\sqrt{2}}{(2k-3)^{3/2}} \left(\frac{Mt^{\delta_2 + \gamma_1 - 1}}{1 + \lambda_k t^{\delta_2}} \right) \\
&\quad + \|h''(\cdot)\| \sum_{k=0}^{\infty} \frac{|h^*| 4\sqrt{2}}{(2k-3)^{3/2}} \left(|E(t)| + |E(0)| \left(\frac{M\lambda_k t^{\delta_2}}{1 + \lambda_k t^{\delta_2}} \right) \right) \\
&\quad + \|h''(\cdot)\| \sum_{k=0}^{\infty} \frac{|h^*| 4\sqrt{2}}{(2k-3)^{3/2}} \int_0^t \left(\frac{M\lambda_k (t-s)^{\delta_2}}{1 + \lambda_k (t-s)^{\delta_2}} \right) |E'(s)| \, ds.
\end{aligned}
$$

We presume $E' \in C[0, T]$ and, considering the Weierstrass M-test, one can see that the series representation of $t^{1-\delta_2-\gamma_1} u(x, t)$ converges uniformly.

Now, for the second term of Eq. (26.1), we have that

$$
\begin{aligned}
&D_{0+}^{(\alpha_1, \beta_1)\mu_1} \frac{\partial}{\partial x} \left[(1 - x^2) u_x \right] \\
&= - \sum_{k=0}^{\infty} \lambda_k D_{0+}^{(\alpha_1, \beta_1)\mu_1} u_k(t) P_k(x) \\
&= - \sum_{k=0}^{\infty} \lambda_k \Big[\psi_k t^{\gamma_2 - \delta_1 - 1} E_{\delta_2, \gamma_2 - \delta_1}(-\lambda_k t^{\delta_2}) \\
&\quad + \varphi_k t^{\delta_2 + \gamma_1 - \delta_1 - 1} E_{\delta_2, \delta_2 + \gamma_1 - \delta_1}(-\lambda_k t^{\delta_2}) \\
&\quad + h_k h^* D_{0+}^{(\alpha_1, \beta_1)\mu_1} E(t) - \lambda_k h_k h^* E(0) t^{\delta_2 - \delta_1} E_{\delta_2, \delta_2 - \delta_1 + 1}(-\lambda_k t^{\delta_2}) \\
&\quad - \lambda_k h_k h^* \int_0^t (t-s)^{\delta_2 - \delta_1} E_{\delta_2, \delta_2 - \delta_1 + 1}[-\lambda_k (t-s)^{\delta_2}] E'(s) ds \Big] P_k(x).
\end{aligned}
$$

First, we integrate by parts in order to get away from singularity and by taking estimate with help of properties of Legendre polynomials and Mittag-Leffler function, we get

$$\left| D_{0+}^{(\alpha_1,\beta_1)\mu_1} \frac{\partial}{\partial x} \left[(1-x^2) u_x(\cdot,t) \right] \right|$$

$$\leq \sum_{k=0}^{\infty} \left[|\lambda_k \psi_k| \left(\frac{Mt^{\gamma_2-\delta_1-1}}{1+\lambda_k t^{\delta_2}} \right) + |\lambda_k \varphi_k| \left(\frac{Mt^{\delta_2+\gamma_1-\delta_1-1}}{1+\lambda_k t^{\delta_2}} \right) \right.$$

$$+ |\lambda_k h_k| |h^* D_{0+}^{(\alpha_1,\beta_1)\mu_1} E(t)| + |\lambda_k h_k h^*| |\lambda_k E(0)| \left(\frac{Mt^{\delta_2-\delta_1}}{1+\lambda_k t^{\delta_2}} \right)$$

$$+ |\lambda_k h_k h^*| |\lambda_k E'(0)| \left(\frac{Mt^{\delta_2+1-\delta_1}}{1+\lambda_k t^{\delta_2}} \right)$$

$$\left. + |\lambda_k^2 h_k h^*| \int_0^t \left(\frac{M(t-s)^{\delta_2+1-\delta_1}}{1+\lambda_k (t-s)^{\delta_2}} \right) |E''(s)| ds \right]$$

$$\leq \sum_{k=0}^{\infty} \frac{\|\psi''(\cdot)\| 4\sqrt{2}}{(2k-3)^{3/2}} \left(\frac{\lambda_k M t^{\gamma_2-\delta_1-1}}{1+\lambda_k t^{\delta_2}} \right) + \sum_{k=0}^{\infty} \frac{\|\varphi''(\cdot)\| 4\sqrt{2}}{(2k-3)^{3/2}} \left(\frac{\lambda_k M t^{\gamma_2-\delta_1-1}}{1+\lambda_k t^{\delta_2}} \right)$$

$$+ \sum_{k=0}^{\infty} \frac{6\sqrt{2}\lambda_k \|h^{(4)}(\cdot)\|}{(2k-7)^{7/2}} \left(|h^* D_{0+}^{(\alpha_1,\beta_1)\mu_1} E(t)| + |E(0)| \left(\frac{M\lambda_k t^{\delta_2-\delta_1}}{1+\lambda_k t^{\delta_2}} \right) \right)$$

$$+ \sum_{k=0}^{\infty} \frac{6\sqrt{2}\lambda_k \|h^{(4)}(\cdot)\|}{(2k-7)^{7/2}} \left(|E'(0)| \left(\frac{M\lambda_k t^{\delta_2+1-\delta_1}}{1+\lambda_k t^{\delta_2}} \right) + \right.$$

$$\left. + |h^*| \int_0^t \left(\frac{M\lambda_k (t-s)^{\delta_2+1-\delta_1}}{1+\lambda_k (t-s)^{\delta_2}} \right) |E''(s)| ds \right).$$

If $\psi'' \in L^2(-1,1)$, $\varphi'' \in L^2(-1,1)$, $h^{(4)} \in L^2(-1,1)$ and $E'' \in L^1(0,T)$, then the series on the right side converge, and according to the Weierstrass M test, the representation of $D_{0+}^{(\alpha_1,\beta_1)\mu_1} \frac{\partial}{\partial x} \left[(1-x^2) u_x \right]$ converges uniformly in Ω.

Finally, in a similar way, one can show that the series $D_{0+}^{(\alpha_1,\beta_1)\mu_1} D_{0+}^{(\alpha_2,\beta_2)\mu_2} u$ is uniformly convergent in Ω. This proves the existence of the solution to the ISP.

Most of the time, certain characteristics of the equations and features of the fractional-order derivative help us to select the approaches to problems. For instance, the series technique can be divided into two problems of solving ordinary differential equations to solve PDEs with any arbitrary order of derivatives.

We selected very strict conditions in the theorem for the given functions to fulfil the considered problem. It is also possible to choose quite simple conditions in terms of spaces for the functions, but this will require extra calculations at points of the singularities.

Acknowledgments The author is grateful to his supervisors for their contributions in terms of improving the quality of this research.

References

1. Stefanov, P., Vasy, A., Zworski, M.: Inverse Problems and Applications. American Mathematical Society, Providence (2014)
2. Blaunstein, N., Yakubov, V.: Electromagnetic and Acoustic Wave Tomography: Direct and Inverse Problems in Practical Applications. Chapman and Hall/CRC (2018)
3. Hendy, A.S., Van Bockstal, K.: On a reconstruction of a solely time-dependent source in a time-fractional diffusion equation with non-smooth solutions. J. Sci. Comput. **90**, 41 (2022)
4. Sakamoto, K., Yamamoto, M.: Initial value/boundary value problems for fractional diffusion-wave equations and applications to some inverse problems. J. Math. Anal. Appl. **382**(1), 426-447 (2011)
5. Kilbas, A.A., Srivastava, H.M., Trujillo, J.J.: Theory and Applications of Fractional Differential Equation. Elsevier, Amsterdam (2006)
6. Karimov, E., Ruzhansky, M., Toshtemirov, B.: Solvability of the boundary-value problem for a mixed equation involving hyper-Bessel fractional differential operator and bi-ordinal Hilfer fractional derivative. Math. Methods Appl. Sci. **46**(1), 54–70 (2023)
7. Toshtemirov, B: On solvability of the non-local problem for the fractional mixed-type equation with bessel operator. Fractional Differ. Calc., **12**(1), 63–76 (2022).
8. Karimov, E., Toshtemirov, B.: On a time-nonlocal boundary value problem for time-fractional partial differential equation. Int. J. Appl. Math. **35**(3), 423–438 (2022)
9. Kaplan, W.: Advanced Calculus, 5th edn. Pearson (2002)

Chapter 27
Very Weak Solution of the Discrete Wave Equation for Harmonic Oscillator

Shyam Swarup Mondal and Abhilash Tushir

Abstract In this paper, we announce some preliminary findings of the joint work with [6] for the wave equation with Schrödinger operator. In particular, we study the semiclassical analogue of the wave equation for the discrete harmonic oscillator on the lattice $\hbar\mathbb{Z}^n$. In the case of regular coefficients, we show that the Cauchy problem is well-posed in the associated Sobolev-type spaces. Moreover, we also show that the Cauchy problem is very weakly well-posed when the coefficients are allowed to have a distributional singularity.

27.1 Introduction

In a variety of contexts, the wave equation with constant coefficients, time-dependent propagation speed or space-dependent potential has received considerable attention from many researchers. We refer to [4, 5] for the work relevant to the wave equation with time-dependent coefficients and [1, 3] for the wave equation with space-dependent potential. The wave equation for operators with purely discrete spectrum was extensively studied by Ruzhansky and Tokmagambetov in [10–12]. Moreover, they allowed the coefficients to have distributional irregularities. In the case of the lattice $\hbar\mathbb{Z}^n$, the discrete wave equation with the harmonic oscillator is a particular case of the wave equation for the Schrödinger operator with unbounded potential. In this work, we provide some preliminary findings of the work [6] for the Schrödinger operator with potential $V(k) = |k|^2$. Moreover, we also allow the coefficients and the source term (in Cauchy problem (27.1)) to have a distributional singularity. In this case, we adapt the notion of very weak solution that was developed by Ruzhansky and Garetto in [9] and later applied in various settings. We refer to [6] for a more thorough examination and detailed proof of the results stated in this article.

S. S. Mondal · A. Tushir (✉)
Department of Mathematics, Indian Institute of Technology, Delhi, Hauz Khas, New Delhi, India

M. Ruzhansky, K. Van Bockstal (eds.), *Extended Abstracts 2021/2022*, Research Perspectives Ghent Analysis and PDE Center 2,
https://doi.org/10.1007/978-3-031-42539-4_27

In this article, we study the semiclassical version of the wave equation with a discrete harmonic oscillator on the lattice

$$\hbar\mathbb{Z}^n = \left\{x \in \mathbb{R}^n : x = \hbar k, k \in \mathbb{Z}^n\right\},$$

where $\hbar > 0$ is a (small) discretisation parameter. Let v_j be the jth basis vector in $\mathbb{Z}^n$, having all zeros except for 1 as the jth component. The discrete harmonic oscillator on $\hbar\mathbb{Z}^n$ denoted by $\mathcal{H}_\hbar$ is defined by

$$\mathcal{H}_\hbar u(k) := \left(-\hbar^{-2}\mathcal{L}_\hbar + |k|^2\right) u(k), \quad k \in \hbar\mathbb{Z}^n,$$

where the discrete Laplacian $\mathcal{L}_\hbar$ is given by

$$\mathcal{L}_\hbar u(k) := \sum_{j=1}^{n} \left(u\left(k + \hbar v_j\right) + u\left(k - \hbar v_j\right)\right) - 2nu(k), \quad k \in \hbar\mathbb{Z}^n.$$

For $s \in \mathbb{R}$, the Sobolev space associated with the harmonic oscillator is given by

$$\mathrm{H}^s_{\mathcal{H}_\hbar} := \left\{u \in \mathrm{H}^{-\infty}_{\mathcal{H}_\hbar} : \left(I + \mathcal{H}_\hbar\right)^{s/2} u \in \ell^2(\hbar\mathbb{Z}^n)\right\},$$

with the norm $\|f\|_{\mathrm{H}^s_{\mathcal{H}_\hbar}} := \|\left(I + \mathcal{H}_\hbar\right)^{s/2} f\|_{\ell^2(\hbar\mathbb{Z}^n)}$, where $\mathrm{H}^{-\infty}_{\mathcal{H}_\hbar}$ is the space of $\mathcal{H}_\hbar$-distributions. Clearly, the Sobolev space $\mathrm{H}^s_{\mathcal{H}_\hbar}$ also includes $\ell^2(\hbar\mathbb{Z}^n)$ as a special case when $s = 0$. For more details, we refer to [6, Section 3].

The non-homogeneous wave equation with discrete harmonic oscillator and time-dependent coefficients can be read as follows:

$$\begin{cases} \partial_t^2 u(t,k) + a(t)\mathcal{H}_\hbar u(t,k) + q(t)u(t,k) = f(t,k), & (t,k) \in (0,T] \times \hbar\mathbb{Z}^n, \\ u(0,k) = u_0(k), & k \in \hbar\mathbb{Z}^n, \\ \partial_t u(0,k) = u_1(k), & k \in \hbar\mathbb{Z}^n. \end{cases} \tag{27.1}$$

The above Cauchy problem is the semiclassical discretisation of the following classical wave equation with the harmonic oscillator

$$\begin{cases} \partial_t^2 u(t,x) + a(t)\mathcal{H}u(t,x) + q(t)u(t,x) = f(t,x), & (t,x) \in (0,T] \times \mathbb{R}^n, \\ u(0,x) = u_0(x), & x \in \mathbb{R}^n, \\ \partial_t u(0,x) = u_1(x), & x \in \mathbb{R}^n, \end{cases}$$

where $\mathcal{H} := -\mathcal{L} + |x|^2$ is the usual harmonic oscillator and $\mathcal{L}$ is the Laplacian on $\mathbb{R}^n$.

The semiclassical version of the wave equation with time-dependent coefficients, Klein-Gordon equation with space-dependent potential and heat equation with irregular thermal conductivity was extensively studied in [2, 7, 8], respectively. They proved that the wave equation and the Klein-Gordan equation are well-posed in $\ell^2(\hbar\mathbb{Z}^n)$ while the heat equation is well-posed in the space of tempered distributions. In contrast to the above equations, the wave equation with the harmonic oscillator is well-posed in the associated Sobolev-type spaces. Specifically, in this work, we announce the results for the well-posedness of Cauchy problem (27.1) with $(u_0, u_1) \in \mathrm{H}^1_{\mathcal{H}_\hbar} \times \mathrm{H}^0_{\mathcal{H}_\hbar} \equiv \mathrm{H}_{\mathcal{H}_\hbar} \times \ell^2(\hbar\mathbb{Z}^n)$. We refer to Sect. 27.3 for the main results related to the well-posedness of the Cauchy problem (27.1).

27.2 Preliminaries

The discrete harmonic oscillator $\mathcal{H}_\hbar : \mathrm{Dom}\left(\mathcal{H}_\hbar\right) \to \ell^2\left(\hbar\mathbb{Z}^n\right)$ is a densely defined, self-adjoint linear operator in $\ell^2\left(\hbar\mathbb{Z}^n\right)$ with domain $\mathrm{Dom}\left(\mathcal{H}_\hbar\right)$ given by

$$\mathrm{Dom}\left(\mathcal{H}_\hbar\right) := \left\{u \in \ell^2\left(\hbar\mathbb{Z}^n\right) : \left(I + \mathcal{H}_\hbar\right) u \in \ell^2\left(\hbar\mathbb{Z}^n\right)\right\}.$$

Furthermore, it also has a purely discrete spectrum consisting of non-negative eigenvalues only, see [6, Theorem 1.2]. Let $\{\lambda_\xi \geq 0 : \xi \in \mathcal{I}_\hbar\}$ be the spectrum of the discrete harmonic oscillator and u_ξ be the eigenfunction associated with the eigenvalue λ_ξ. The $\mathcal{H}_\hbar$-Fourier transform $\widehat{f}$ of $f : \hbar\mathbb{Z}^n \to \mathbb{C}$ is defined as

$$\widehat{f}(\xi) := \sum_{k \in \hbar\mathbb{Z}^n} f(k)\overline{u_\xi(k)}, \quad \xi \in \mathcal{I}_\hbar.$$

The $\mathcal{H}_\hbar$-Plancherel formula is as follows

$$\sum_{k \in \hbar\mathbb{Z}^n} |f(k)|^2 = \sum_{\xi \in \mathcal{I}_\hbar} |\widehat{f}(\xi)|^2.$$

The $\mathcal{H}_\hbar$-Fourier transform of the harmonic oscillator is given by

$$\left(\mathcal{F}_{\mathcal{H}_\hbar}\mathcal{H}_\hbar f\right)(\xi) = (\mathcal{H}_\hbar f, u_\xi) = (f, \mathcal{H}_\hbar u_\xi) = (f, \lambda_\xi u_\xi) = \lambda_\xi \widehat{f}(\xi), \quad \xi \in \mathcal{I}_\hbar. \tag{27.2}$$

Combining Plancherel's identity with (27.2), we have

$$\|f\|_{\mathrm{H}_{\mathcal{H}_\hbar}} := \left\|(I + \mathcal{H}_\hbar)^{1/2} f\right\|_{\ell^2(\hbar\mathbb{Z}^n)} = \left(\sum_{\xi \in \mathcal{I}_\hbar} \left(1 + \lambda_\xi\right) |\widehat{f}(\xi)|^2\right)^{\frac{1}{2}}.$$

In order to simplify the notation, we write $A \lesssim B$, if there exists a constant C independent of the appearing parameter such that $A \leq CB$. We write $a \in L_1^\infty([0,T])$ whenever $a \in L^\infty([0,T])$ is differentiable and also $\partial_t a \in L^\infty([0,T])$.

27.3 Main Results

This section is devoted to the main results of this paper. First, we state our well-posedness results in the discrete Sobolev space $\mathrm{H}_{\mathcal{H}_\hbar}$ for the Cauchy problem (27.1) with regular coefficients as well as irregular coefficients. We start with the following well-posedness theorem for the Cauchy problem (27.1) with coefficients a, q and the source term f are regular:

Theorem 27.1 (Classical Solution) *Let $f \in L^2([0,T]; \ell^2(\hbar\mathbb{Z}^n))$. Assume that $a \in L_1^\infty([0,T])$ satisfies $\inf_{t\in[0,T]} a(t) = a_0 > 0$ and $q \in L^\infty([0,T])$. If the initial Cauchy data $(u_0, u_1) \in H_{\mathcal{H}_\hbar} \times \ell^2(\hbar\mathbb{Z}^n)$, then the Cauchy problem* (27.1) *has a unique solution $u \in C([0,T]; H_{\mathcal{H}_\hbar}) \cap C^1([0,T]; \ell^2(\hbar\mathbb{Z}^n))$ satisfying the estimate*

$$\|u(t,\cdot)\|_{H_{\mathcal{H}_\hbar}}^2 + \|u_t(t,\cdot)\|_{\ell^2(\hbar\mathbb{Z}^n)}^2 \lesssim \|u_0\|_{H_{\mathcal{H}_\hbar}}^2 + \|u_1\|_{\ell^2(\hbar\mathbb{Z}^n)}^2 + \|f\|_{L^2([0,T];\ell^2(\hbar\mathbb{Z}^n))}^2,$$

for all $t \in [0,T]$.

Now, we permit the source term f and the coefficients a and q to have singularities in the time variable. In this situation, it is possible that the Cauchy problem (27.1) does not have a meaningful solution in the usual distributional sense. Thus, we need to take into account the notion of very weak solution.

Let a be a compactly supported distribution. The Friedrichs-mollifier, i.e., $\psi \in C_0^\infty(\mathbb{R})$, $\psi \geq 0$ such that $\int_{\mathbb{R}} \psi = 1$, allows one to construct the families of smooth functions $(a_\varepsilon)_\varepsilon$ given by

$$a_\varepsilon(t) := \left(a * \psi_{\omega(\varepsilon)}\right)(t),$$

where $\psi_{\omega(\varepsilon)}(t) = \frac{1}{\omega(\varepsilon)}\psi\left(\frac{t}{\omega(\varepsilon)}\right)$ for some $\omega(\varepsilon) \geq 0$ such that $\omega(\varepsilon) \to 0$ as $\varepsilon \to 0$.

A distribution a is said to be positive, if $\langle a, \psi\rangle \geq 0$ for all $\psi \in C^\infty(\mathbb{R})$ satisfying $\psi \geq 0$. Furthermore, a distribution a is said to be strictly positive if there exists a positive constant α such that $a - \alpha$ is a positive distribution.

In order to define the notion of very weak solution, we will make use of the following notion of moderateness and negligibility:

Definition 27.1

(i) A net of functions $(a_\varepsilon)_\varepsilon \in L^\infty_m(\mathbb{R})^{(0,1]}$ is said to be L^∞_m-moderate if for all $K \Subset \mathbb{R}$, there exist $N \in \mathbb{N}_0$ and $c > 0$ such that

$$\left\| \partial^k a_\varepsilon \right\|_{L^\infty(K)} \leq c\varepsilon^{-N-k}, \qquad \text{for all } k = 0, 1, \ldots, m,$$

for all $\varepsilon \in (0, 1]$.

(ii) A net of functions $(a_\varepsilon)_\varepsilon \in L^\infty_m(\mathbb{R})^{(0,1]}$ is said to be L^∞_m-negligible if for all $K \Subset \mathbb{R}$ and $q \in \mathbb{N}_0$, there exists $c > 0$ such that

$$\left\| \partial^k a_\varepsilon \right\|_{L^\infty(K)} \leq c\varepsilon^{q}, \qquad \text{for all } k = 0, 1, \ldots, m,$$

for all $\varepsilon \in (0, 1]$.

(iii) A net $(u_\varepsilon)_\varepsilon \in L^2([0, T]; \mathrm{H}_{\mathcal{H}_\hbar})^{(0,1]}$ is said to be $L^2([0, T]; \mathrm{H}_{\mathcal{H}_\hbar})$-moderate if there exist $N \in \mathbb{N}_0$ and $c > 0$ such that

$$\|u_\varepsilon\|_{L^2([0,T];\mathrm{H}_{\mathcal{H}_\hbar})} \leq c\varepsilon^{-N},$$

for all $\varepsilon \in (0, 1]$.

(iv) A net $(u_\varepsilon)_\varepsilon \in L^2([0, T]; \mathrm{H}_{\mathcal{H}_\hbar})^{(0,1]}$ is said to be $L^2([0, T]; \mathrm{H}_{\mathcal{H}_\hbar})$-negligible if for all $q \in \mathbb{N}_0$, there exists $c > 0$ such that

$$\|u_\varepsilon\|_{L^2([0,T];\mathrm{H}_{\mathcal{H}_\hbar})} \leq c\varepsilon^{q},$$

for all $\varepsilon \in (0, 1]$.

We may now introduce the notion of a very weak solution for the Cauchy problem (27.1) as follows:

Definition 27.2 Let $(u_0, u_1) \in \mathrm{H}_{\mathcal{H}_\hbar} \times \ell^2(\hbar\mathbb{Z}^n)$. We say that the net of functions $(u_\varepsilon)_\varepsilon \in L^2([0, T]; \mathrm{H}_{\mathcal{H}_\hbar})^{(0,1]}$ is a very weak solution to the Cauchy problem (27.1) if there exist

1. a L^∞_1-moderate regularisation a_ε of the coefficient a;
2. a L^∞-moderate regularisation q_ε of the coefficient q; and
3. a $L^2([0, T]; \ell^2(\hbar\mathbb{Z}^n))$-moderate regularisation f_ε of the source term f,

such that $(u_\varepsilon)_\varepsilon$ solves the regularised problem

$$\begin{cases} \partial_t^2 u_\varepsilon(t, k) + a_\varepsilon(t)\mathcal{H}_\hbar u_\varepsilon(t, k) + q_\varepsilon(t) u_\varepsilon(t, k) = f_\varepsilon(t, k), & (t, k) \in (0, T] \times \hbar\mathbb{Z}^n, \\ u_\varepsilon(0, k) = u_0(k), \quad k \in \hbar\mathbb{Z}^n, \\ \partial_t u_\varepsilon(0, k) = u_1(k), \quad k \in \hbar\mathbb{Z}^n, \end{cases}$$

for all $\varepsilon \in (0, 1]$, and is $L^2([0, T]; \mathrm{H}_{\mathcal{H}_\hbar})$-moderate.

The following theorem concerns the existence of a very weak solution to the Cauchy problem (27.1) with distributional coefficients.

Theorem 27.2 (Existence) *Let a and q be distributions with supports contained in $[0, T]$ such that $a \geq a_0 > 0$ for some positive constant a_0. Let the source term $f(\cdot, k)$ be a distribution with support contained in $[0, T]$, for all $k \in \hbar\mathbb{Z}^n$. If the initial Cauchy data $(u_0, u_1) \in H_{\mathcal{H}_\hbar} \times \ell^2(\hbar\mathbb{Z}^n)$, then the Cauchy problem (27.1) has a very weak solution.*

The notion of uniqueness of a very weak solution for the Cauchy problem (27.1) can be understood as: the family of very weak solutions affects negligibility by the negligible changes in the approximations of the coefficients and the source term.

Definition 27.3 A very weak solution of the Cauchy problem (27.1) is said to be $L^2([0, T]; \mathrm{H}_{\mathcal{H}_\hbar})$-unique, if

1. for all L_1^∞-moderate nets $a_\varepsilon, \tilde{a}_\varepsilon$ such that $(a_\varepsilon - \tilde{a}_\varepsilon)_\varepsilon$ is L_1^∞-negligible,
2. for all L^∞-moderate nets $q_\varepsilon, \tilde{q}_\varepsilon$ such that $(q_\varepsilon - \tilde{q}_\varepsilon)_\varepsilon$ is L^∞-negligible; and
3. for all $L^2([0, T]; \ell^2(\hbar\mathbb{Z}^n))$-moderate nets $f_\varepsilon, \tilde{f}_\varepsilon$ such that $(f_\varepsilon - \tilde{f}_\varepsilon)_\varepsilon$ is $L^2([0, T]; \ell^2(\hbar\mathbb{Z}^n))$-negligible,

the net $(u_\varepsilon - \tilde{u}_\varepsilon)_\varepsilon$ is $L^2([0, T]; \mathrm{H}_{\mathcal{H}_\hbar})$-negligible, where $(u_\varepsilon)_\varepsilon$ and $(\tilde{u}_\varepsilon)_\varepsilon$ are the families of solutions corresponding to the ε-parametrised problems

$$\begin{cases} \partial_t^2 u_\varepsilon(t, k) + a_\varepsilon(t)\mathcal{H}_\hbar u_\varepsilon(t, k) + q_\varepsilon(t) u_\varepsilon(t, k) = f_\varepsilon(t, k), & (t, k) \in (0, T] \times \hbar\mathbb{Z}^n, \\ u_\varepsilon(0, k) = u_0(k), \quad k \in \hbar\mathbb{Z}^n, \\ \partial_t u_\varepsilon(0, k) = u_1(k), \quad k \in \hbar\mathbb{Z}^n, \end{cases}$$

and

$$\begin{cases} \partial_t^2 \tilde{u}_\varepsilon(t, k) + \tilde{a}_\varepsilon(t)\mathcal{H}_\hbar \tilde{u}_\varepsilon(t, k) + \tilde{q}_\varepsilon(t) \tilde{u}_\varepsilon(t, k) = \tilde{f}_\varepsilon(t, k), & (t, k) \in (0, T] \times \hbar\mathbb{Z}^n, \\ \tilde{u}_\varepsilon(0, k) = u_0(k), \quad k \in \hbar\mathbb{Z}^n, \\ \partial_t \tilde{u}_\varepsilon(0, k) = u_1(k), \quad k \in \hbar\mathbb{Z}^n, \end{cases}$$

respectively.

The following theorem establishes the uniqueness of the very weak solution to the Cauchy problem (27.1) in the sense of Definition 27.3.

Theorem 27.3 (Uniqueness) *Let a and q be distributions with supports contained in $[0, T]$ such that $a \geq a_0 > 0$ for some positive constant a_0, also let the source term $f(\cdot, k)$ be a distribution with support contained in $[0, T]$, for all $k \in \hbar\mathbb{Z}^n$. If the initial Cauchy data $(u_0, u_1) \in H_{\mathcal{H}_\hbar} \times \ell^2(\hbar\mathbb{Z}^n)$, then the very weak solution obtained in Theorem 27.2 is $L^2([0, T]; H_{\mathcal{H}_\hbar})$-unique.*

The following theorem illustrates the consistency of the very weak solution in Theorem 27.2 with the classical solution obtained in Theorem 27.1.

Theorem 27.4 (Consistency) *Let* $f \in L^2([0,T], \ell^2(\hbar\mathbb{Z}^n))$. *Assume that* $a \in L_1^\infty([0,T])$ *satisfies* $\inf_{t\in[0,T]} a(t) = a_0 > 0$ *and* $q \in L^\infty([0,T])$. *If the initial Cauchy data* $(u_0, u_1) \in H_{\mathcal{H}_\hbar} \times \ell^2(\hbar\mathbb{Z}^n)$, *then the very weak solution* $(u_\varepsilon)_\varepsilon$ *converges to the classical solution in* $L^2([0,T]; H_{\mathcal{H}_\hbar})$ *as* $\varepsilon \to 0$.

Acknowledgments We would like to thank Prof. Michael Ruzhansky and Prof. Aparajita Dasgupta for their insightful reviews and suggestions. S. S. Mondal is supported by Core Research Grant, RP03890G, Science and Engineering Research Board (SERB), DST, India. A.Tushir is supported by the institute assistantship from Indian Institute of Technology Delhi, India.

References

1. Altybay, A., Ruzhansky, M., Sebih, M.E., Tokmagambetov, N.: Fractional Klein-Gordon equation with singular mass. Chaos Solitons Fractals **143**, 110579 (2021)
2. Chatzakou, M., Dasgupta, A., Ruzhansky, M., Tushir, A.: Discrete heat equation with irregular thermal conductivity and tempered distributional data. *Proceedings of the Royal Society of Edinburgh Section A: Mathematics*, 1–24 (2023). https://doi.org/10.1017/prm.2023.84
3. Chatzakou, M., Ruzhansky, M., Tokmagambetov, N.: Fractional Klein-Gordon equation with singular mass. II: Hypoelliptic case. Complex Var. Elliptic Equ. **67**(3), 615–632 (2022)
4. Cicognani, M., Colombini, F.: A well-posed Cauchy problem for an evolution equation with coefficients of low regularity. J. Differ. Equ. **254**(8), 3573–3595 (2013)
5. Colombini, F., De Giorgi, E., Spagnolo, S.: Sur les équations hyperboliques avec des coefficients qui ne dépendent que du temps. Ann. Scuola Norm. Sup. Pisa Cl. Sci. **6**(3), 511–559 (1979)
6. Dasgupta, A., Mondal, S.S., Ruzhansky, M., Tushir, A.: Discrete time-dependent wave equation for the Schrödinger operator with unbounded potential (2023). Preprint https://doi.org/10.48550/arXiv.2306.02409
7. Dasgupta, A., Ruzhansky, M., Tushir, A.: Discrete time-dependent wave equations I. Semiclassical analysis. J. Differ. Equ. **317**, 89–120 (2022)
8. Dasgupta, A., Ruzhansky, M., Tushir, A.: Discrete time-dependent wave equations II. Semiclassical fractional Klein-Gordon equation (2022). https://doi.org/10.48550/arXiv.2205.05360
9. Garetto, C., Ruzhansky, M.: Hyperbolic second order equations with non-regular time dependent coefficients. Arch. Ration. Mech. Anal. **217**(1), 113–154 (2015)
10. Ruzhansky, M., Tokmagambetov, N.: Very weak solutions of wave equation for Landau Hamiltonian with irregular electromagnetic field. Lett. Math. Phys. **107**(4), 591–618 (2017)
11. Ruzhansky, M., Tokmagambetov, N.: Wave equation for operators with discrete spectrum and irregular propagation speed. Arch. Ration. Mech. Anal. **226**(3), 1161–1207 (2017)
12. Ruzhansky, M., Tokmagambetov, N.: Wave equation for 2D Landau Hamiltonian. Appl. Comput. Math. **18**(1), 69–78 (2019)

Chapter 28
An Estimate for the Multivariate Mittag-Leffler Function

Frederick Maes and Karel Van Bockstal

Abstract In this contribution, we discuss Maes and Van Bockstal (Fract Calc Appl Anal 26(4) 2023, Lemma 5). This lemma provides a bound on the multivariate Mittag-Leffler function, which may be valuable in the analysis of multiterm time-fractional wave equations. It is an extension of Li and Yamamoto (Appl Math Comput 257:381–397, 2015, Lemma 3.2) used to analyse multiterm fractional diffusion equations.

28.1 Introduction

The solution to single-term time-fractional diffusion and wave equations can often be explicitly represented by the two-parameter Mittag-Leffler function, see e.g. [1] or [2]. This function is defined by Erdélyi et al. [3]

$$E_{\alpha,\beta}(z) = \sum_{k=0}^{\infty} \frac{z^k}{\Gamma(\alpha k + \beta)}, \qquad z \in \mathbb{C}, \quad \alpha > 0, \quad \beta \in \mathbb{R},$$

where Γ denotes the Gamma function.

When analysing the regularity of the solution (by establishing stability estimates) to single-term equations, the following lemma obtained by Podlubny is crucial, see [4, Theorem 1.6].

F. Maes
Research Group NaM^2, Department of Electronics and Information Systems, Ghent University, Ghent, Belgium
e-mail: frederick.maes@ugent.be

K. Van Bockstal (✉)
Ghent Analysis & PDE Center, Department of Mathematics: Analysis, Logic and Discrete Mathematics, Ghent University, Ghent, Belgium
e-mail: karel.vanbockstal@ugent.be

M. Ruzhansky, K. Van Bockstal (eds.), *Extended Abstracts 2021/2022*,
Research Perspectives Ghent Analysis and PDE Center 2,
https://doi.org/10.1007/978-3-031-42539-4_28

Lemma 28.1 *Let $\alpha \in (0, 2)$ and $\beta \in \mathbb{R}$. Assume that μ is such that $\alpha\pi/2 < \mu < \min\{\alpha\pi, \pi\}$. Then, there exists a constant $C > 0$ such that*

$$\left|E_{\alpha,\beta}(z)\right| \leq \frac{C}{1+|z|}, \qquad \mu \leq |\arg(z)| \leq \pi, \quad |z| \geq 0.$$

Corollary 28.1 *Let $\alpha \in (0, 2)$ and $\beta \in \mathbb{R}$. Then, there exists a constant $C > 0$ such that*

$$\left|E_{\alpha,\beta}(-x)\right| \leq \frac{C}{1+x} \leq C, \qquad x \geq 0.$$

Explicit solutions to multiterm equations may be expressed in terms of the multivariate Mittag-Leffler function, which is defined by Hadid and Luchko [5, Eq. (3.16)]

$$E_{(\alpha_1,\ldots,\alpha_m),\beta}(z_1,\ldots,z_m) = \sum_{k=0}^{\infty} \sum_{\substack{k_1+\cdots+k_m=k \\ k_j \geq 0}} \binom{k}{k_1,\ldots,k_m} \frac{\prod_{j=1}^m z_j^{k_j}}{\Gamma(\beta + \sum_{j=1}^m \alpha_j k_j)}, \tag{28.1}$$

where $\binom{k}{k_1,\ldots,k_m}$ is the multinomial coefficient, $z_i \in \mathbb{C}$ $(1 \leq i \leq m)$, $\alpha_i > 0$ $(1 \leq i \leq m)$ and $\beta > 0$.

In [6, Lemma 3.2], the authors provide the following bound on (28.1).

Lemma 28.2 *Let $0 < \beta < 2$ and $0 < \alpha_m < \cdots < \alpha_1 < 1$. Assume that $\alpha_1\pi/2 < \mu < \min\{\alpha_1\pi, \pi\}$, $\mu \leq |\arg(z_1)| \leq \pi$ and there exists $K > 0$ such that $-K \leq z_j < 0$ for $j = 2, \ldots, m$. Then, there exists a constant $C > 0$ depending only on μ, K, α_j $(j = 1, \ldots, m)$ and β such that*

$$\left|E_{(\alpha_1,\alpha_1-\alpha_2,\alpha_1-\alpha_m),\beta}(z_1,\ldots,z_m)\right| \leq \frac{C}{1+|z_1|}.$$

This lemma has been used as a crucial tool in analysing multiterm time-fractional diffusion equations.

28.2 Main Result

In [7, Lemma 5], we have extended [6, Lemma 3.2] such that it is applicable in the analysis of multiterm time-fractional wave equations. A similar estimate has been obtained for a special case of the multivariate Mittag-Leffler function in [8, Lemma

3] assuming $0 < \beta < 1 + \alpha_1$. However, our result presented in the next theorem is more generally valid.

Theorem 28.1 *Let $\beta > 0$ and $\alpha_1 > \alpha_2 > \cdots > \alpha_m > 0$ be given with $\alpha_1 \in (0, 2)$. Assume that $\alpha_1\pi/2 < \mu < \min\{\alpha_1\pi, \pi\}$, $\mu \leq |\arg(z_1)| \leq \pi$ and there exists $K > 0$ such that $-K \leq z_j < 0$ for $j = 2, \ldots, m$. Then, there exists a constant $C > 0$ depending only on μ, K, α_j $(j = 1, \ldots, m)$ and β such that*

$$\left|E_{(\alpha_1,\ldots,\alpha_m),\beta}(z_1, \ldots, z_m)\right| \leq \frac{C}{1 + |z_1|}.$$

We also note the following results obtained in [7, Lemma 3] and [7, Lemma 4], which will be useful in the next section. The first result is a slight generalisation of [6, Lemma 3.3], which we recover by setting $\gamma = \beta$. The second result is a generalisation of the property $E_{\alpha,\beta}(z) = zE_{\alpha,\beta+\alpha}(z) + \frac{1}{\Gamma(\beta)}$ for the two-parameter Mittag-Leffler function and [6, Lemma 3.1] to its multivariate version.

Theorem 28.2 *Let $\alpha_1, \ldots, \alpha_m$ be positive constants and $q_1, \ldots, q_m, \beta$ be real. Then, for real γ, we have that*

$$\begin{aligned}
&\frac{d}{dt}\left[t^{\gamma} E_{(\alpha_1,\ldots,\alpha_m),\beta+1}\left(q_1 t^{\alpha_1}, \ldots, q_m t^{\alpha_m}\right)\right] \\
&\quad = t^{\gamma-1} E_{(\alpha_1,\ldots,\alpha_m),\beta}\left(q_1 t^{\alpha_1}, \ldots, q_m t^{\alpha_m}\right) \\
&\qquad + (\gamma - \beta) t^{\gamma-1} E_{(\alpha_1,\ldots,\alpha_m),\beta+1}\left(q_1 t^{\alpha_1}, \ldots, q_m t^{\alpha_m}\right).
\end{aligned}$$

Theorem 28.3 *Let $\alpha_1, \ldots, \alpha_m$ be positive constants, $z_1, \ldots, z_m \in \mathbb{C}$ and $\beta > 0$ be fixed. Then*

$$\sum_{j=1}^{m} z_j E_{(\alpha_1,\ldots,\alpha_m),\beta+\alpha_j}(z_1, \ldots, z_m) + \frac{1}{\Gamma(\beta)} = E_{(\alpha_1,\ldots,\alpha_m),\beta}(z_1, \ldots, z_m).$$

This result remains valid in the limit case $\beta \to 0$ as $\lim_{\beta \searrow 0} 1/\Gamma(\beta) = 0$.

28.3 Example

We consider the following fractional differential equation for $t \in [0, \mathcal{T}]$:

$$\partial_t^{\alpha} T' + aT' + b\partial_t^{\alpha} T + cT = 0, \qquad a, b \in \mathbb{R}, \quad c > 0, \quad \alpha \in (0, 1), \tag{28.2}$$

where ∂_t^α is the classical Caputo fractional derivative of order α [9] defined by

$$\partial_t^\alpha = \frac{1}{\Gamma(1-\alpha)} \int_0^t (t-s)^{-\alpha} u'(s) ds.$$

Additionally, we take the following initial conditions

$$T(0) = c_1, \quad T'(0) = c_2, \tag{28.3}$$

where $c_i \in \mathbb{R}$ for $i = 1, 2$. We will now obtain the solution to the initial value problem (28.2–28.3) in terms of the multivariate Mittag-Leffler function by means of the Laplace transform $\mathcal{L}$. Taking the Laplace transform of (28.2), we find that

$$\mathcal{L}[T(t)](s) = T(0)\left(s^\alpha + bs^{\alpha-1} + a\right) g(s) + T'(0) s^{\alpha-1} g(s),$$

where by Podlubny [4, Eqs. (5.31)–(5.33)] the function $g(s)$ is given by

$$g(s) = \sum_{m=0}^{\infty} (-1)^m c^m \sum_{k=0}^{m} \binom{m}{k} \left(\frac{b}{c}\right)^k \frac{s^{\alpha k - m - 1}}{(s^\alpha + a)^{m+1}},$$

and forms a Laplace pair with

$$G(t) = \sum_{m=0}^{\infty} \frac{1}{m!} (-c)^m \sum_{k=0}^{m} \binom{m}{k} \left(\frac{b}{c}\right)^k \times t^{(\alpha+1)(m+1)-\alpha k-1} E^{(m)}_{\alpha,\alpha+1+m-\alpha k}\left(-at^\alpha\right). \tag{28.4}$$

Using properties of the Laplace transform related to fractional derivatives [4, §2.8], we find that

$$\mathcal{L}\left[\partial_t^\alpha G(t)\right](s) = s^\alpha g(s) \quad \text{and} \quad \mathcal{L}\left[\int_0^t \partial_t^\alpha G(z) \mathrm{d}z\right](s) = s^{\alpha-1} g(s), \tag{28.5}$$

noting that $G(0) = 0$. Following the approach from [10], we can rewrite (28.4) by means of multivariate Mittag-Leffler functions. After relabelling, we obtain that

$$G(t) = t^\alpha E_{(\alpha+1,1,\alpha),\alpha+1}\left(-ct^{\alpha+1}, -bt, -at^\alpha\right). \tag{28.6}$$

Employing (28.5), the solution $T(t)$ to the problem (28.2–28.3) takes the form

$$T(t) = c_1 T_1(t) + c_2 T_2(t),$$

with

$$T_1(t) = \partial_t^\alpha G(t) + b\int_0^t \partial_t^\alpha G(s)\mathrm{d}s + aG(t),$$

$$T_2(t) = \int_0^t \partial_t^\alpha G(s)\mathrm{d}s,$$

where $G(t)$ is given by (28.6). The results in [11, Theorem 2.3] show that

$$\partial_t^\alpha G(t) = E_{(\alpha+1,1,\alpha),1}\left(-ct^{\alpha+1}, -bt, -at^\alpha\right),$$

$$\int_0^t \partial_t^\alpha G(s)\mathrm{d}s = tE_{(\alpha+1,1,\alpha),2}\left(-ct^{\alpha+1}, -bt, -at^\alpha\right).$$

Hence, we find that

$$\begin{aligned} T_1(t) = E_{(\alpha+1,1,\alpha),1}&\left(-ct^{\alpha+1}, -bt, -at^\alpha\right) \\ &+ btE_{(\alpha+1,1,\alpha),2}\left(-ct^{\alpha+1}, -bt, -at^\alpha\right) \\ &\qquad + at^\alpha E_{(\alpha+1,1,\alpha),\alpha+1}\left(-ct^{\alpha+1}, -bt, -at^\alpha\right) \end{aligned}$$

and

$$T_2(t) = tE_{(\alpha+1,1,\alpha),2}\left(-ct^{\alpha+1}, -bt, -at^\alpha\right).$$

From Theorem 28.3, we have that

$$\begin{aligned} &E_{(\alpha+1,1,\alpha),1}\left(-ct^{\alpha+1}, -bt, -at^\alpha\right) \\ &\quad = 1 - ct^{\alpha+1}E_{(\alpha+1,1,\alpha),\alpha+2}\left(-ct^{\alpha+1}, -bt, -at^\alpha\right) \\ &\qquad - btE_{(\alpha+1,1,\alpha),2}\left(-ct^{\alpha+1}, -bt, -at^\alpha\right) \\ &\qquad\quad - at^\alpha E_{(\alpha+1,1,\alpha),\alpha+1}\left(-ct^{\alpha+1}, -bt, -at^\alpha\right). \end{aligned}$$

Hence, T_1 simplifies to

$$T_1(t) = 1 - ct^{\alpha+1}E_{(\alpha+1,1,\alpha),\alpha+2}\left(-ct^{\alpha+1}, -bt, -at^\alpha\right).$$

From Theorem 28.1, it follows that there exist constants $C_i > 0$ such that

$$\max_{t\in[0,\mathcal{T}]} |T_i(t)| \leq C_i, \qquad i = 1, 2.$$

Using Theorem 28.2, we obtain that

$$T_1'(t) = -ct^\alpha E_{(\alpha+1,1,\alpha),\alpha+1}\left(-ct^{\alpha+1}, -bt, -at^\alpha\right),$$

$$T_1''(t) = -ct^{\alpha-1} E_{(\alpha+1,1,\alpha),\alpha}\left(-ct^{\alpha+1}, -bt, -at^\alpha\right),$$

$$T_2'(t) = E_{(\alpha+1,1,\alpha),1}\left(-ct^{\alpha+1}, -bt, -at^\alpha\right),$$

$$T_2''(t) = t^{-1} E_{(\alpha+1,1,\alpha),0}\left(-ct^{\alpha+1}, -bt, -at^\alpha\right).$$

Note that T_2'' can be rewritten by Theorem 28.3 as

$$T_2''(t) = -ct^\alpha E_{(\alpha+1,1,\alpha),\alpha+1}\left(-ct^{\alpha+1}, -bt, -at^\alpha\right)$$
$$-bE_{(\alpha+1,1,\alpha),1}\left(-ct^{\alpha+1}, -bt, -at^\alpha\right) - at^{\alpha-1} E_{(\alpha+1,1,\alpha),\alpha}\left(-ct^{\alpha+1}, -bt, -at^\alpha\right).$$

Therefore, from Theorem 28.1, it follows that there exist constants $\dot{C}_i > 0$ such that

$$\max_{t\in[0,\mathcal{T}]} \left|T_i'(t)\right| \leq \dot{C}_i, \qquad i = 1, 2.$$

Moreover, there exist constants $\ddot{C}_i > 0$ such that

$$\left|T_1''(t)\right| \leq \ddot{C}_1 t^{\alpha-1}$$

and

$$\left|T_2''(t)\right| \leq \ddot{C}_2\left(1 + t^{\alpha-1}\right).$$

To summarise, the obtained estimates imply that the solution T to problem (28.2–28.3) satisfies $T \in C^1([0, \mathcal{T}])$ and $T'' \in L^1(0, \mathcal{T})$.

Acknowledgments K. Van Bockstal was supported by the Methusalem programme of Ghent University Special Research Fund (BOF) (Grant Number 01M01021).

References

1. Sakamoto, K., Yamamoto, M.: Initial value/boundary value problems for fractional diffusion-wave equations and applications to some inverse problems. J. Math. Anal. Appl. **382**(1), 426–447 (2011)
2. Luchko, Y.: Initial-boundary-value problems for the one-dimensional time-fractional diffusion equation. Fract. Calc. Appl. Anal. **15**(1), 141–160 (2012)
3. Erdélyi, A., Magnus, W., Oberhettinger, F., Tricomi, F.G.: Higher transcendental functions. Vol. III. Bateman Manuscript Project, California Institute of Technology, vol. XVII. McGraw-Hill Book, New York (1955)
4. Podlubny, I.: Fractional Differential Equations: An Introduction to Fractional Derivatives, Fractional Differential Equations, to Methods of Their Solution and Some of Their Applications. Mathematics in Science and Engineering. Elsevier, Amsterdam (1998)
5. Hadid, S.B., Luchko, Yu.F.: An operational method for solving fractional differential equations of an arbitrary real order. Panam. Math. J **6**(1), 57–73 (1996)
6. Li, Z., Liu, Y., Yamamoto, M.: Initial-boundary value problems for multi-term time-fractional diffusion equations with positive constant coefficients, Appl. Math. Comput. **257**, 381–397 (2015)
7. Maes, F., Van Bockstal, K.: Existence and uniqueness of a weak solution to fractional single-phase-lag heat equation. Fract. Calc. Appl. Anal. **26**(4) (2023). https://doi.org/10.1007/s13540-023-00177-w
8. Sin, C.-S., Rim, J.-U., Choe, H.-S.: Initial-boundary value problems for multi-term time-fractional wave equations. Fract. Calc. Appl. Anal. **25**(5), 1994–2019 (2022)
9. Caputo, M.: Linear models of dissipation whose Q is almost frequency independent-II. Geophys. J. R. Astron. Soc. **13**(5), 529–539 (1967)
10. Li, C., Li, C.: The fractional Green's function by Babenko's approach. Tbil. Math. J. **13**(3), 19–42 (2020)
11. Bazhlekova, E.: Completely monotone multinomial Mittag–Leffler type functions and diffusion equations with multiple time-derivatives. Fract. Calc. Appl. Anal. **24**(1), 88–111 (2021)

Part III
Mathematical Modelling

Chapter 29
Mathematical Modelling of the Lomb–Scargle Method in Astrophysics

Yeskendyr Ashimov

Abstract In astrophysics, the Lomb–Scargle method is widely used to analyse time series observations of stellar objects. The method allows us to detect periodic variations in the light intensity of a star, which may be due to its rotation, pulsations or interaction with a companion. In this paper, we describe the mathematical modelling of the Lomb–Scargle method and its applications in astrophysics. The Lomb–Scargle method is based on a spectral analysis of the time series of the star's light intensity. In particular, it uses a periodogram, a function of the power of the spectrum of the time series, which reflects the contribution of each frequency to the total variance of the series. The periodogram can be obtained from the time series using a Fourier transform.

29.1 Introduction

In astrophysics, one of the most important tasks is the detection and analysis of periodic signals in time series data. These signals can originate from various sources, such as pulsating stars, variable stars, and exoplanets. The Lomb–Scargle method is a widely used technique for analyzing such periodic signals. It was first introduced by Lomb in 1976 and later extended by Scargle in 1982. In this article, we will discuss the Lomb–Scargle method and its application in astrophysics. The Lomb–Scargle method is a powerful tool for detecting and analyzing periodic signals in time series data. It is based on fitting a sinusoidal function to the data and calculating a power spectrum that measures the strength of the periodic signal at different frequencies. The power spectrum can be used to identify the dominant frequency or frequencies in the data and to estimate the amplitude and phase of the signal [1–3].

Y. Ashimov (✉)
Al-Farabi Kazakh National University, Almaty, Kazakhstan

M. Ruzhansky, K. Van Bockstal (eds.), *Extended Abstracts 2021/2022*,
Research Perspectives Ghent Analysis and PDE Center 2,
https://doi.org/10.1007/978-3-031-42539-4_29

The Lomb–Scargle method is particularly useful when the data is unevenly sampled or contains missing data points. In these cases, traditional Fourier analysis techniques cannot be used because they require evenly spaced data [4, 5]. The Lomb–Scargle method, on the other hand, can handle irregularly sampled data and can interpolate missing data points using a weighted average.

Mathematically, the Lomb–Scargle method involves fitting a sinusoidal function of the form:

$$f(t) = A\sin(2\pi f t + \phi) + C$$

to the data, where A is the amplitude, f is the frequency, ϕ, is the phase, and C is a constant offset. The Lomb–Scargle periodogram is then calculated by summing the squared residuals of the data around the fitted sinusoidal function at each frequency:

$$P(f) = \frac{1}{2\sigma^2}\left|\frac{\sum_{i=1}^{N} y_i \cos(wt_i)^2}{\sum_{i=1}^{N} \cos^2(wt_i)} + \frac{(\sum_{i=1}^{N} y_i \sin(wt_i)^2)}{\sum_{i=1}^{N} \sin^2(wt_i)}\right|,$$

where N is the number of data points, $y(i)$ is the value of the data at time $t(i)$, $\omega = 2\pi f$, and σ is the standard deviation of the data [6]. The Lomb–Scargle periodogram measures the power of the periodic signal at each frequency, and the frequency with the highest power corresponds to the dominant frequency in the data. The amplitude and phase of the signal at this frequency can be calculated by fitting a sinusoidal function to the data using the parameters obtained from the Lomb–Scargle periodogram.

29.2 Application in Astrophysics

The Lomb–Scargle method is a powerful mathematical tool widely used in astrophysics to analyse periodic signals. It was developed by Lance Lomb and Nicholas Skargle in 1976 and has since become an indispensable tool for studying cosmic objects such as stars, galaxies and pulsars. The principle of the Lomb–Scargle method is to look for periodic oscillations in a data set. This could be the oscillation frequency of a star or the rotation period of a galaxy. The Lomb-Scargle method allows such periods to be found, even if they are not constant or are lost in the noise of the data. Using the Lomb–Scargle method, astrophysicists can study many different objects. One of the most interesting examples is Cepheid-type variable stars.

These stars have regular periodic variations in their luminosity, which can be used to determine their distance. This is very important for measuring distances in space, as there is currently no other way to accurately measure distances to distant objects. However, the Lomb–Scargle method can be used for more than just measuring distances. It could be applied to find periodic oscillations in other objects, such

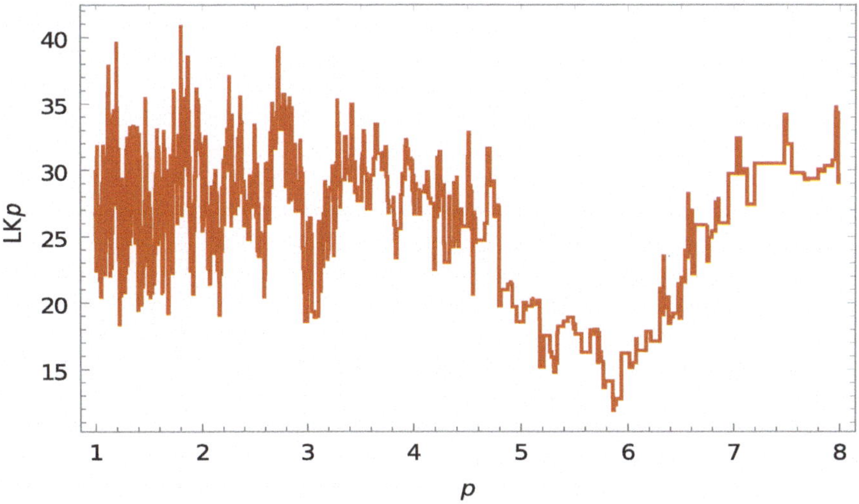

Fig. 29.1 Observed light curve from Horne object

as active galaxies, pulsars and space stations. It can help scientists better understand these objects and predict their future behaviour. The Lomb–Scargle method also has a number of limitations. For example, it may not work with data that contain non-periodic oscillations. As an example of a typical application of this method, consider the data shown in Fig. 29.1: this is a non-uniformly sampled time series showing one object from the Horne study [7], with unfiltered magnitude measured 300 times within 8 years. p is the time for the intensity of the object (LKp). It can be seen by eye that the brightness of the object varies with time in a range spanning about 0.8 magnitude, but what is not immediately clear is that this change is periodic in time. Also, it may not always be able to accurately determine the period, especially if it is very close to other periods in the data [8]. In general, the Lomb–Scargle method is an important tool for astrophysicists to investigate periodic oscillations in space objects. It is widely used in modern studies of astrophysical objects, and its application helps scientists better understand the evolution of the universe and its components.

The Lomb–Scargle method can be particularly useful for analysing data obtained from space-based observations. Space observatories such as the Hubble Space Telescope and Chandra X-ray Observatory collect a huge amount of data that can be used to study cosmic objects. However, these data can contain noise, which can significantly affect the results of analysis. The Lomb–Scargle method eliminates noise from the data and detects periodic oscillations that may be related to physical processes in the object. The Lomb–Scargle method can also be used to study exoplanets, which are planets outside the solar system [9–11]. These planets can have periodic oscillations in their motion, which can be detected by the Lomb-Scargle method. This helps scientists determine the parameters of the planet,

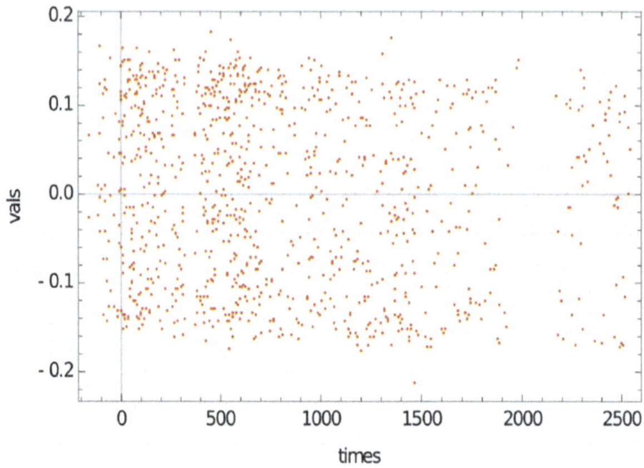

Fig. 29.2 The influence of noise on these stars is shown

such as its mass, radius and orbit. Before we begin to explore the Lomb-Skargle periodogram in more detail, it is worth briefly considering the broader context of detection methods and the characterization of periodicity in time series. First, it is important to note that there are many different modes of observing time series. The point observation shown in Fig. 29.2 is typical of optical astronomy: values (often with uncertainty) are measured at discrete times (vals) that can be spaced equally or unequally apart. Other modes of observation–for example, timestamped events, binned event data, time-spilled events, and so on–are common in high-energy astronomy and other fields. We will not consider such event-driven data modes here, but note that some interesting research has been done on unified statistical processing of all the above observation modes.

The Lomb-Scargle method is thus a powerful mathematical tool in astrophysics that allows periodic oscillations in different observational objects to be detected. This can be useful for studying the properties of these objects, such as mass, radius, orbital period, luminosity and other parameters. Another example of the use of the Lomb–Scargle method is the analysis of magnetic fields on the surface of stars. When a magnetic field is on the surface of a star, it can create periodic fluctuations in stellar brightness [12, 13]. Using the Lomb-Scargle method, scientists can detect these oscillations and analyse them to study the properties of the magnetic fields on the surface of the star. The Lomb–Scargle method can also be used to detect periodic oscillations in galaxies and quasars. Galaxies can have periodic variations in stellar velocity, which can be detected by spectroscopy. Quasars may have periodic variations in their brightness, which can be detected by photometry. Using the Lomb–Scargle method, scientists can analyse these oscillations and study the properties of these objects.

During mathematical modeling on the wolfram program, the results of this method were obtained. We compared with other methods, we got the results and they are shown in Fig. 29.3. By the distribution of the phase, we see the periodicity of the

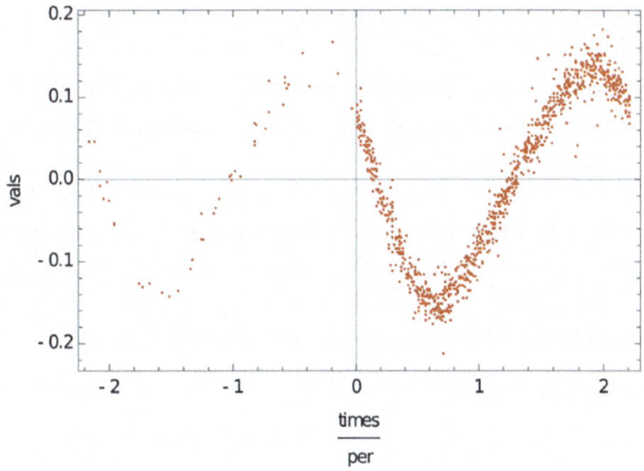

Fig. 29.3 The results of the Lomb-Skargle method are shown

sinusoid, which proves the accuracy of this method. When a classical periodogram is applied to uniformly sampled Gaussian white noise, the values of the resulting periodogram are x-square distributed. This property becomes quite useful in practice when the periodogram is used in the context of classical hypothesis testing to distinguish between periodic and non-periodic objects. The Lomb–Scargle method works well with data that comes with different cosmic noises. When constructing a periodogram, it complements the absence of amplitude values, which makes it possible to calculate full periods.

29.3 Conclusion

Various reviews have been undertaken to compare the effectiveness and efficiency of available methods. Schwarzenberg-Czerny [14] focuses on the statistical properties of methods. Based on smooth model fitting, which is not phase binning. While Graham et al. [15] instead use an empirical approach. When considering performance on real datasets, no sufficiently efficient algorithm outperforms others in light of the variety of methods available. First, the Lomb-Skargle periodogram is perhaps the best-known method for calculating the periodicity of unevenly spaced data in astronomy and other fields, and is therefore the first tool that many will turn to when looking for periodic content in a signal. Secondly; that the Lomb-Skargle method occupies a unique niche: it is based on Fourier analysis, but it can also be considered as a method of least squares. It can be derived from the principles of Bayesian probability theory and has been shown in some circumstances to be closely related to bin-based phase convolution methods [16]. Thus, the Lomb-Skargle periodogram occupies a unique point of correspondence between many classes of methods and thus provides a focus for discussing the considerations associated with all of these methods.

The Lomb–Scargle method is a powerful tool for detecting and analyzing periodic signals in time series data, particularly in cases where the data is unevenly sampled or contains missing data points. It has numerous applications in astrophysics, including the study of pulsating stars, exoplanets, variable stars, pulsar timing, and gravitational waves. The Lomb–Scargle method involves fitting a sinusoidal function to the data and calculating a power spectrum that measures the strength of the periodic signal at different frequencies. The power spectrum can be used to identify the dominant frequency or frequencies in the data and to estimate the amplitude and phase of the signal. The Lomb–Scargle method has been widely used in astrophysics and has led to numerous discoveries and insights into the properties and behavior of various astrophysical objects and phenomena.

References

1. Lomb, N.R.: Least-squares frequency analysis of unequally spaced data. Astrophys. Space Sci. **39**(2), 447–462 (1976)
2. Scargle, J.D.: Studies in astronomical time series analysis. II-Statistical aspects of spectral analysis of unevenly spaced data. Astrophys. J. **263**, 835–853 (1982)
3. Al-Ani, M., Tarczynski, A.: Evaluation of Fourier transform estimation schemes of multidimensional signals using random sampling. Signal Process. **92**(10), 2484–2496 (2012)
4. Zhunussova, Zh.Kh., Ashimov, Ye.K., Dosmagulova, K.A., Zhunussova, L.Kh.: Optimal packing of two disks on torus. Appl. Math. Inform. Sci. **16**(4), 549–554 (2022)
5. Mityushev, V., Rylko, N., Zhunussova, Z., Ashimov, Y.: Inverse conductivity problem for spherical particles. In: Mechanics and Physics of Structured Media: Asymptotic and Integral Equations Methods of Leonid Filshtinsky, pp. 109–121 (2022)
6. Brown, T.M.: Kepler mission design, realized photometric performance, and early science. Publ. Astronom. Soc. Pac. **125**(931), 103–114 (2013)
7. Horne, J.H., Baliunas, S.L.: A prescription for period analysis of unevenly sampled time series. Astrophys. J. **302**, 757–763 (1986)
8. Dosmagulova, K., Mityushev, V., Zhunussova, Zh.: Optimal conductivity of packed two-dimensional dispersed composites. SIAM J. Appl. Math. **83**(3), 152766, 985–999 (2023)
9. Anglada-Escudé, G., et al.: A terrestrial planet candidate in a temperate orbit around Proxima Centauri. Nature, **536**(7617), 437–440 (2016)
10. Babu, P., Stoica, P.: Spectral analysis of nonuniformly sampled data—a review. Digit. Signal Process. **20**(2), 359–378 (2010)
11. Baluev, R.V.: Assessing the statistical significance of periodogram peaks. Mon. Not. R. Astronom. Soc. **385**(3), 1279–1285 (2008)
12. Baluev, R.V.: Detecting multiple periodicities in observational data with the multifrequency periodogram—I. Analytic assessment of the statistical significance. Mon. Not. R. Astronom. Soc. **436**(1), 807–818 (2013)
13. Mortier, A., Faria, J.P., Correia, C.M., Santerne, A., Santos, N.C.: BGLS: a Bayesian formalism for the generalised Lomb–Scargle periodogram. Astron. Astrophys. **573**, A101 (2015)
14. Schwarzenberg- czerny, A.: Optimum period search: quantitative analysis. Astrophys. J. **516**, 315–323 (1999)
15. Aguti, B., Walters, R., Wills, G.: A framework for evaluating the effectiveness of blended e-learning within universities. In: Society for Information Technology & Teacher Education International Conference, pp. 1982–1987 (2013)
16. Swingler, D.N.: Astrophys. J. **97**, 280 (1989)

Chapter 30
The Application of Physics Informed Networks to Solve Hyperbolic Partial Differential Equations with Nonconvex Flux Function and Diffusion Term

Yedilkhan Amirgaliyev and Timur Merembayev

Abstract In this paper, we study the possibility of using neural networks to solve rare derivatives, in particular transport problems in a porous medium. Neural networks can approximate the solution of differential equations, particularly multivariate partial differential equations (PDEs). We use physics-informed neural networks (PINN) for the classic hyperbolic model problem, namely the Buckley–Leverett. The experiment shows fairly accurate results; the error is RSME = 3.7356e-01. However, there is the open question of whether a solution to the Buckley–Leverett problem with a nonconvex flow function can be learnt by deep neural networks without the aid of artificial physical constraints.

30.1 Introduction

Deep learning enables neural networks consisting of multiple layers of computation to acquire knowledge about raw input data at various levels of complexity. These networks excel in tasks that involve supervised learning, where the availability of abundant labelled data is crucial for their successful application [1]. However, in numerous engineering applications, the process of gathering data is often excessively costly, resulting in a scarcity of labelled data [2–4]. This scarcity is particularly evident in computational problems that revolve around modelling the dynamics of subsurface flow, where specific site data is often limited.

Neural networks can approximate the solution of differential equations, particularly multivariate partial differential equations (PDEs). One of the most promising approaches to efficiently solving nonlinear partial differential equations is physics-informed neural networks (PINN). PINNs can solve supervised learning problems

Y. Amirgaliyev · T. Merembayev (✉)
Institute of Information and Computational Technologies, Almaty, Kazakhstan
e-mail: amir_ed@mail.ru

M. Ruzhansky, K. Van Bockstal (eds.), *Extended Abstracts 2021/2022*, Research Perspectives Ghent Analysis and PDE Center 2,
https://doi.org/10.1007/978-3-031-42539-4_30

with PDE constraints, such as continuum fluid and solid mechanics conservation laws.

Multiscale modelling of subsurface flow and transport is crucial for various applications, such as petroleum recovery, nuclear waste disposal, CO_2 sequestration, groundwater remediation, and contaminant plume migration in porous media. However, direct numerical simulations of these processes can be computationally demanding due to the fine-scale characterisation of subsurface formations. In the research [5], the author developed an efficient multiscale framework that reduces the computational burden associated with fine-scale properties while maintaining accuracy in quantities of interest, including mass balance, pressure, velocity, and concentration.

In the paper [6, 7], authors investigated the effect of the post-processed pressure on the velocity approximation in the interface of subdomains within the framework of the Enhanced Velocity Mixed Finite Element Method. Specifically, the focus is on incompressible Darcy flow in a non-matching multiblock grid setting. Multiple numerical results substantiate that incorporating the post-processed pressure leads to an improved approximation of the interface velocity. These findings have significant implications for enhancing our understanding of velocity approximation and provide a robust foundation for posterior error analysis, such as recovery-based estimates.

In the research [8], authors considered an acoustic problem concerning the propagation of waves through a discontinuous medium. The problem is formulated as a dissipative wave equation with distributional dissipation. A very weak solution to this problem is introduced and analysed, shedding light on its properties. Numerical simulations are performed to approximate the solutions of the full dissipative model for a piecewise continuous synthetic medium, illustrating the theoretical results. The concept of very weak solutions overcomes these difficulties and establishes well-posedness results for equations with singular coefficients.

The idea of PINN is to train the neural network with automatic differentiation by minimising the error of the objective function, including initial and boundary constraints [9, 10]. PINNs can serve as efficient simulators of physical processes described by differential equations. Once the PINN model has been trained, it has the potential to run faster and make more accurate predictions than standard numerical models of complex real-world phenomena.

30.2 Multi-Phase Transport Problem

We use the PINN approach to solve the classic hyperbolic model problem, namely the Buckley–Leverett [11] equation. The Buckley–Leverett equation with a non-convex flow function is an excellent benchmark for testing the general potential of PINN in solving hyperbolic PDEs. During the experiment, it was shown that PINNs effectively capture the propagation of the shock wave front and are able to provide high-quality solutions for the coefficients of mobility inside the convex body of the training facility. PINNs can provide smooth and accurate shock fronts without explicitly introducing additional constraints or residual dissipation due to an artificial diffusion term or expansion of spatial derivatives.

Consider the problem of the flow of two immiscible fluids through a horizontal porous medium. There is a limitation in that the fluids and the system as a whole are incompressible, i.e. there is no pressure. The Buckley–Leverett [12] equation describes the saturation of the wetting phase (with water) with a diffusion time by changing in time and space. Let $u : \mathbf{R}_0^+ \times \mathbf{R}_0^+ \to [0, 1]$ be the solution of the Buckley–Leverett equation:

$$\begin{aligned} \frac{\partial u}{\partial t}(x,t) + \frac{\partial f}{\partial x}(x,t) = \epsilon \frac{\partial^2 u}{\partial x^2}(x,t) & \\ u(x,t) = 0, \forall x, t = 0 \quad & \text{initial condition,} \\ u(x,t) = 1, x = 0 \quad & \text{border condition,} \end{aligned}$$

where u is the saturation of the wetting phase, $\epsilon > 0$ is the scalar diffusion coefficient inverse to the Peclet number, Pe is the ratio of the characteristic scattering time to the characteristic convection time. If the Peclet number is large then diffusion effects are negligible and convection predominates. The f is a partial flow function and it is represented as an S-figurative flow function and we consider a convex and nonconvex function:

$$f_w(u) = \frac{u}{u + \frac{1-u}{M}} \qquad \text{convex function,}$$

$$f_w(u) = \frac{u^2}{u^2 + \frac{(1-u)^2}{M}} \qquad \text{nonconvex function,}$$

where M is the mobility coefficient of the two liquid phases and is characterized by the mobility coefficient.

30.3 Physics-Informed Neural Networks

The PINN implementation approach is shown in Fig. 30.1, which shows the concept scheme to solve the objective function optimisation problem.

To solve this problem, a neural network architecture was built with the following topology: input 2 values $[x, t]$, 7 hidden layers of 9 neurons, output 1 of size $[u]$, Fig. 30.2. For this, the Tensorflow v.2.10 framework was used.

The forecasting performance of machine learning models is estimated by one indicator, the root mean squared error (RSME) score [13].

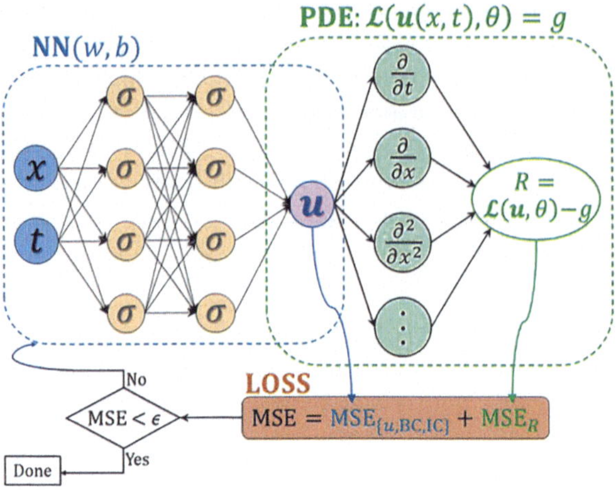

Fig. 30.1 Schematic of a physics-informed neural network, where the loss function of neural network contains a mismatch in the given data on the state variables or boundary and initial conditions [10]

```
Net(
  (hidden_layer1): Linear(in_features=2, out_features=9, bias=True)
  (hidden_layer2): Linear(in_features=9, out_features=9, bias=True)
  (hidden_layer3): Linear(in_features=9, out_features=9, bias=True)
  (hidden_layer4): Linear(in_features=9, out_features=9, bias=True)
  (hidden_layer5): Linear(in_features=9, out_features=9, bias=True)
  (hidden_layer6): Linear(in_features=9, out_features=9, bias=True)
  (hidden_layer7): Linear(in_features=9, out_features=9, bias=True)
  (output_layer): Linear(in_features=9, out_features=1, bias=True)
)
```

Fig. 30.2 The architecture of neural network

30.4 Result

After training 10,000 epochs, the neural network has RSME = 3.7356e-01 and duration = 01:50 min.

Comparison of PINN results with the analytical solution provided in [9]. The analytical solution to this problem contains a shock wave and a rarefaction wave and is constructed as follows:

$$u(x,t) = \begin{cases} 0,\ \frac{x}{t} > f'_w(u_{shock}) \\ u(\frac{x}{t}),\ f'_w(u_{shock}) \geq \frac{x}{t} \geq f'_w(u=1) \\ 1,\ f'_w(u=1) \geq \frac{x}{t} \end{cases}$$

where u denotes the position of the shock wave, which is determined by the Rankine-Hugoniot condition. Figure 30.3 shows the result of the prediction of the

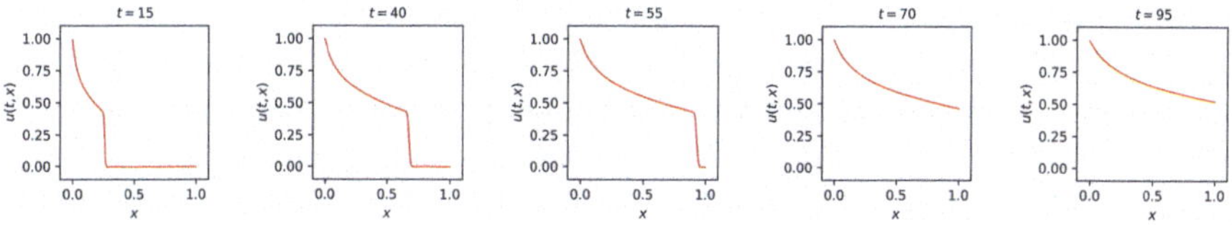

Fig. 30.3 PINN solution for nonconvex and with diffusion term condition in BL problem

Fig. 30.4 PINN solution for nonconvex and with diffusion term condition for $t = 0.55$ period

saturation flow without diffusion using the PINN model. It can be seen that the neural network is quite good at predicting saturation over time.

Figure 30.4 shows the increased scale of the predicted saturation in the period $t = 0.55$. There is no precision prediction of the shock effect, the line should be more linear.

This first-order hyperbolic equation is interesting as its solution can display smooth solutions (rarefactions) and sharp fronts (shocks). Whether deep neural networks can learn the solution to the BL problem with nonconvex flux function without the aid of artificial physical constraints remains an open question.

30.5 Conclusion

In this study, we provide a solution to hyperbolic equations using neural networks using the example of the Buckley–Leverett problem. The experiment showed that PINN solves the Buckley–Leverett problem well for the case with a convex partial flow function and diffusion, the experiment shows fairly accurate results, the error is RSME = 3.7356e-01.

The problem of solving a first-order hyperbolic equation is interesting because its solution can have both smooth solutions (rare fractions) and sharp fronts (jumps). It remains an open question whether a solution to the Buckley–Leverett problem with a nonconvex flow function can be learnt by deep neural networks without the aid of artificial physical constraints.

The computation time for the neural network approach is an order of value less than for the finite element method, and the efficiency increases as the size and number of discrete elements increase. This property makes neural networks very attractive when it comes to solving numerical problems with poor complexity scaling (such as reservoir modelling). Additionally, no labelled data is required for training, which can significantly speed up the learning process.

In our further research, we intend to concentrate on the prediction of the propagation of waves through a discontinuous medium using the PINN approach [14].

Acknowledgments The work was supported by the program-targeted funding projects of the Ministry of Education and Science of the Republic of Kazakhstan AP14871625.

References

1. Mukhamediev, R., Popova, Y., Kuchin, Y., Zaitseva, E., Kalimoldayev, A., Symagulov, A., Levashenko, V., Abdoldina, F., Gopejenko, V., Yakunin, K., et al.: Review of artificial intelligence and machine learning technologies: classification, restrictions, opportunities and challenges. Mathematics **10**, 2552 (2022)
2. Merembayev, T., Kurmangaliyev, D., Bekbauov, B., Amanbek, Y.A.: Comparison of machine learning algorithms in predicting lithofacies: case studies from Norway and Kazakhstan. Energies **14**, 1896 (2021)
3. Merembayev, T., Bekkarnayev, K., Amanbek, Y.: The identification models of the copper recovery using supervised machine learning algorithms for the geochemical data. In: 55th US Rock Mechanics/Geomechanics Symposium (2021)
4. Yeleussinov, A., Amirgaliyev, Y., Cherikbayeva, L.: Improving OCR Accuracy for Kazakh handwriting recognition using GAN models. Appl. Sci. **13**, 5677 (2023)
5. Amanbek, Y.: A New Adaptive Modeling of Flow and Transport in Porous Media Using an Enhanced Velocity Scheme. The University of Texas at Austin (2018)
6. Amanbek, Y., Singh, G., Wheeler, M.: Recovery of the interface velocity for the incompressible flow in enhanced velocity mixed finite element method. In: Computational Science–ICCS 2019: 19th International Conference, Faro, Portugal, June 12–14, 2019, Proceedings, Part IV 19, pp. 510–523 (2019)
7. Amanbek, Y., Singh, G., Pencheva, G., Wheeler, M.: Error indicators for incompressible Darcy flow problems using enhanced velocity mixed finite element method. Comput. Methods Appl. Mech. Eng. **363**, 112884 (2020)
8. Muñoz, J., Ruzhansky, M., Tokmagambetov, N.: Wave propagation with irregular dissipation and applications to acoustic problems and shallow waters. Journal De Mathematiques Pures Et Appliquees **123**, 127–147 (2019)
9. Fuks, O., Tchelepi, H.: Limitations of physics informed machine learning for nonlinear two-phase transport in porous media. J. Mach. Learn. Model. Comput. **1** (2020)
10. Meng, X., Li, Z., Zhang, D., Karniadakis, G.: PPINN: parareal physics-informed neural network for time-dependent PDEs. Comput. Methods Appl. Mech. Eng. **370**, 113250 (2020)
11. LeVeque, R., Leveque, R.: Numerical Methods for Conservation Laws. Springer, Berlin (1992)
12. Buckley, S., Leverett, M.: Mechanism of fluid displacement in sands. Trans. AIME **146**, 107–116 (1942)
13. Grandini, M., Bagli, E., Visani, G.: Metrics for multi-class classification: an overview (2020). ArXiv Preprint ArXiv:2008.05756
14. Munoz, J., Ruzhansky, M., Tokmagambetov, N.: Acoustic and shallow water wave propagation with irregular dissipation. Funct. Anal. Appl. **53**, 153–156 (2019)

Chapter 31
Fractional Differential Equations: A Primer for Structural Dynamics Applications

Shashank Pathak

Abstract This paper presents a very brief overview of the framework for the application of fractional calculus (FC) in structural dynamics with the target audience from two scholarly communities, namely, structural dynamists and applied mathematicians. For the former, this paper provides a quick and simplified introduction to fractional differential equations (FDE) to explore the dynamics of structures consisting of rate-dependent materials. A systematic link has been shown between conventional structural dynamics (CSD) and its fractional generalisation. Whereas, for the latter, it highlights the powerful applications of fractional equations in the broad field of structural dynamics with some potential problems that may need their attention in the near future.

31.1 Introduction

In the community of mathematicians, FC is well known since the time of Leibniz and Euler (a few centuries ago). However, engineering interest started to grow only a few decades ago. Machado et al. [1] prepared an exhaustive list of special issues in the journals, conferences, books, and software devoted to FC for the period 1974–2011. After that, in 2018, Sun et al. [2] summarised various powerful applications of FC in engineering problems related to the transport phenomena in heterogeneous gaseous media, diffusion in viscoelastic biological liquids, wave propagation and attenuation in viscoelastic polymers and porous media, nonlocal elasticity and viscoelasticity of nanostructures, fractional viscoelasticity and thermoelasticity, dynamics of viscoelastic structures, soil-structure interaction considering the viscoelastic nature of soils, time-dependent deformation of materials depicting the strain softening behaviour, fractional order controllers, computer vision and image

S. Pathak (✉)
School of Civil and Environmental Engineering, Indian Institute of Technology Mandi, Mandi, Himachal Pradesh, India
e-mail: shashank@iitmandi.ac.in

M. Ruzhansky, K. Van Bockstal (eds.), *Extended Abstracts 2021/2022*,
Research Perspectives Ghent Analysis and PDE Center 2,
https://doi.org/10.1007/978-3-031-42539-4_31

processing, resolution of microscopic images, microfluidic devices, non-Newtonian fluid mechanics, a constitutive model of temperature-dependent shape-memory polymers, etc. Quite recently, in 2022, Diethelm et al. [3] thoroughly analysed the bibliographic information of more than 6500 documents related to FC to decipher the current trends and future scope of research.

Structural dynamics is the field of engineering that discusses the vibratory motion of civil, mechanical, and aerospace structures under the influence of external excitations such as wind, earthquakes, blasts, impacts, sea waves, turbulence, fatigue, etc. With technological advancements such as 3-D printing, robotic constructions, and material advancements, modern structures are also being constructed using advanced materials such as shape-memory alloys, viscoelastic materials, metamaterials, carbon nanotubes, etc. Such materials exhibit unconventional constitutive behaviour, including hereditary and viscoelastic properties, when subjected to dynamic stresses. The proven capabilities of FC to capture nonlocal and hereditary behaviour shows that there is a tremendous scope for its application in understanding the dynamics of such modern structures. Shitikova [4] and Li [5] are the most recent and relevant reviews on the application of FC in the dynamics of viscoelastic solids.

There are well-established and standard approaches in CSD that are regularly used to understand the dynamics of structures made up of conventional materials such as concrete, steel, aluminium, fibres, etc. However, the problems of modern-material-based structures can be solved more efficiently using FC. Thus, this paper provides a quick and simplified introduction to FDE to explore the dynamics of structures consisting of rate-sensitive materials. A systematic link between the CSD and fractional structural dynamics (FSD) is highlighted, indicating that the standard approaches existing in CSD can be generalised to the FSD.

31.2 Material Response

It is well established that several engineering materials demonstrate rate-dependent viscoelastic behaviour in the form of stress relaxation and frequency-dependent damping and stiffness [6–9]. Torvik and Bagley [8] discussed a generalized five parameter fractional-order constitutive model as follows:

$$\sigma(t) + V_\sigma \frac{d^\beta \sigma(t)}{dt^\beta} = E_\epsilon \epsilon(t) + V_\epsilon \frac{d^\gamma \epsilon(t)}{dt^\gamma}, \tag{31.1}$$

where $\sigma(t)$ and $\epsilon(t)$ are stress and strain time-histories, respectively. E_ϵ and V_ϵ are strain coefficients related to elasticity and viscosity, respectively. V_σ is the viscosity-related stress coefficient. d^β/dt^β and d^γ/dt^γ (also may be indicated as D^β or D^γ) represent fractional derivatives of the order β and $\gamma \in [0, 1]$. Equation (31.1) may be regarded as a generalised material model because: (1) $V_\sigma = V_\epsilon = 0$ represents Hooke's law, (2) $V_\sigma = E_\epsilon = 0, \gamma = 1$ represents the Newtonian fluid, (3) $V_\sigma =$

$E_\epsilon = 0$ represents the Scott Blair model, and several other models can also be obtained as a special case of Eq. (31.1).

For engineering applications, the 5-parameters $(V_\sigma, \gamma, E_\epsilon, V_\epsilon, \beta)$ can be determined by appropriately fitting the experimental data. Although in the current paper, these parameters are assumed to be deterministic, but the material parameters may have significant uncertainty in their behaviour owing to ambient conditions, testing procedures, inherent variability, limited and incomplete test data, etc. Thus, in real-life situations, all five parameters will be random variables and can be represented by a joint probability density function (PDF). For all practical purposes, these parameters can be assumed to be statistically independent and their joint distribution would be multiplication of their respective marginal PDFs.

In addition to viscoelasticity, problems involving extreme transient loads such as earthquakes, shock and explosions, wind-storms, and space-debris impacts require the material models which include the dependency on stress and/or strain rates (e.g. [10]). Thus, there may be suitable applications of fractional derivatives in those problems as well. In such cases, E_ϵ, V_ϵ, and V_σ may be arbitrary functions of the time derivatives of ϵ and σ.

31.3 Equation of Motion

Consider a single-degree-of-freedom (sdof) oscillator of total mass m represented by the generalised material model of Eq. (31.1). The Fourier transform (FT) of Eq. (31.1) leads to (a discussion on FT of fractional derivatives is presented later)

$$\frac{\tilde{\sigma}(\omega)}{\tilde{\epsilon}(\omega)} = \frac{E_\epsilon + V_\epsilon (i\omega)^\gamma}{1 + V_\sigma (i\omega)^\beta}, \tag{31.2}$$

where $i = \sqrt{-1}$, $\tilde{[.]}$ is FT of [.], and ω is frequency (domain). The left-hand side of Eq. (31.2) represents the modulus of the material $E(\omega)$ and is related to the equivalent stiffness $(k_{eq}(\omega))$ as $k_{eq}(\omega) = pE(\omega)$ where p is the constant of proportionality depending on the structural geometry.

If the mass m of equivalent stiffness k_{eq} is in a state of rest at the equilibrium position (i.e., initial conditions are zero) and is subjected to an external forcing function $f(t)$ then the equation of motion is

$$m\ddot{x}(t) + k_{eq}x(t) = f(t). \tag{31.3}$$

It may be noted that $\dot{[.]}$ and $\ddot{[.]}$ represent the single and double time-derivatives of [.], respectively, and $x(t)$ is displacement of the mass. The FT of Eq. (31.3) gives

$$[-m\omega^2 + k_{eq}]\tilde{x}(\omega) = \tilde{f}(\omega). \tag{31.4}$$

Substituting Eq. (31.2) in Eq. (31.4) and using $k_{eq}(\omega) = pE(\omega)$ leads to

$$\left[-m\omega^2 + p\frac{E_\epsilon + V_\epsilon(i\omega)^\gamma}{1 + V_\sigma(i\omega)^\beta}\right]\tilde{x}(\omega) = \tilde{f}(\omega),$$

which further simplifies to

$$\left[m(i\omega)^2 + mV_\sigma(i\omega)^{\beta+2} + pV_\epsilon(i\omega)^\gamma + pE_\epsilon\right]\tilde{x}(\omega)$$
$$= \tilde{f}(\omega) + V_\sigma(i\omega)^\beta\tilde{f}(\omega). \qquad (31.5)$$

By observation, it can be seen that Eq. (31.5) corresponds to the following time-domain equation

$$m\left(\ddot{x} + V_\sigma\frac{d^{\beta+2}x}{dt^{\beta+2}}\right) + (pV_\epsilon)\left(\frac{d^\gamma x}{dt^\gamma}\right) + (pE_\epsilon)x(t) = f(t) + V_\sigma\frac{d^\beta f}{dt^\beta},$$

which in the operator format looks like

$$[m(1 + V_\sigma D^\beta)D^2 + pV_\epsilon D^\gamma + pE_\epsilon]x(t) = [1 + V_\sigma D^\beta]f(t). \qquad (31.6)$$

The operator $(1 + V_\sigma D^\beta)$ gives rise to fractional inertia and fractional excitation terms. With $V_\sigma = 0$, the well-known inertia term $(m\ddot{x})$ and the excitation term $(f(t))$ are obtained. Since the physical interpretation of fractional inertia and fractional excitation is not known to the author, the rest of the discussion in this paper will be for $V_\sigma = 0$. The term pV_ϵ is the viscosity-related term that is recognised as the viscous damping coefficient c in CSD. The third term pE_ϵ is the elastic part of the complex modulus and represents the spring coefficient k in CSD. Thus, Eq. (31.6) can be rewritten as

$$m\ddot{x} + c\overset{\gamma}{x} + kx = f,$$

where $\overset{\gamma}{x}$ is the time-derivative of x of fractional-order γ.

It is worth noting that in the cases of material and geometric nonlinearity, the stiffness may be dependent on time and deformation and its derivatives. Therefore, the generalised way would be to represent k as a summation of a linear elastic component and a deformation-dependent nonlinear component as: $k = m(\omega_n^2 + g(t, x, \overset{\alpha}{x}, \dot{x}))$, where ω_n is the natural frequency of the linear component of the oscillator and g is an arbitrary function explaining the nonlinear component of the stiffness, $\overset{\alpha}{x}$ is the time derivative of x of fractional order α. Thus, the generalized equation of motion can be written as

$$m\ddot{x} + c\overset{\gamma}{x} + m\omega_n^2 x + mg(t, x, \overset{\alpha}{x}, \dot{x}) = f(t). \qquad (31.7)$$

Dividing Eq. (31.7) by m gives the well-known physically intuitive form

$$\ddot{x} + \eta \overset{\gamma}{x} + \omega_n^2 x + g(t, x, \overset{\alpha}{x}, \dot{x}) = a(t), \tag{31.8}$$

where $a(t) = f(t)/m$ is the external acceleration applied to the mass and $\eta = c/m$. In CSD, $\eta = 2\zeta\omega_n$, where ζ is a positive nondimensional damping ratio. By conducting the dimensional analysis of Eq. (31.8), it is noted that all terms on the left-hand side should have a dimension $[L^1T^{-2}]$ (where L= dimension of length and T = dimension of time). Thus, the dimension of η can be back-calculated as $[L^0T^{(\gamma-2)}]$ which indicates that, for Eq. (31.8), $\eta = 2\zeta\omega_n^{(2-\gamma)}$. Thus, Eq. (31.8) can be further written as

$$\ddot{x} + 2\zeta\omega_n^{(2-\gamma)} \overset{\gamma}{x} + \omega_n^2 x + g(t, x, \overset{\alpha}{x}, \dot{x}) = a(t). \tag{31.9}$$

Interestingly, Eq. (31.9) converts to the well-known CSD equation for $\gamma = 1$ and linear oscillators ($g \equiv 0$)

$$\ddot{x} + 2\zeta\omega_n\dot{x} + \omega_n^2 x = a(t). \tag{31.10}$$

Thus, in view of the above discussion, Eq. (31.6) can be regarded as the complete-form of the fractional generalization of the equation of motion of a sdof oscillator, whereas, Eq. (31.9) can be seen as the simplified-form of the generalized equation of motion.

31.4 Frequency Response Function

The structural response $x(t)$ when the structure is subjected to the unit harmonic excitation is popularly known as the frequency response function (represented as $H(\omega)$ or $G(s)$) and provides important information for the estimation of response in the frequency domain (s-Laplace domain or ω-Fourier domain).

The Laplace transform (LT, represented by $\mathcal{L}$) of nth integer-order derivative of $x(t)$ can be written as

$$\mathcal{L}[D^n x(t)] = s^n \mathcal{L}[x(t)] - \sum_{k=0}^{n-1} s^{n-1-k} [D^k x(t)]|_{t=0}.$$

Here, $[D^k x(t)]|_{t=0}$ physically represents the initial conditions. For example, in the case of CSD (i.e., $n = 2$ then $k = 0, 1$), the initial conditions are: initial displacement $x(0)$ and initial velocity $\dot{x}(0)$. Using the Caputo's definition of fractional derivative [11] (represented as ${}^C D^r x(t)$), LT of rth real-order derivative of $x(t)$ can be naturally generalised to

$$\mathcal{L}[{}^{C}D^{r}x(t)] = s^{r}\mathcal{L}[x(t)] - \sum_{k=0}^{n-1} s^{r-1-k}[D^{k}x(t)]|_{t=0}$$

where $r \in [n-1, n]$. It is straightforward to note that the initial conditions are conveniently taken care of in the LT of the Caputo's derivative. However, if one uses the Riemann-Liouville's definition [12]) of fractional derivative (represented as ${}^{RL}D^{r}x(t)$) then LT includes the fractional integral/derivative terms as follows:

$$\mathcal{L}[{}^{RL}D^{r}x(t)] = s^{r}\mathcal{L}[x(t)] - \sum_{k=0}^{n-1} s^{k}[{}^{RL}D^{r-1-k}x(t)]|_{t=0}.$$

Thus, one gets the initial conditions in the form of fractional derivatives. For example, if $n = 2$ and let $r = 1/2$, then the initial conditions are: ${}^{RL}D^{-1/2}x(t)|_{t=0}$ and ${}^{RL}D_{t}^{-3/2}x(t)|_{t=0}$ which do not make any physical sense. Using the Riemann-Liouville's definition of the fractional integral of order r, i.e.

$${}^{RL}D^{-r}x(t) = \frac{1}{\Gamma(r)}\int_{0}^{t}(t-u)^{(r-1)}x(u)du.$$

If the above integral vanishes in the limit $t \to 0$ then for zero initial conditions (i.e., initially the oscillator is at equilibrium and in the state of rest), both (Riemann-Liouville and Caputo) the definitions lead to the same LT, i.e., $s^{r}\mathcal{L}[x(t)]$. This implies, in principle, that both definitions will lead to the identical response of the oscillator when initial conditions are zero, whereas, if the oscillator vibrates with non-zero initial displacement or velocity then Riemann-Liouville's definition may not be able to capture the correct physics of the problem and, in such a case, Caputo's definition would be the best choice.

Using the above discussion, one may easily write the frequency response function $H(\omega) = \tilde{x}(\omega)/\tilde{a}(\omega)$ or $G(s) = \tilde{x}(s)/\tilde{a}(s)$ as (for the case of linear oscillators, i.e. $g \equiv 0$ in Eq. (31.9))

$$G(s) = \frac{1}{\omega_n^2 + 2\zeta\omega_n^{(2-\gamma)}s^{\gamma} + s^2}.$$

Since the Laplace and Fourier domains are related as $s = i\omega$, $H(\omega)$ can be directly written as

$$H(\omega) = \frac{1}{\omega_n^2 + 2\zeta\omega_n^{(2-\gamma)}(i\omega)^{\gamma} - \omega^2},$$

which reduces to the following very well-known form of CSD when $\gamma = 1$:

$$H(\omega) = \frac{1}{\omega_n^2 + 2i\zeta\omega_n\omega - \omega^2}.$$

For non-integer values of γ, i^γ can be evaluated using DeMoivre's Theorem which will lead to multiple complex-conjugate solutions. Once, $H(\omega)$ is known, the inverse Fourier transform of $H(\omega) \times \tilde{a}(\omega)$ provides the time-domain response $x(t)$.

31.5 Superposition Principle

In CSD, Duhamel's convolution integral is used to solve the linear problems which work on the superposition principle of the impulse responses occurring at different time intervals. Suarez and Shokooh [13] demonstrated the extension of Duhamel's integral to the case of FSD with $\gamma = 1/2$. When external excitation is a unit impulse load defined as $a(t) = \delta(t)$ or $\tilde{a}(\omega) = 1$ (Dirac-Delta function), the inverse transform approach gives the impulse solution $h(t)$ of the following form:

$$h(t) = 2\sum_{i=1}^{4} \frac{\lambda_i e^{\lambda_i^2 t}}{\prod_{k=1,k\neq i}^{4} (\lambda_i - \lambda_k)} + \frac{2\eta}{\pi} \int_0^\infty \frac{u^2 e^{-u^2 t}}{(u^4 + \omega_n^2)^2 + \eta^2 u^2} du. \tag{31.11}$$

The λ's are the solutions of the polynomial $\lambda^4 + 2\zeta\omega_n^{3/2}\lambda + \omega_n^2 = 0$. The first term on the right-hand side of Eq. (31.11) is the oscillatory part of the solution and as shown by Suarez and Shokooh [13], it has a functional form like [decaying exponential] × [sinusoidal], whereas, the second term is only a decaying function and it seems that finding an exact closed-form solution to this term may be quite complicated or not possible. Nevertheless, one may always use the appropriate numerical integration techniques and find a solution for the second term. Once the complete solution of $h(t)$ is known, the following convolution integral can be used to obtain the response $x(t)$ for any arbitrary excitation $a(t)$ of total duration t_0:

$$x(t) = \int_0^{t_0} a(t - \tau)h(\tau)d\tau.$$

It may be noted that Eq. (31.11) is applicable only when $\gamma = 1/2$. If γ is of the form p/q such that p and q are positive integers ($q \geq p$) then λ's will be the solutions of: $\lambda^{2q} + 2\zeta\omega_n^{(2-p/q)}\lambda^p + \omega_n^2 = 0$ and accordingly Eq. (31.11) will be modified.

31.6 State-Space Approach

The state-space approach is quite popular in control engineering where the dynamists deal with the sensors and actuators (see Chapter-9 of [14]). Mathematically, the state-space approach converts the second-order ordinary differential equations (ODEs) to the system of first-order ODEs which are computationally efficient to solve. Consider the case of Eq. (31.10). Using the transformation $x_1 = x$ and $x_2 = \dot{x}$, one can write

$$\dot{X} = AX + Ba, \tag{31.12}$$

where $X = [x_1 \quad x_2]^T$, $A = \begin{bmatrix} 0 & 1 \\ -\omega_n^2 & -2\zeta\omega_n \end{bmatrix}$, and $B = [0 \quad 1]^T$. The eigenvalues (λ's) of the system of Eqs. (31.12) can be obtained by solving the characteristic equation $|\lambda I - A| = 0$, where I is 2×2 identity matrix and, subsequently, the modal matrix Φ can be obtained such that the ith column of Φ corresponds to the ith eigenvalue λ_i. It can be shown that the operation $\Phi^{-1}A\Phi$ transforms the matrix A to the diagonal matrix $\Lambda = diag(\lambda_i)$. This orthogonalization property can be used to obtain the decoupled system of equations

$$\dot{X}' = \Lambda X' + B'a,$$

where $X' = \Phi^{-1}X$ and $B' = \Phi^{-1}B$. Now, it is straightforward to obtain the desired solution $x(t)$.

Suarez and Shokooh [15] presented the extension of the state-space approach to the FDE of the form given by Eq. (31.8) with $g \equiv 0$. Their main idea was to convert the FDE to the set of semi-differential equations [16] which can provide the solution (in some cases the closed-form solution) very easily. Considering Eq. (31.8) with $g \equiv 0$ and $\gamma = 1/2$ and applying the transformation $x_1 = x$, $x_2 = \overset{\gamma}{x}$, $x_3 = \overset{2\gamma}{x}$, and $x_4 = \overset{3\gamma}{x}$, following matrix format equation can be written

$$\overset{\gamma}{X} = AX + Ba, \tag{31.13}$$

where $X = [x_1 \quad x_2 \quad x_3 \quad x_4]^T$, $A = \begin{bmatrix} \mathbf{0}_{3\times1} & \mathbf{I}_{3\times3} \\ -\omega_n^2 & -\eta\mathbf{I}_\eta \end{bmatrix}$, $\mathbf{0}_{3\times1}$ is 3×1 zero-vector, $\mathbf{I}_{3\times3}$ is 3×3 identity matrix, $\mathbf{I}_\eta = [1 \quad \mathbf{0}_{1\times2}]$, and $B = [\mathbf{0}_{1\times3} \quad 1]^T$. It may be noted that as $\gamma = 1/2$, $\overset{4\gamma}{x_4}$ becomes $\ddot{x}$ and therefore, four state-variables $\{X\}$ appear in the problem. If γ is of the form $2/m$ such that m is an integer (≥ 2) then there will be m state-variables. For example, in the case discussed above m is equal to 4. In the general case, $X = [x_1 \quad x_2 \quad \ldots \quad x_m]^T$, $A = \begin{bmatrix} \mathbf{0}_{(m-1)\times(m-1)} & \mathbf{I}_{(m-1)\times(m-1)} \\ -\omega_n^2 & -\eta\mathbf{I}_\eta \end{bmatrix}$, $\mathbf{I}_\eta = [1 \quad \mathbf{0}_{1\times(m-2)}]$, and $B = [\mathbf{0}_{1\times(m-1)} \quad 1]^T$. The characteristic equation $|\lambda I - A| = 0$

(for Eq. (31.13)) will lead to an mth order polynomial in λ which upon solving gives the m eigenvalues and the corresponding m eigenvectors. The rest of the procedure is quite similar to the one for the linear case (Eq. (31.12)). Suarez and Shokooh [15] demonstrated the procedure for $m = 4$ and it seems that this procedure may also be extended to the other values of m as well. However, the author is not sure if there is an approach available in the literature when γ is not of the form $2/m$.

31.7 Nonlinear Fractional Structural Dynamics

So far, in the discussion, we assumed that $g(t, x, \overset{\alpha}{x}, \dot{x}) \equiv 0$. In CSD applications, the equivalent linearisation techniques are quite popular in solving the nonlinear oscillators [17]. Some recent papers [18, 19] present the equivalent linearisation methodology for solving the nonlinear fractional oscillators. It may be noted that equivalent linearisation is just an approximation and provides reasonable solutions as long as the assumptions used in developing the linearisation are valid. For example, one may rewrite Eq. (31.9) in the following form:

$$\ddot{x} + 2\zeta\omega_n\dot{x} + g_0(t, x, \overset{\gamma}{x}, \overset{\alpha}{x}, \dot{x}) = a(t),$$

such that $g_0(t, x, \overset{\gamma}{x}, \overset{\alpha}{x}, \dot{x}) = 2\zeta\omega_n^{(2-\gamma)}\overset{\gamma}{x} + \omega_n^2 x + g(t, x, \overset{\gamma}{x}, \overset{\alpha}{x}, \dot{x}) - 2\zeta\omega_n\dot{x}$. The target is to obtain an equivalent ODE of the form

$$\ddot{x} + \eta_{eq}\dot{x} + \omega_{eq}^2 x = a(t), \tag{31.14}$$

such that η_{eq} and ω_{eq}^2 represent the equivalent damping and equivalent stiffness terms (normalized with mass). The equation error Δ can be written as

$$\Delta = (2\zeta\omega_n - \eta_{eq})\dot{x} + (g_0 - \omega_{eq}^2)x$$

As demonstrated in [20], the mean square of the error $E[\Delta^2]$ can be minimised to obtain η_{eq} and ω_{eq}^2.

Similarly, one may visualise another problem where Eq. (31.9) (with $g \equiv 0$) is converted to the equivalent ODE (Eq. (31.14)) such that

$$\Delta = 2\zeta\omega_n^{(2-\gamma)}\overset{\gamma}{x} - \eta_{eq}\dot{x} + (\omega_n^2 - \omega_{eq}^2)x. \tag{31.15}$$

However, the feasibility and accuracy of such an equivalence can be a question for further research. Once the equivalently linearised oscillator is obtained, then the approaches discussed with $g \equiv 0$ can be applied directly.

31.8 Concluding Remarks

Equations (31.6) and (31.9) are the two generalised forms of equations of motion that need the attention of applied mathematicians. Solutions of these equations will be of significant use in the field of structural dynamics. In addition to this, further investigations on the superposition principle and state-space approach for $\gamma \neq 1/2$ and developing equivalent ordinary differential equations by minimising Δ of Eq. (31.15) may lead to enhanced applications of fractional differential equations in the field of structural dynamics.

Acknowledgments This paper is motivated by the lecture ("Differential Equations and Structural Dynamics") delivered by the author at the Ghent Analysis and PDE centre (GAPC), Ghent University in January 2023 and subsequent discussions with the researchers at the GAPC. The visit of the author to Ghent University was partially supported by IIT Mandi and Ghent University.

References

1. Machado, J.T., Kiryakova, V., Mainardi F.: Recent history of fractional calculus. Commun. Nonlinear Sci. Numer. Simul. **16**(3), 1140–1153 (2011)
2. Sun, H., Zhang, Y., Baleanu, D., Chen, W., Chen, Y.: A new collection of real world applications of fractional calculus in science and engineering. Commun. Nonlinear Sci. Numer. Simul. **1**(64), 213–231 (2018)
3. Diethelm K., Kiryakova V., Luchko Y., Machado J.T., Tarasov, V.E.: Trends, directions for further research, and some open problems of fractional calculus. Nonlinear Dyn. **107**(4), 3245–3270 (2022)
4. Shitikova, M.V.: Fractional operator viscoelastic models in dynamic problems of mechanics of solids: a review. Mech. Solids **57**, 1–33 (2022)
5. Li, M.: Theory of fractional engineering vibrations. In: Theory of Fractional Engineering Vibrations. De Gruyter (2021)
6. Bagley, R.L., Torvik, P.J.: A theoretical basis for the application of fractional calculus to viscoelasticity. J. Rheol. **27**(3), 201–210 (1983)
7. Bagley, R.L., Torvik, P.J.: On the fractional calculus model of viscoelastic behavior. J. Rheol. **30**(1), 133–155 (1986)
8. Torvik, P.J., Bagley, R.L.: On the appearance of the fractional derivative in the behavior of real materials. J. Appl. Mech. **51**(2) 294–298 (1984)
9. Bagley, R.L., Torvik, P.J.: Fractional calculus in the transient analysis of viscoelastically damped structures. AIAA J. **23**(6), 918–925 (1985)
10. Armstrong, R.W., Walley, S.M.: High strain rate properties of metals and alloys. Int. Mater. Rev. **53**(3), 105–128 (2008)
11. Caputo, M.: Linear models of dissipation whose Q is almost frequency independent—II. Geophys. J. Int. **13**(5), 529–539 (1967)
12. Podlubny, I.: Fractional Differential Equations. Mathematics in Science and Engineering, vol. 198. Academic Press, California (1999)
13. Suarez, L., Shokooh, A.: Response of systems with damping materials modeled using fractional calculus. Appl. Mech. Rev. **48**(11S), S118–S126 (1995)
14. Preumont, A.: Vibration Control of Active Structures: An Introduction. Springer, Berlin (2011)

15. Suarez, L.E., Shokooh, A.: An eigenvector expansion method for the solution of motion containing fractional derivatives. J. Appl. Mech. **64**(3), 629–635 (1997)
16. Oldham, K.B., Spanier, J.: The Fractional Calculus. Academic Press, New York (1974)
17. Hieu, D.V., Hai, N.Q., Hung, D.T.: The equivalent linearization method with a weighted averaging for solving undamped nonlinear oscillators. J. Appl. Math. **2018**, 7487851 (2018)
18. Fragkoulis, V.C., Kougioumtzoglou, I.A., Pantelous, A.A., Beer, M.: Non-stationary response statistics of nonlinear oscillators with fractional derivative elements under evolutionary stochastic excitation. Nonlinear Dyn. **97**, 2291–2303 (2019)
19. El-Dib, Y.O.: Immediate solution for fractional nonlinear oscillators using the equivalent linearized method. J. Low Freq. Noise. Vib. Active Control **41**(4), 1411–1425 (2022)
20. Roberts, J.B., Spanos, P.D.: Random Vibration and Statistical Linearization. Courier Corporation (2003)

Chapter 32
Text Matching as Time Series Matching

Xuechao Wang

Abstract Text matching plays a fundamental and vital role in many aspects of natural language processing (NLP), where *multi-level* text matching is the most challenging task. Recently, the prevailing use of deep neural networks for text matching to fuse the multi-granularity semantic features has been witnessed. In this study, we adopt a two-stage text matching frame from rough to fine, which combines the recall stage and the ranking stage to extract different levels of matching information. In particular, the word embeddings of *Query* and *Document* are novel as time series in the ranking stage, and in response to this idea, we propose a simple and efficient neural network to extract matching information of signal trend patterns, which is built on **R**esidual connection module, spatial **A**ttention mechanism module and **M**ulti-scale convolution module, namely RAM-CNN. The extensive experimental results show that the efficiency of the RAM-CNN model exceeds those of shallow models. And we finally demonstrate the effectiveness of the two-stage text matching model on a practical book-matching task.

32.1 Introduction

Text matching, as the core ingredient of natural language processing, has been widely used in many fields, including, but not limited to, questioning and answering [1], machine translation [2] and document retrieval [3]. In general, it refers to a model that takes two text sequences as input and predicts a category or relevance score indicating the relationship between them.

With the development of deep learning, many text matching works [4–6] focus on how to build the neural networks such as CNN, RNN [7] and Transformer [8] to automatically capture the rich text semantic features. In addition, the appropriate

X. Wang (✉)
Department of Mathematics: Analysis, Logic and Discrete Mathematics, Ghent University, Gent, Belgium
e-mail: xuechao.wang@ugent.be

M. Ruzhansky, K. Van Bockstal (eds.), *Extended Abstracts 2021/2022*, Research Perspectives Ghent Analysis and PDE Center 2,
https://doi.org/10.1007/978-3-031-42539-4_32

increase of attention mechanism module can effectively capture the interactive information between text pairs, such as counterparts, which is crucial for the model to determine the degree of matching between text pairs. Recently, due to the accumulation of text corpus and the huge amount of parameters of traditional neural networks, the pre-training models, e.g., ELMo, GPT and BERT are proposed to train the model parameters in advance, and then fine-tune them to reduce training time and improve robustness at the same time. Depending on the order of transformation and interaction, these text matching models can be divided into two categories: transformation-based models [9] and interaction-based models [6, 10].

One of the intrinsic challenges for the text matching model is how to capture rich *multi-level* matching information in order to more accurately identify the similarity. To interpret this process, we consider the following example:

- Q : *Apple Inc. is a technology-led company.*
- D_1 : *Iron man is a popular movie.*
- D_2 : *Apple is a kind of fruit that is beneficial to the human body.*
- D_3 : *Apple mobile phones sell well all over the world.*
- D_4 : *Apple Inc. is a company that makes money through technology.*

Here, Q and D_1,D_2,D_3,D_4 are sentences sampled from the *Query* and *Document* datasets, respectively, where Q and D_4 is the best match. As we can see, the similarity of Q and D_1,D_2,D_3,D_4 can be measured from different levels—literally and semantically. First, taking the literal-level information to match, the words *"Apple"* in Q is same as *"Apple"* in D_2,D_3,D_4, so D_1 is removed to quickly narrow the scope of matches. The further similarity of Q and D_2,D_3,D_4 can be measured from semantical-level, including word information, phrase information and sentence information. Obviously, for Q and D_2, their overall and local semantic scenario are different at word information; for Q and D_3, the *"Apple"* local semantics scenario is same at word information, but *"technology-led company"* and *" mobile phones"* local semantics is different at phrase information; for Q and D_4, their overall and local semantic scenario are same at word, phrase and sentence information. So we can see that it is always a non-trivial problem to simultaneously identify their matching information at different levels.

In response to the above problems, we adopt a two-stage text matching model, including recall stage and ranking stage, to extract literal-level and semantical-level matching information between text pairs, respectively. In the recall stage, an unsupervised statistical model is adopted to quickly recall the part of *Document* dataset that is similar to *Query* based on literal-level matching information, which can effectively ensure a recall rate and reduce the computational consumption at the same time. In the ranking stage, we novelty employ the idea of a signal processing field to transform the text data into a time series form; then, we propose a simple and effective neural network RAM-CNN model based on this idea to extract the matching patterns between time series pairs, so as to *Query* is performed more precise text matching in the part of *Document* dataset. The main contributions of this paper are summarized as follows:

- We adopt a two-stage text matching model from rough to fine, and gradually extract different levels of matching information between text pairs.
- The text data is novelly transformed into a time series form, and the corresponding RAM-CNN model is concisely designed, which is built on residual connection, spatial attention mechanism and multi-scale convolution.

The remainder of this paper is organised as follows. We present the proposed text matching model in Sect. 32.2. Section 32.3 demonstrates the effectiveness of text matching model on the book-matching task with an extensive experimental comparison. Finally, Sect. 32.4 concludes this paper.

32.2 Proposed Method

The *multi-level* matching information plays a crucial role in text matching task. In this study, the two-stage text matching framework consisting of a recall stage and a ranking stage is adopted to quickly capture the wealthy literal-level and semantical-level matching information. In particular, we propose a straight and effective neural network model RAM-CNN in the ranking stage, which novelly uses time series matching idea to extract the semantical-level matching information of text pairs. In order to facilitate the explanation, let $T_Q = \{w_1, w_2, \cdots, w_m\}$ and $T_D = \{v_1, v_2, \cdots, v_n\}$ be a pair of *Query* and *Document*, where w_i (v_j) is the i-th (j-th) word in T_Q (T_D) and m (n) represents the length of the T_Q (T_D). Formally, the matching similarity of them is typically defined as,

$$match(T_Q, T_D) = F([\Psi(T_Q), \Psi(T_D)]),$$

where function Ψ maps each text sequence T to a feature vector $\Psi(T)$, function F measures the similarity of the vector pair, $match(T_Q, T_D)$ is the corresponding similarity score. As shown in Fig. 32.1, the architecture of the whole scheme consists of two key components:

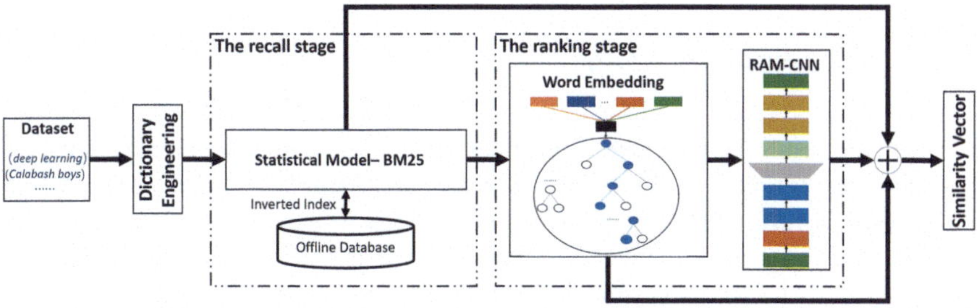

Fig. 32.1 The architecture of two-stage text matching model

1. **The recall stage** is composed of the statistical model and the offline database to quickly extract literal-level matching information between text pairs.
2. In **the ranking stage**, the word embedding model and the RAM-CNN model are adopted to capture the semantical-level matching information.

32.2.1 The Recall Stage

The main function of the recall stage is to capture the literal-level matching information of text pairs, so as to perform rough but fast text matching and recall similar part of data in *Document* dataset. Given the input text pair (T_Q, T_D), we employ the BM25 model to encode the first similarity vector as:

$$BM25(T_Q, T_D) = \sum_{i=1}^{m} \frac{f_{iD}(k+1)}{f_{iD} + k(1 - b + b\frac{dl_D}{avgdl})} \times \log \frac{N - n_i + 0.5}{n_i + 0.5},$$

where m is the length of T_Q, f_{iD} represents the frequency of the i-th word in T_D, dl_D is the length of T_D, $avgdl$ represents the average length of texts in *Document* dataset, N represents the amount of texts in $Document$ dataset, n_i represents the number of texts containing the i-th word in $Document$ dataset, k and b are hyperparameters. At the same time, the Inverted Index technology is adopted to extract the statistical information of each word w_i in $T_Q = \{w_1, w_2, \cdots, w_m\}$, thereby building the offline database. For example, the statistical information corresponding to w_i is $[\{size\}, D_1, \cdots, D_{size}]$, where $\{size\}$ represents the amount of texts containing w_i in $Document$ dataset and D_j represents the specific statistical information of the j-th text which includes the word w_i. More specifically, $D_j = (Text\_id_j, Frequency_j, Length_j)$ has three weights:

1. $Text\_id_j$ is index of the text D_j in *Document* dataset.
2. $Frequency_j$ represents the number of w_i in D_j.
3. $Length_j$ represents the length of D_j.

32.2.2 The Ranking Stage

Based on the results of the recall stage, the ranking stage implements more accurate text matching for $Query$ on the recalled part of *Document* dataset and extracts the semantical-level matching information between text pairs. There are two key components. First, the semantic encoder word2vec is used to learn the vectorised representation of text pairs and encode the second similarity vector by cosine distance. Then, based on the multiple observation that Consistent Trend and Pearson Correlation between vector pairs, we find that the similarity between text pairs is proportional to the similarity between the corresponding vector pairs. So we treat

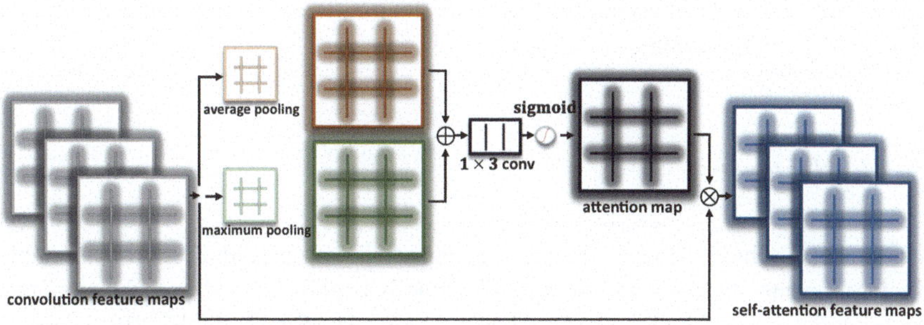

Fig. 32.2 Spatial attention module

them as time series and propose RAM-CNN, as shown in Fig. 32.1, on these new time series to extract the third similarity vector. Especially, we try to replace the convolutional layer in each module by a spatial attention module [11] to increase representation ability, thereby giving a new module, called SAM, as shown in Fig. 32.2, throughout the paper. This spatial attention module simulates the visual attention mechanism in the spatial domain. Here, it is used to emphasise important features and weaken the unnecessary ones. As shown in Fig. 32.2, given an input feature map $F \in R^{1\times1\times W}$, we first aggregate channel information of a feature map by using two pooling operations: average pooling (average-pooling gathers important clue about uniform object features) and maximum pooling (max-pooling gathers another important clue about distinctive object features), obtain two 1D maps: $F_{avg} \in R^{1\times1\times W}$ and $F_{max} \in R^{1\times1\times W}$. Then, we concatenate them into a 2D map and a standard convolution layer is adopted to convulse it in order to get a spatial attention map. The spatial attention map $M(F)$ can be expressed as:

$$M(F) = \sigma(f^{1\times3}([AvgPool(F); MaxPool(F)])),$$

where σ denotes the sigmoid function and $f^{1\times3}$ represents a convolution operation with the filter size of 1×3. Finally, we multiply every element in F with its corresponding element in $M(F)$ and get a feature map F with attention mechanism. The overall attention process can be summarized as:

$$F = M(F) \otimes F.$$

32.3 Experiments

In this section, we introduce the datasets and the setup used to validate the parameters of our methods. Then, we evaluate the performance of our text matching model. The results are analysed together with the limitations of our model and future directions of improvement.

32.3.1 Datasets

The datasets include "Query" dataset, "Document" dataset and a relationship table which represents the corresponding relationship between "Query" and "Document". The book dataset is divided into four basic categories: "Single book", "Bulk book", "Set book", and "Other book". We use "Single book" for experiments, and other categories are only used for pre-trained word embedding models. The unregistered word dictionary is used to alleviate the out-of-vocabulary(OOV) problem, the synonym dictionary is used to deal with the artificial acronyms problem and the stop word dictionary is used to supplement the Jieba word segmentation database.

32.3.2 The Priori Experiments

Due to the fact that our dataset is rather different from that in the original word2vec algorithm, we here retrain the word embedding model with the data of "Bulk book", "Set book", and "Other book" and then the model is fine-tuned with the "Single book" data. The length of sliding window is 3, the number of iterations is set to 400 and the fixed-length feature vectors are 8,16,32,64 dimensions respectively. By analysing the trend characteristics and Pearson correlation coefficient of the representation of vector sequence, we find that the similarity of these two features is proportional to the similarity of their corresponding texts.

As shown in Fig 32.3, the black line in the figure are dissimilar vector, and the rest are similar vectors. Taking the 16-dim space as an example, we find that four similar vectors have similar characteristics that are obviously different from the black line in certain intervals, such as the same trend or similar values. In addition, we calculate the Pearson correlation coefficient between these vectors according to the following formula:

$$PCCs = \frac{\mathrm{cov}(X, Y)}{\sigma_X \sigma_Y} = \frac{\sum_{i=1}^{n} (x_i - \overline{x})(y_i - \overline{y})}{\sqrt{\sum_{i=1}^{n} (x_i - \overline{x})^2} \sqrt{\sum_{i=1}^{n} (y_i - \overline{y})^2}}.$$

From the above, we prove that the vectorized representation of text sequence obtained by word2vec contains matching patterns, such as trend feature and

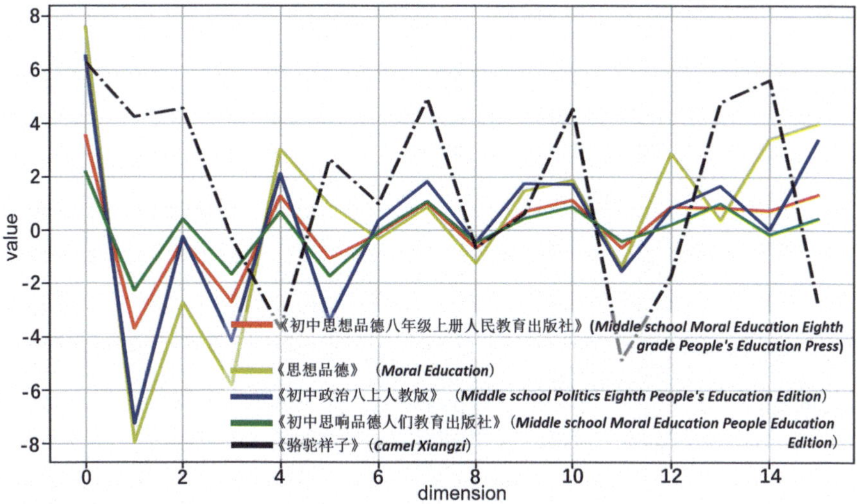

Fig. 32.3 Consistent Trend. The vectorized representation of text sequences in 16-dim (16 dimensions). The "X" axes is the number of dimension, and "Y" axes is the value. The red line represents the standard text sequence, and the remaining lines show the results are degraded by some factors such as "keyword abbreviations" (yellow line), "synonym" substitutions (blue line), "typos" (green line) and "unrelated" (black line)

Pearson correlation feature. So we concatenate a pair of vectorised representations horizontally into a time series, then use the high fitting ability [12] of RAM-CNN to learn matching patterns.

32.3.3 *Main Experiments*

To facilitate the explanation, the recall stage and the ranking stage settings are described, separately. First, in the recall stage, we search for the optimal combination of k and b hyperparameters in the statistical model with extensive experiments. Second, in the ranking stage, we prove the effectiveness of residual connection module, spatial attention mechanism module and multi-scale convolution module in RAM-CNN in 8,16,32,64 dimensions, respectively. Finally, combining three similarity vectors proves the effectiveness of our text matching model.

32.3.3.1 BM25

As shown in formula (2), there are two hyperparameters k and b in the model, where k adjusts the effect of word frequency on the similarity between text pairs, and b adjusts the effect of document length on the similarity between text pairs. As shown

Table 32.1 Hyperparameter combination

		$b = 0.0$	$b = 0.25$	$b = 0.5$	$b = 0.75$	$b = 1.0$
Recall-50	$k = 1.0$	0.9132	0.9248	0.9303	0.9345	0.9374
	$k = 1.5$	0.9130	0.9263	0.9320	0.9371	0.9383
	$k = 2.0$	0.9128	0.9267	0.9331	0.9374	0.9396
	$k = 2.5$	0.9128	0.9272	0.9338	0.9379	0.9394
	$k = 3.0$	0.9127	0.9277	0.9341	0.9378	**0.9399**
Recall-100	$k = 1.0$	0.9355	0.9434	0.9479	0.9506	0.9528
	$k = 1.5$	0.9355	0.9444	0.9494	0.9520	0.9542
	$k = 2.0$	0.9353	0.9454	0.9498	0.9526	0.9545
	$k = 2.5$	0.9351	0.9457	0.9501	0.9533	**0.9546**
	$k = 3.0$	0.9351	0.9459	0.9506	0.9534	0.9543
Recall-200	$k = 1.0$	0.9520	0.9570	0.9598	0.9618	0.9630
	$k = 1.5$	0.9519	0.9576	0.9606	0.9627	0.9641
	$k = 2.0$	0.9519	0.9582	0.9610	0.9631	0.9644
	$k = 2.5$	0.9519	0.9583	0.9614	0.9636	0.9652
	$k = 3.0$	0.9519	0.9586	0.9616	0.9638	**0.9653**
Recall-500	$k = 1.0$	0.9666	0.9697	0.9706	0.9718	0.9734
	$k = 1.5$	0.9665	0.9698	0.9712	0.9727	0.9742
	$k = 2.0$	0.9665	0.9700	0.9715	0.9733	0.9749
	$k = 2.5$	0.9665	0.9701	0.9716	0.9734	**0.9750**
	$k = 3.0$	0.9665	0.9701	0.9717	0.9736	0.9749

in Table 32.1, We conduct experiments on the hyperparameters k and b, and record the best combination of k and b when recalling the top 50, 100, 200, and 500 books most similar to Q in *Document* dataset. The recall rate is used as the evaluation indicator.

32.3.3.2 RAM-CNN

In this section, we report the performance of the proposed model. Among them, the experiment is about the influence of spatial attention mechanism module, residual connection module, and multi-scale convolution module on network performance. The accuracy rate is used as the evaluation indicator. According to the model with different modules, we built four neural networks, namely RAM-CNN-I, RAM-CNN-II, RAM-CNN-III and RAM-CNN-IV. RAM-CNN-I is the elementary network that does not contain any modules; the "stem" block of RAM-CNN-II includes the spatial attention mechanism module; in RAM-CNN-III, we add the spatial attention mechanism module to "stem" block and the residual connection module to "BM-1 and BM-2" block, respectively; based on RAM-CNN-III, we replace the maximum pooling operation with the multi-scale convolution module in "Reduction" block, here we call it RAM-CNN-IV. Moreover, in order to prevent

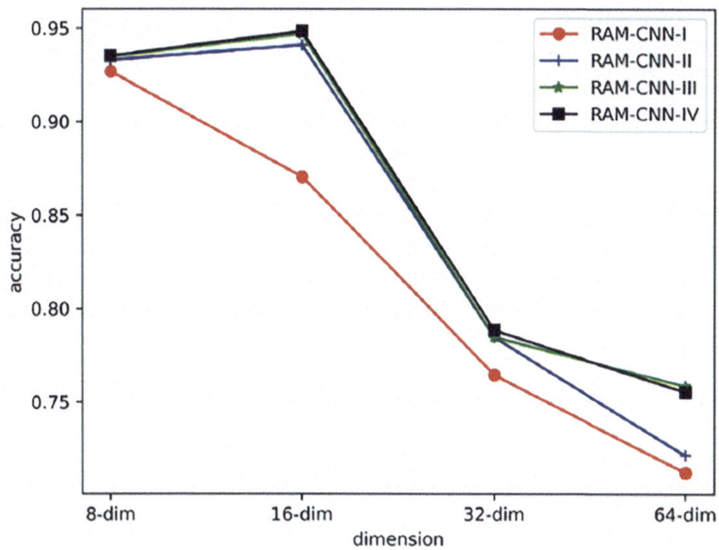

Fig. 32.4 The performance of RAM-CNN

the model from overfitting, we add "Early Stopping" operation. During the training process, if the decrease of the loss function is less than 0.05 during the following 6 iterations, the model training is stopped. And in order to stabilize the model as soon as possible, gradual warm-up [13] operation is adopted and the initial learning rate is set as 1e-3. We use the Cross-Entropy Function as the loss function. For the optimisation strategy, the adaptive optimisation algorithm Adam is adopted. In 8-dim, 16-dim, 32-dim, and 64-dim, we train and test the performance of each model, separately. As illustrated in Fig 32.4, we find that:

- The spatial attention mechanism module, the residual connection module and the multi-scale convolution module have improved the performance of RAM-CNN in different dimensions.
- From RAM-CNN-I to RAM-CNN-IV, the overall improvement effect is 0.81%, 6.77%, 2.37%, 4.30% in 8-dim, 16-dim, 32-dim, 64-dim vector space, respectively.
- The matching patterns extracted from 16-dim vector space is the most suitable. And from the curve trend, it shows that the matching information contained in the matching features may be more redundant in higher-dimensional hidden spaces.

32.3.3.3 The Two-Stage Matching Model

As presented in Table 32.2, in the recall stage, the data volume of the *Document* dataset to be matched is reduced to 50, 100, 200, 500 respectively, and the optimal

Table 32.2 The result of two-stage matching model. The best results are in bold

	Top-5	Top-10	Top-20
Recall-50 ($k = 3, b = 1$)	0.840	0.880	**0.916**
Recall-100 ($k = 2.5, b = 1$)	0.831	0.881	0.912
Recall-200 ($k = 3, b = 1$)	0.840	0.880	0.915
Recall-500 ($k = 2.5, b = 1$)	0.841	0.880	0.913

combination of the corresponding hyperparameters k and b is adopted; in the ranking stage, a 16-dim word embedding vector is selected and converted into a time series form, and then RAM-CNN is used to capture matching patterns between time series.

32.4 Conclusion

In this paper, we adopt the two-stage text matching model to capture literal-level and semantical-level matching patterns. In particular, we propose a simple and effective neural network RAM-CNN based on the spatial attention mechanism module, residual connection module, and multi-scale convolution module which novelly treats the spatial vector representation of text pairs as time series. In the process of the data processing, we adopt that:

- By making artificial dictionaries, we introduce book domain knowledge to standardize the raw datasets.
- With "Dislocation" idea, high-cost manual labeling is avoided.
- We adopt straight and effective models to keep computational complexity limited during the online and shift the bulk of computations to the offline through pre-training and offline database.

Moreover, we implement RAM-CNN on an actual data set with various practical optimisations, e.g., "Early Stopping" operation and gradual warm-up operation. Extensive experiment results demonstrate that each module in RAM-CNN improves the performance of the model. In the future research, we will try other more advanced word embedding models, e.g. ALBERT and XLNet. In addition, it is desirable to design more effective interaction methods to improve the performance of network, such as introducing the knowledge graph.

References

1. Xue, X., Jeon, J., Croft, W.B.: Retrieval models for question and answer archives. In: Proceedings of the 31st Annual International ACM SIGIR Conference on Research and Development in Information Retrieval, pp. 475–482 (2008)

2. Brown, P.F., Della Pietra, S.A., Della Pietra, V.J., et al.: The mathematics of statistical machine translation: parameter estimation. Comput. Ling. **19**(2), 263–311 (1993)
3. Li, H., Xu, J.: Semantic matching in search. Found. Trends Inform. Retr. **7**(5), 343–469 (2014)
4. Huang, P.-S., He, X., Gao, J., Deng, L., Acero, A., Heck, L.: Learning deep structured semantic models for web search using clickthrough data. In: Proceedings of the 22nd ACM International Conference on Information & Knowledge Management, pp. 2333–2338 (2013)
5. Hu, B., Lu, Z., Li, H., Chen, Q.: Convolutional neural network architectures for matching natural language sentences. In: Advances in Neural Information Processing Systems, pp. 2042–2050 (2014)
6. Pang, L., Lan, Y., Guo, J., Xu, J., Wan, S., Cheng, X.: Text matching as image recognition. In: Proceedings of the AAAI Conference on Artificial Intelligence, vol. 30 (2016)
7. Chen, Z., Cheng, X., Dong, S., Dou, Z., Guo, J., Huang, X., Lan, Y., Li, C., Li, R., Liu, T.-Y., et al.: Information retrieval: a view from the Chinese ir community. Front. Comput. Sci. **15**(1), 1–15 (2021)
8. Vaswani, A., Shazeer, N., Parmar, N., Uszkoreit, J., Jones, L., Gomez, A.N., Kaiser, L., Polosukhin, I.: Attention is all you need. In: Advances in Neural Information Processing Systems, pp. 5998–6008 (2017)
9. Shen, Y., He, X., Gao, J., Deng, L., Mesnil, G.: A latent semantic model with convolutional-pooling structure for information retrieval. In: Proceedings of the 23rd ACM International Conference on Conference On Information and Knowledge Management, pp. 101–110 (2014)
10. Dai, Z., Xiong, C., Callan, J., Liu, Z.: Convolutional neural networks for soft-matching n-grams in ad-hoc search. In: WSDM 2018: The Eleventh ACM International Conference on Web Search and Data Mining Marina Del Rey CA USA February, 2018, pp. 126–134 (2018)
11. Woo, S., Park, J., Lee, J.Y., et al.: CBAM: convolutional block attention module. In: Proceedings of the European Conference on Computer Vision (ECCV), pp. 3–19 (2018)
12. Hornik, K., Stinchcombe, M., White, H.: Multilayer feedforward networks are universal approximators. Neural Netw. **2**(5), 359–366 (1989)
13. He, K., Zhang, X., Ren, S., Sun, J.: Deep residual learning for image recognition. In: Proceedings of the IEEE Conference on Computer Vision and Pattern Recognition, pp. 770–778 (2016)

Chapter 33
Performing Particle Image Segmentation on an Extremely Small Dataset

Marianna Chatzakou, Junqing Huang, Bogdan V. Parakhonskiy, Michael Ruzhansky, Andre G. Skirtach, Junnan Song, and Xuechao Wang

Abstract Image segmentation is one of the typical computer vision tasks that has received great success with the recent advance of deep-learning methods. However, it is still a challenging problem, particularly when encountering limited data. In this paper, we present a new strategy for particle image segmentation, which relies on extensive data augmentation methods to reuse the available annotated samples for more effective performance. The procedure consists of the K-nearest neighbour (KNN) matting to fine-tune manually annotated boundaries and the use of data augmentations to expand available annotated data. After that, we employ the U-net architecture to train the model based on a small dataset consisting of twenty images. The results showed that the proposed strategy could effectively extract particle boundary features, thereby obtaining accurate segmentation results based on a very limited source of images.

33.1 Introduction

In recent years, deep learning methods for image segmentation have witnessed significant growth across various scientific fields, including classical image processing, hyper-spectral image segmentation, autonomous driving application, and biomedical and chemical image analysis, and so on [1, 2]. Despite their impressive achievements, these methods often rely heavily on large and high-quality training datasets. For instance, Mask R-CNN [3] is trained on the Microsoft COCO dataset [4], which contains 2.5 million labelled instances in 328,000 images. However,

M. Chatzakou · J. Huang · M. Ruzhansky · X. Wang (✉)
Department of Mathematics: Analysis, Logic and Discrete Mathematics, Ghent University, Ghent, Belgium
e-mail: xuechao.wang@ugent.be

B. V. Parakhonskiy · A. G. Skirtach · J. Song
Department of Biotechnology, Faculty of Bioscience Engineering, Ghent University, Ghent, Belgium

M. Ruzhansky, K. Van Bockstal (eds.), *Extended Abstracts 2021/2022*, Research Perspectives Ghent Analysis and PDE Center 2,
https://doi.org/10.1007/978-3-031-42539-4_33

such a huge amount of data is always unavailable in many practical applications, especially in the context of biomedical and chemical image analysis. Consequently, there is an ongoing need to explore deep learning methods for image segmentation that can effectively utilise limited-sized datasets, both in theory and practice.

Image segmentation can be posed as a pixel classification problem with semantic labels (semantic segmentation), or a partition problem of individual objects (instance segmentation), or both (panoptic segmentation) [2]. The early work on using deep learning for image segmentation can be dated back to the introduction of the fully convolutional network (FCN) [5] for deep learning-based semantic image segmentation. After that, a number of deep learning methods have been proposed joint with the development of computer vision community, for example, convolutional neural networks, recurrent neural networks and long short-term memory, encoder-decoder and autoencoder models, generative adversarial networks, and so on. Among these techniques, CNN-based methods have gained prominence as mainstream and widely used architectures for image segmentation. One key advantage of CNN-based approaches is weight sharing across layers, resulting in a significantly reduced parameter count compared to FCN-based methods. Several notable CNN architectures have made substantial contributions to image segmentation, including U-net [6], Dilated CovNet [3], Mask-rcnn [7], and so on.

In this paper, we focus on a type of image segmentation for microscope image analysis in the biomedical field. Specifically, we process the image segmentation for scanning electron microscope (SEM) images provided by the widely-used system in Fig. 33.1. SEM image analysis aims to provide a clear observation of objects' surface morphologies, including the size, shape, and even their interaction (dispersed well or aggregation extensively), which would dominate their pathway of metabolism and guide their application exploration [8]. Fluorescence

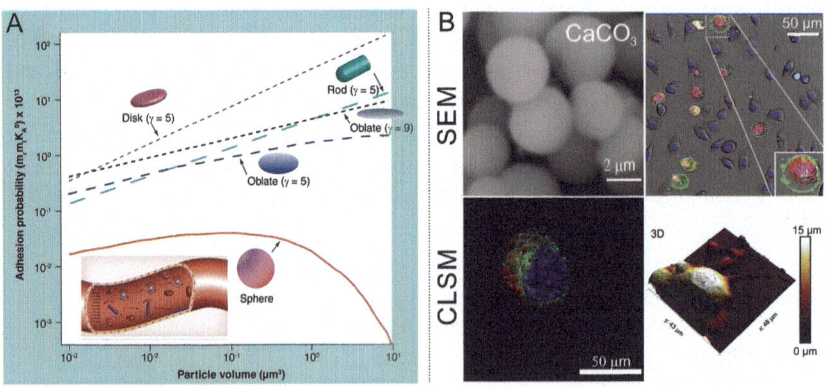

Fig. 33.1 (**a**) Effect of shape on micro-nano particles adhesion probability, reproduced from Liu and Toy's work [13, 14], with the permission from Europe PM. (**b**) SEM images of $CaCO_3$ micro particles; FM, CLSM and AFM images of cells reproduced from Van der Meeren, L's [11], with the permission from Elsevier

microscopy (FM) has made it down-to-earth to visualise interactions among cells and molecules using specific probe dyes. This technique can also be used to analyse the absorption, transport, distribution, and localisation of small molecules in cells. This process is known as "in situ hybridisation" [9]. Confocal laser scanning microscope (CLSM) upgrades the 2D surface micrograph of (FM) to 3D detection, dramatically increasing the observation accuracy [10]. Atomic force microscope (AFM) furthermore provides the real surface 3D image with high resolution in the atomic level and allows for manipulation and observation in the normal physical environment. Besides the surface properties like surface roughness, particle size, average gradient, pore structure and pore size distribution, AFM analysis could also output mechanical properties like electrical, magnetic and viscoelastic properties [11, 12]. We, therefore, began with the most obvious and essential parameters, dealing with the shape obtained from SEM images, anticipating in the future to accomplish the vision of the high throughput images automatically and reliably analysis based on multi-dimensional mutual confirmation including size, shape, colour and so on.

For this purpose, a new Particle Image Segmentation (PIS) framework based on the classical U-net architecture was proposed in our study. The overall scheme is illustrated in Fig. 33.2. The framework mainly expands the available dataset by fusing different data preprocessing strategies and then learns the local information of particle images through the U-net architecture. Firstly, a combination of manual annotation and algorithmic fine-tuning strategy is employed to obtain the corresponding high-quality annotation information for the limited available particle images. Secondly, the corresponding data augmentation methods are widely used to expand the available images according to the different properties of the particle images. Finally, we get the training data by randomly cropping local patches over the expanded dataset and then retraining the U-net model for particle extraction. The combination of manual annotation and algorithm fine-tuning in this framework ensures high-quality annotation for supervised learning in case of low-quality manual annotation in practical tasks. The experimental results further demonstrate that the proposed strategy along with the U-net architecture can effectively extract particle boundary features, thereby effectively obtaining accurate segmentation results.

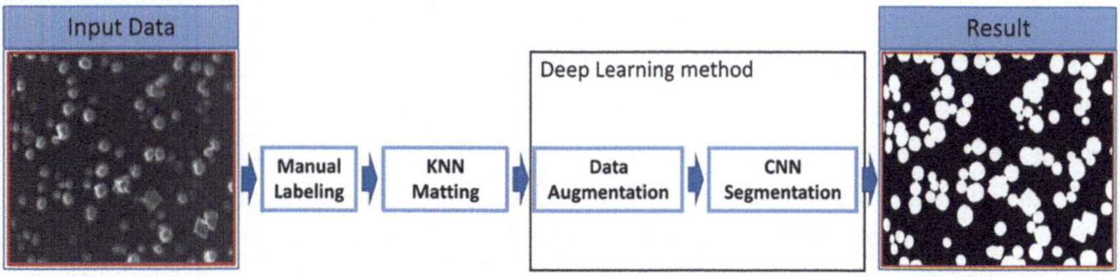

Fig. 33.2 Schematic of the overall strategy, including manual annotation boundaries, KNN matting [15] to improve annotation quality, data augmentation to expand the available annotation data, and segmentation of the entire particle image for training the neural network model

The rest of this paper is organised as follows. The proposed particle segmentation pipeline is presented in Sect. 33.2. Section 33.3 demonstrates the effectiveness of U-net for the particle segmentation task through an extensive experimental comparison. In Sect. 33.4, we draw the concluding remarks and future directions.

33.2 Proposed Method

This section elaborates on the particle image segmentation strategy, which consists of KNN matting to fine-tune manually annotated boundaries and data augmentations to expand available annotated data. Subsequently, the CNN based on the U-net architecture was adopted to learn particle boundary information in the extremely small dataset.

33.2.1 KNN Matting

Image matting is a typical image problem aiming to decompose an image into foreground and background counterparts. The basic image matting is to solve a matting equation (Eq. 33.1),

$$I = \alpha \times F + (1 - \alpha) \times B \qquad (33.1)$$

where I, F, and B are composite colour, foreground colour, and background colour, respectively, and α is a fusion coefficient. KNN matting [15] is a nonlocal matting method, and the goal is to solve the limitation of matting Laplacian by allowing the values of α to propagate in nonlocal pixels. For our particle segmentation task, as shown in Fig. 33.3, manual labelling inevitably causes labelling errors at the boundary of particles due to the limited contrast around the boundaries, in which cases the KNN matting can help to improve the quality of annotation by fine-tuning the manual labelling results.

33.2.2 Data Augmentation

One of the main challenges of particle segmentation is the lack of sufficient data. The current measures that can be taken to solve this challenge are generally to combine multiple data augmentation techniques or adopt pre-trained transfer learning strategies. In the present work, data augmentation techniques are employed to significantly increase the diversity of particle samples, which can be roughly classified into the following three categories:

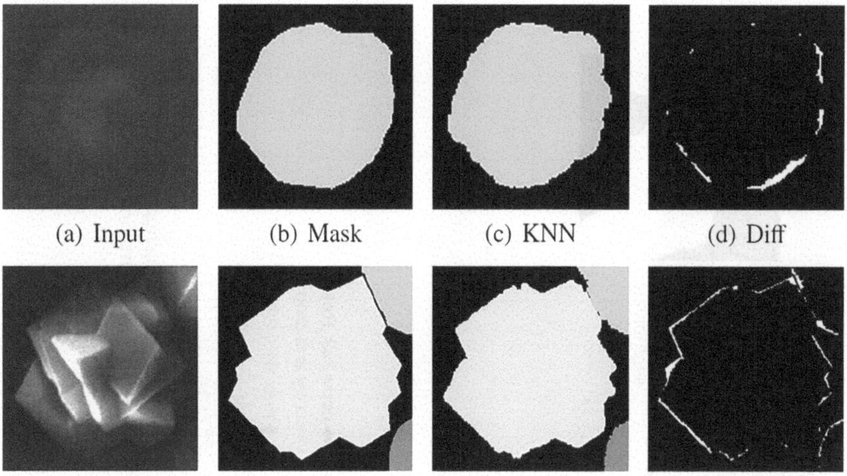

(a) Input (b) Mask (c) KNN (d) Diff

(e) Input (f) Mask (g) KNN (h) Diff

Fig. 33.3 Comparison of the mask before and after fine-tuning. (**a**) and (**e**) represent two types of particle images, the corresponding manual annotations are denoted by (**b**) and (**f**), (**c**) and (**g**) are the result of KNN matting [15], (**d**) and (**h**) denote the difference between the corresponding manual annotations and KNN matting fine-tuning results

- Flipping: flipping produces mirror data, where the particle image and the corresponding mask are flipped horizontally and vertically, respectively.
- Rotation: the data are rotated by a given angle, such as 90, 180, or 270°, where the particle image and the corresponding mask are rotated around the center point.
- Illumination: illumination can be implemented by adjusting the colour map (RGB values), where random values are added to the R, G, and B channels of the particle image.

It is worth noting that the entire particle image has finally been split into uniformly sized (256×256) patch datasets for training the model.

33.2.3 Network Architecture

The network architecture consists of a contracting path (left side) and an expansive path (right side), as illustrated in Fig. 33.4. The contracting path follows the typical architecture of a convolutional network. It consists of the repeated application of two 3×3 convolutions (unpadded convolutions), each followed by a rectified linear unit (ReLU) [16] and a 2×2 max pooling operation with stride 2 for sampling. At each downsampling step, we double the number of feature channels. Every step in the expansive path consists of an up-sampling of the feature map followed by a 2×2 convolution ("up-convolution"), a concatenation with the correspondingly cropped

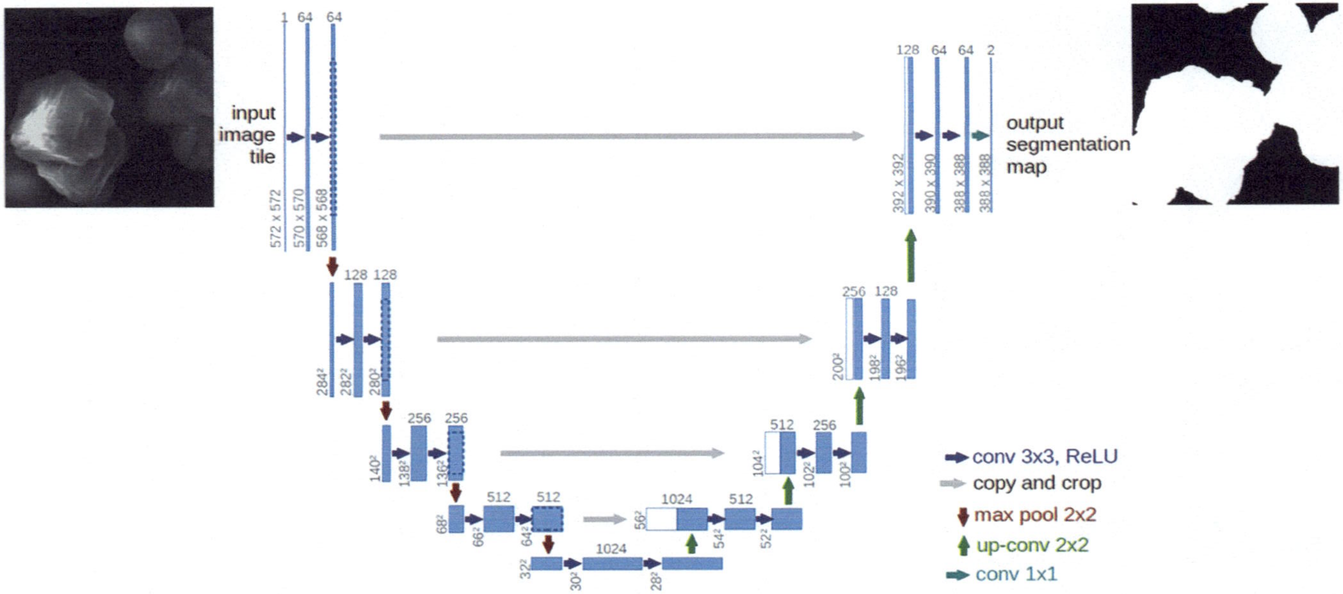

Fig. 33.4 Schematic of the U-net model architecture, where there is a large number of symmetric feature channels in the up-sampling part, which allows the network to propagate semantic information to higher resolution layers. Therefore, the expansion path is symmetrical to the contraction path and produces a U-shaped structure

feature map from the contracting path, and two 3 × 3 convolutions, each followed by a ReLU [16].

33.3 Experiments

In this section, we evaluate the effectiveness of the proposed PIS framework on the extremely small dataset and show the segmentation performance of the trained model on new particle images. Model training and testing were carried out on a PC that has an Intel(R) Core(TM) i7-11700K CPU with 3.60 GHZ (8 CPU), 32 GB RAM and an NVIDIA GEFORCE RTX3070Ti graphics card with 8 GB memory.

The dataset contains a total of 20 original particle images. 21,660 training images are obtained through the pre-processing step of the PSI framework, of which 80% are used for training and 20% for testing. We train the CNN based on the U-net architecture using CrossEntropy loss with label smoothing, Adam optimizer [17], an initial learning rate $lr = 0.0001$, and a batch size of 32. In addition, intersection over union (IOU) [18] is used to evaluate the segmentation performance of the model. The training results of the model on the training set and the test set are shown in Fig. 33.5a, b and c, d, respectively. It can be seen that with the increase of the number of iterations, IOU and Loss tend to be stable at the end of the training and eventually converge to satisfactory performance. In addition, the performance is finally verified with the segmentation of a 1902 × 1500 image taking less than a second on a desktop GPU platform. A representative image segmentation result is shown in Fig. 33.6. It can be seen that our model can have high efficiency and accurately segment the different particles.

33.4 Conclusion

This paper has described a new particle segmentation framework that can obtain accurate segmentation results based on a very limited source of images. One of the drawbacks of our segmentation framework is the difficulty of distinguishing the boundaries between overlapping particles, which is still to be investigated in the future.

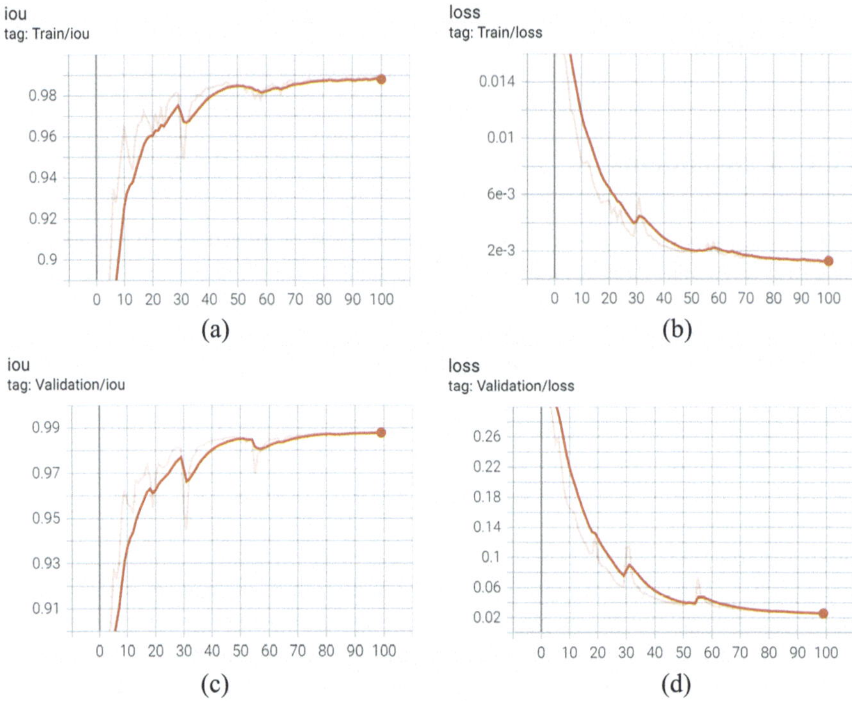

Fig. 33.5 The IOU and the Loss curves of the U-net model on the training set and the test set, respectively, where the abscissa represents the number of iterations, and the ordinate represents the metrics. (**a**) Train-IOU. (**b**) Train-loss. (**c**) Test-IOU. (**d**) Test-Loss

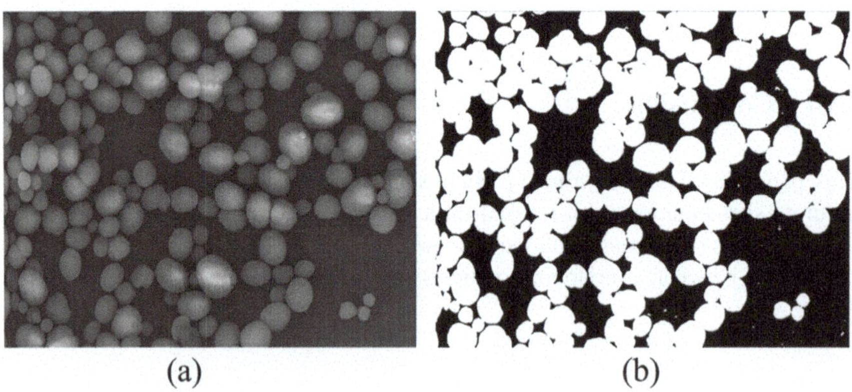

Fig. 33.6 Results of particle segmentation. (**a**) is the input images of the test dataset, and (**b**) is the corresponding segmentation results of the U-net model

Acknowledgments A.S. and B.P. and J.S. thank the Special Research Fund (BOF) of Ghent University (01IO3618), FWO-Vlaanderen (G043322N; I002620N), and EOS of FWO-F.N.R.S. (project # 40007488) for support. J.S. acknowledges the support of the China Research Council (CSC, No. 202006150025). M.R., J.H., X.W. and M.C. were also partially supported by the FWO Odysseus 1 grant G.0H94.18N: Analysis and Partial Differential Equations, and the Methusalem programme of the Ghent University Special Research Fund (BOF) (Grant number 01M01021). M.R. is also supported by EPSRC grant EP/R003025/2. M.C. is a postdoctoral fellow of the Research Foundation – Flanders (FWO) under the postdoctoral grant No 12B1223N.

References

1. Ghosh, S., Das, N., Das, I., Maulik, U.: Understanding deep learning techniques for image segmentation. ACM Comput. Surv. **52**(4), 1–35 (2019)
2. Minaee, S., Boykov, Y., Porikli, F., Plaza, A., Kehtarnavaz, N., Terzopoulos, D.: Image segmentation using deep learning: a survey. IEEE Trans. Pattern Anal. Mach Intell. **44**(7), 3523–3542 (2021)
3. He, K., Gkioxari, G., Dollár, P., Girshick, R.: Mask R CNN. In: Proceedings of the IEEE International Conference on Computer Vision, pp. 2961–2969 (2017)
4. Lin, T.-Y., Maire, M., Belongie, S., Hays, J., Perona, P., Ramanan, D., Dollár, P., Zitnick, C.L.: Microsoft coco: common objects in context. In: Computer Vision–ECCV 2014: 13th European Conference, Zurich, Switzerland, September 6–12, 2014, Proceedings, Part V 13, pp. 740–755. Springer, Berlin (2014)
5. Long, J., Shelhamer, E., Darrell, T.: Fully convolutional networks for semantic segmentation. In: Proceedings of the IEEE Conference on Computer Vision and Pattern Recognition, pp. 3431–3440 (2015)
6. Ronneberger, O., Fischer, P., Brox, T.: U-net: convolutional networks for biomedical image segmentation. In: Medical Image Computing and Computer-Assisted Intervention–MICCAI 2015: 18th International Conference, Munich, Germany, October 5–9, 2015, Proceedings, Part III 18, pp. 234–241. Springer, Berlin (2015)
7. Yu, F., Koltun, V., Funkhouser, T.: Dilated residual networks. In: Proceedings of the IEEE Conference on Computer Vision and Pattern Recognition, pp. 472–480 (2017)
8. Song, J., Vikulina, A.S., Parakhonskiy, B.V., Skirtach, A.G.: Hierarchy of hybrid materials. Part II: the place of organics-on-inorganics in it, their composition and applications. Front. Chem. **11**, 1078840 (2023)
9. Lichtman, J.W., Conchello, J.-A.: Fluorescence microscopy. Nat. Methods **2**(12), 910–919 (2005)
10. Alvarez-Román, R., Naik, A., Kalia, Y., Fessi, H., Guy, R.H.: Visualization of skin penetration using confocal laser scanning microscopy. Eur. J. Pharm. and Biopharm. **58**(2), 301–316 (2004)
11. Van der Meeren, L., Verduijn, J., Krysko, D.V., Skirtach, A.G.: AFM analysis enables differentiation between apoptosis, necroptosis, and ferroptosis in murine cancer cells. Iscience **23**(12), 101816 (2020)
12. Van der Meeren, L., Verduijn, J., Krysko, D.V., Skirtach, A.G.: High-throughput mechano-cytometry as a method to detect apoptosis, necroptosis, and ferroptosis. Cell Proliferation **56**(6), e13445 (2023)
13. Liu, Y., Tan, J., Thomas, A., Ou-Yang, D., Muzykantov, V.R.: The shape of things to come: importance of design in nanotechnology for drug delivery. Ther. Delivery **3**(2), 181–194 (2012)
14. Toy, R., Peiris, P.M., Ghaghada, K.B., Karathanasis, E.: Shaping cancer nanomedicine: the effect of particle shape on the in vivo journey of nanoparticles. Nanomedicine **9**(1), 121–134 (2014)
15. Chen, Q., Li, D., Tang, C.-K.: KNN matting. IEEE Trans. Pattern Anal. Mach. Intell. **35**(9), 2175–2188 (2013)

16. Glorot, X., Bordes, A., Bengio, Y.: Deep sparse rectifier neural networks. In: Proceedings of the Fourteenth International Conference on Artificial Intelligence and Statistics, pp. 315–323. JMLR Workshop and Conference Proceedings (2011)
17. Kingma, D.P., Ba, J.: Adam: a method for stochastic optimization (2014). arXiv preprint arXiv:1412.6980
18. Rezatofighi, H., Tsoi, N., Gwak, J., Sadeghian, A., Reid, I., Savarese, S.: Generalized intersection over union: a metric and a loss for bounding box regression. In: Proceedings of the IEEE/CVF Conference on Computer Vision and Pattern Recognition, pp. 658–666 (2019)

Chapter 34
Two-Dimensional Dispersed Composites on a Square Torus

Zhanat Zhunussova

Abstract The article is devoted to the recently established connection between the problem of packing disks on a torus and the effective conductivity of composites with circular inclusions. The packing problem is usually investigated by geometric considerations, the conductivity problem by means of elliptic functions. An algorithm has been developed for determining the optimal arrangement of three disks on a torus formed by a hexagonal lattice. The corresponding minimization function is constructed in terms of expressions consisting of elliptic functions with unknown arguments. This makes it possible to optimize packaging for a hexagonal screen. The numerically found roots coincide with the previously established optimal points, as well as with a purely geometric study.

34.1 Introduction

One of the typical problems in geometry and physics is an optimal design problem. Most of the works were devoted to the problem for elliptic functions, for example, the square lattice, the hexagonal lattice, and problems with the fundamental translation vectors [1–3]. Note that the vectors are periods for the square lattice, and there are two fundamental translation vectors for the hexagonal lattice [4–7]. Moreover, all these combinations are linear. It is considered that the periods can be continued and shifted.

We are going to use the Eisenstein structural sums for multivariable functions. It should be noted that they are applied to compete for a property of the composite fibre materials [8, 9]. The random process is called isotropy. It is possible that there are a lot of disks. Regarding this, some structural sums corresponding to the set of

Z. Zhunussova (✉)
Al-Farabi Kazakh National University, Almaty, Kazakhstan

M. Ruzhansky, K. Van Bockstal (eds.), *Extended Abstracts 2021/2022*,
Research Perspectives Ghent Analysis and PDE Center 2,
https://doi.org/10.1007/978-3-031-42539-4_34

the points $a = \{a_1, a_2, \ldots, a_N\}$ are introduced. Applying the Eisenstein function $E_p(z)$ and the complex conjugate operator, we obtain the following formula:

$$e_2 = \frac{1}{N^2} \sum_{k=1}^{N} \sum_{m=1}^{N} E_2(a_k - a_m).$$

In the case that two disks are investigated, the value a_3 is equal to zero. Hereinafter we write the structural sum e_2. Thus, the optimal centers of the disks $(x_1, y_1), (x_2, y_2), \ldots, (x_N, y_N)$ are studied.

34.2 Main Results

Consider the torus, the hexagonal lattice, with the fundamental translation vectors. As well as we consider the Weierstrass functions. It is well-known that the problem became sophisticated under consideration of complex variable functions [7, 8].

Consider the hexagonal lattice with the fundamental translation vectors

$$\omega_1 = \sqrt[4]{\frac{4}{3}};\ \omega_2 = \sqrt[4]{\frac{4}{3}} e^{\frac{i\pi}{3}};$$

In addition, we pay attention to the invariants of the Weierstrass functions. Let a_2 be taken in the form $a_2 = \frac{1}{2}(\omega_1 + \omega_2)$. An isotropy condition gives the complex equation:

$$\wp(a_1 - \frac{1}{2}(\omega_1 + \omega_2)) = 0,$$

where the isotropy conditions are defined as $e_2^{(2)}(a, r) = \pi$ and $E_1(z) = \wp(z) + S_2,\ S_2 = \pi$.

The series is used under consideration of the Weierstrass invariants. A supplementary Weierstrass' function is introduced. The function is used for the simulation of the disks. A square is chosen from the interval -1/2 till 1/2. Therefore, it is verified as an imposition of the disks. In the case of imposition, the disks are thrown out. The process is repeated for all disks.

More precisely, we investigate the following constrained minimum problem. Given the number of disks N and their radii r. It is required to find the minimal value

$$e_{22}^{(N)}(r) = \min_{\mathbf{a}} e_{22}^{(N)}(\mathbf{a}, r). \tag{34.1}$$

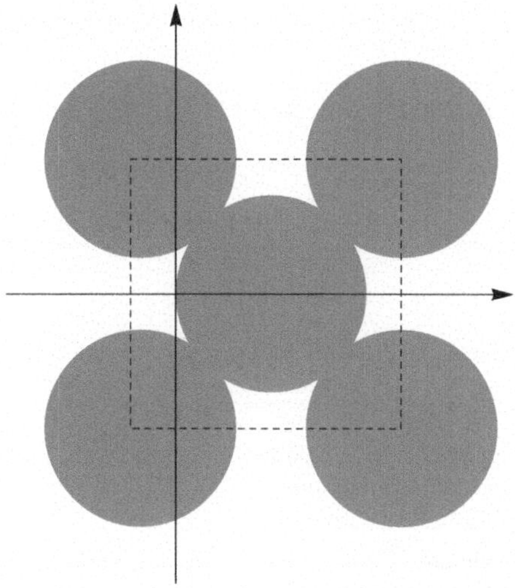

Fig. 34.1 $N = 2$. Optimal centers (34.4) minimising $e_{22}^{(5)}(\mathbf{a})$ with the isotropy condition $e_2^{(5)}(\mathbf{a}) = \pi$ optimal packing points (gray)

The set $\mathbf{a}$ belongs to $Q(\omega_1, \omega_2)$, satisfies the isotropy condition

$$e_2^{(N)}(\mathbf{a}, r) = \pi \qquad (34.2)$$

and the geometrical constraints

$$|a_k - a_m| \geq r \quad \text{for } a_k \neq a_m \quad (k, m = 1, 2, \ldots, N). \qquad (34.3)$$

$\mathbf{a}$ can be arbitrary fixed in $Q(\omega_1, \omega_2)$.

The optimal location for the problem (34.1–34.3) depends on the radius. We obtain the centers (Fig. 34.1)

$$(-0.157, 04615), (0.804, 0.4615), (-0.157, -0.463), \\ (0.804, -0.463), (0.3495, 0) \qquad (34.4)$$

for $r = 0.34$.

We obtain the centers for 3 disks on torus

$$(-0.1152, 0.4406), (0.3444, 0.66), (0.804, 0.463), \\ (0.6108, 0), (0.08325, 0), (-0.1152, -0.4406), \\ (0.3444, -0.66), (0.804, -0.463) \qquad (34.5)$$

for $r = 0.27$.

We obtain the centers for 5 disks on torus (Fig. 34.2)

$$(0.3862, 0.66), (-0.1257, 0.4615), (-0.1257, -0.4615),$$
$$(-0.345, -0.056), (0.2504, 0.2787), (0.068, -0.092),$$
$$(0.6212, 0.1116), (0.4436, -0.2749), (0.2922, -0.6562),$$
$$(0.8092, 0.4667), (0.8144, -0.463) \tag{34.6}$$

for $r = 0.21$ (Fig. 34.3).

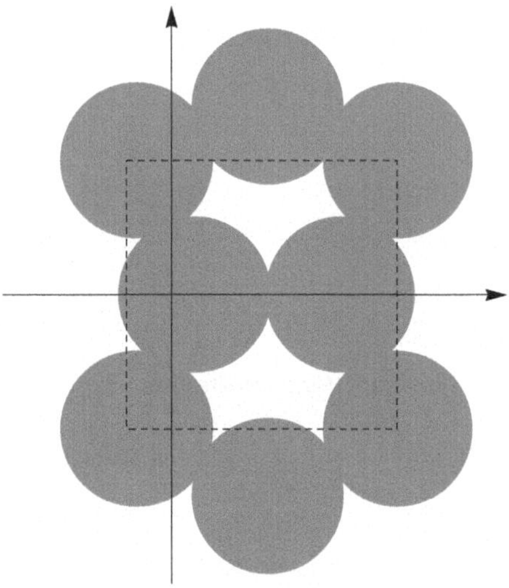

Fig. 34.2 $N = 3$. Optimal centers (34.5) minimizing $e_{22}^{(5)}(\mathbf{a})$ with the isotropy condition $e_2^{(5)}(\mathbf{a}) = \pi$ optimal packing points (gray)

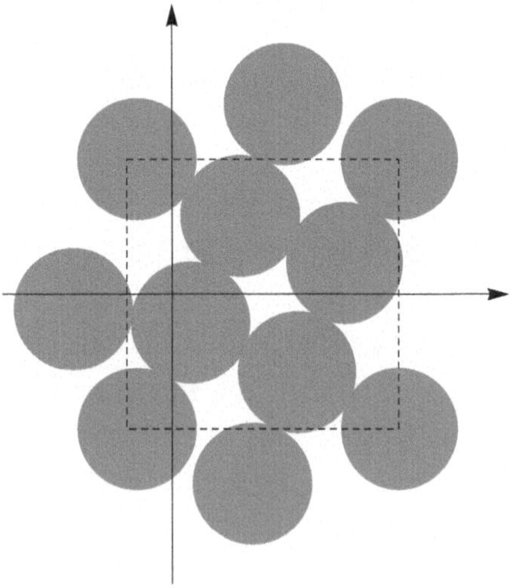

Fig. 34.3 $N = 5$. Optimal centers (34.6) minimizing $e_{22}^{(5)}(\mathbf{a})$ with the isotropy condition $e_2^{(5)}(\mathbf{a}) = \pi$ optimal packing points (gray)

34.3 Conclusion

In the general case, the series associated to periodic analytical functions, Eisenstein's series, is considered. The series is used under consideration of the Weierstrass invariants. A supplementary Weierstrass' function is introduced. The function is used for the simulation of the disks. Similarly procedures on the axis Oy. Therefore, it is verified as an imposition of the disks. In the case of the disks, imposition is thrown out.

The process is repeated for all disks. We are going to use the Eisenstein structural sums for multivariable functions. It should be noted that they are applied to verify a property of composite fibre materials. The random process is called isotropy. It is possible that there are a lot of disks.

Regarding this, we have to simulate and assume that the second disk is known. It is calculated with one unknown point. By minimising the function, we can find the optimal packing of the disks.

References

1. Weil, A.: Elliptic Functions according to Eisenstein and Kronecker. Berlin, Springer (1976)
2. Mityushev, V., Rylko, N.: Optimal distribution of the non-overlapping conducting disks. Multiscale Model. Simul. **10**(1), 180–190 (2012)
3. Andrianov, I., Gluzman, S., Mityushev, V. (eds.): Mechanics and Physics of Structured Media: Asymptotic and Integral Methods of Leonid Filshtinsky. Academic Press, London (2022)
4. Mityushev, V.: R-linear problem on torus and its application to composites. Complex Variables **50**(7–10), 621–630 (2005)
5. Castelaz, A., Dickinson, W., Guillot, D., Xhumari, S.: Optimal Packings of up to 6 Equal Circles on a Triangular Flat Torus, Student Summer Scholars Paper 12 (2008). http://scholarworks.gvsu.edu/sss/12
6. Mityushev, V., Zhunussova, Z.: Optimal random packing of spheres and extremal effective conductivity. Symmetry **13**(6), 1063 (2021)
7. Zhunussova, Zh.Kh., Ashimov, Ye.K., Dosmagulova, K.A., Zhunussova, L.Kh. Optimal packing of two disks on torus. Appl. Math. Inf. Sci. **16**(4), 549-554 (2022)
8. Dosmagulova, K., Mityushev, V., Zhunussova, Zh.: Optimal conductivity of packed two-dimensional dispersed composites. SIAM J. Appl. Math. **83**(3), 985–999 (2023)
9. Zari, I., Gul, T., Dosmagulova, K., Khan, T.S., Haq, S.: Heat transfer analysis of Radiative-Marangoni Convective flow in nanofluid comprising Lorentz forces and porosity effects. Adv. Theory Nonlinear Anal. Appl. **7**(1), 61–81 (2023)

The manufacturer's authorised representative in the EU is Springer Nature Customer Service Centre GmbH, Europaplatz 3, 69115 Heidelberg, Germany. If you have any concerns regarding our products, please contact ProductSafety@springernature.com

Printed and bound by CPI Group (UK) Ltd, Croydon, CR0 4YY
29/07/2026
02180497-0001